MICROBIAL SAFETY *of* MINIMALLY PROCESSED FOODS

MICROBIAL SAFETY *of* MINIMALLY PROCESSED FOODS

Edited by

John S. Novak • Gerald M. Sapers • Vijay K. Juneja

CRC Press
Taylor & Francis Group
Boca Raton London New York

CRC Press is an imprint of the
Taylor & Francis Group, an **informa** business

CRC Press
Taylor & Francis Group
6000 Broken Sound Parkway NW, Suite 300
Boca Raton, FL 33487-2742

First issued in hardback 2018

ISBN 13: 978-1-138-42652-8 (hbk)
ISBN 13: 978-1-58716-041-7 (pbk)

Library of Congress Cataloging-in-Publication Data

Microbial safety of minimally processed foods / edited by John S. Novak, Gerald M. Sapers, and Vijay K. Juneja.
p. cm.
Includes bibliographical references and index.
ISBN 1-58716-041-2 (alk. paper)
1. Food—Microbiology. 2. Food-Safety measures. I. Novak, John S. II. Sapers, Gerald M. III. Juneja, Vijay K., 1956-

QR115 .M458 2002
664'.001'579—dc21 2002073796

Library of Congress Card Number 2002073796

Visit the Taylor & Francis Web site at
http://www.taylorandfrancis.com

and the CRC Press Web site at
http://www.crcpress.com

Preface

There has been a concerted industry effort in recent years toward providing consumers with foods that are more wholesome and natural, prepared with minimal preservatives, or less detrimental to flavor freshness and consumer time constraints. A food shopper venturing into any U.S. supermarket will discover a fresh produce section stocked with exotic fruits and vegetables native to a wide variety of world climates and growing seasons, fresh meat and seafood that is expected to be fresh "farm to fork," and a deli section filled with preprepared entrees, as well as full-course meals. All are expected to be portioned and minimally processed to balance the naturalness of unaltered foods with a concern for safety. Yet the responsibility for proper food preparation and handling remains with the naïve modern consumer, who may be less adept in food preparations than his or her less sophisticated ancestors. As a result, diseases from improperly handled foods are anticipated to escalate.

The U.S. Centers for Disease Control and Prevention (CDC) warn of more than 200 known diseases transmitted through food consumption. The latest current estimates report that food-borne diseases are responsible for approximately 76 million illnesses, 325,000 hospitalizations, and 5,000 deaths in the U.S. annually — where one would expect to find the safest food supply in the world. Many of the chapters in this text deal with preventative solutions to these alarming incidents from the varied perspectives of the producer, handler, consumer, and food preparer, as well as the food inspector and researcher.

This compilation provides the reader with the latest research and insight into assessing the microbial safety of red meats, poultry, fish, vegetables, fruits, and bakery products receiving less-than-stringent sterilizing preparations. In-depth evaluations of hazard analysis critical control point (HACCP) regulations and risk assessments of those minimally processed foods are also provided. The methods used for detection of the pathogens are explored and described along with strategies for prevention of future pathogen occurrences in minimally processed foods. State-of-the-art novel technologies are discussed, such as irradiation, modified atmosphere packaging, biological control measures, and nonthermal preservations, which have potential applications in the enhancement of microbiological safety of minimally processed foods without sacrificing the natural untreated visual appearance and sensory properties.

It is expected that the topics presented here will stimulate thought and future technological research advances toward providing microbiologically safer foods that fulfill a consumer's desire for unadulterated freshness. Toward this goal, the authors address students of food science, industry personnel involved in the safety of minimally processed foods, and government agencies involved in establishment of food safety guidelines. The reader is asked to view the ideas expressed as pliable toward developing improvements to provide increasingly safer foods now and into the future.

The Editors

John S. Novak received his B.S. in biochemical science from the University of Vermont in 1981, M.S. degrees in microbiology and exercise physiology from Syracuse University in 1984 and 1985, respectively, and his Ph.D. in microbiology from the State University of New York College of Environmental Science and Forestry in 1993. Following postdoctoral research at the Ohio State University, Dr. Novak joined the Food Group at American Air Liquide in 1998 as a consulting microbiologist working on research and development in food sanitation, preservation, and packaging technologies. He was hired as a microbiologist for the USDA Agricultural Research Service in 1999. Current research includes stress adaptation studies and virulence gene expression in food-borne pathogens. Dr. Novak has authored 15 scientific papers, 6 trade-journal articles, 2 book chapters, and 1 patent. He is the 2002–2003 chair of the Eastern Food Science Conference planning committee and a member of the editorial board for the *Journal of Food Protection.*

Gerald M. Sapers received his Ph.D. in food technology from MIT in 1961. He joined the USDA's Eastern Regional Research Center (ERRC) in 1968, after 2 years at the U.S. Army Natick Laboratories and 6 years in private industry. He has conducted research on dehydrated potato stability, apple volatiles, safety of home-canned tomatoes, utilization of natural pigments, pigmentation of small fruits, cherry dyeing, control of enzymatic browning in minimally processed fruits and vegetables, mushroom washing, and microbiological safety of fresh produce. He has been a lead scientist at ERRC since 1991. Dr. Sapers has published more than 100 scientific papers, 3 book chapters, and 5 patents. He is an active member of IFT's Fruit and Vegetable Products Division.

Vijay K. Juneja is supervisory microbiologist and lead scientist in the Microbial Food Safety Research Unit at the Eastern Regional Research Center (ERRC) of the Agricultural Research Service (ARS) branch of the U.S. Department of Agriculture (USDA) in Wyndmoor, Pennsylvania. In 1978, Dr. Juneja received his B.V.Sc. and A.H. (D.V.M.) from G. B. Pant University of Agriculture and Technology, India; he received his M.S. and Ph. D. degrees in food technology and science from the University of Tennessee in 1988 and 1991, respectively. Soon after receiving his Ph.D., he was appointed microbiologist at the ERRC–USDA.

Dr. Juneja has developed a nationally and internationally recognized research program on food-borne pathogens, with emphasis on microbiological safety of minimally processed foods and predictive microbiology. He is co-editor of *Control of Foodborne Microorganisms* and serves on the editorial board of the *Journal of Food Protection*. Dr. Juneja is recipient of several awards, including the Agricultural Research Service Early Career Research Scientist, North Atlantic Area Scientist of the Year, 1998; Gold Medalist "Technical Accomplishment," Federal Executive Board (FEB) 1998, 2000; ARS–FSIS Cooperative Research Award, 1998; and USDA–ARS Certificate of Merit for Outstanding Performance, 2002. Currently, Dr. Juneja is a group leader for a multidisciplinary research project concerning the assurance of microbiological safety of processed foods. He develops strategies for research plans, oversees projects, reports results to user groups, and advises regulators (FDA, FSIS, etc.) on technical matters such as research needs and emerging issues. His research interests include intervention strategies for control of food-borne pathogens and predictive modeling. Dr. Juneja's research program has been highly productive, generating more than 180 research articles, book chapters, and abstracts, primarily in the areas of food safety and predictive microbiology.

Contributors

John W. Austin
Microbiology Research Division
Bureau of Microbial Hazards
Ottawa, Ontario

Elizabeth A. Baldwin
USDA, ARS, Citrus and Subtropical Products Lab
Winter Haven, Florida

M. Margaret Barth
Redi-Cut Foods, Inc.
Franklin Park, Illinois

Yuhuan Chen
Rutgers University (SUNJ)
New Brunswick, New Jersey

William S. Conway
USDA, ARS, Produce Quality and Safety Lab
Beltsville, Maryland

Daphne Phillips Daifas
McGill University
Quebec, Ontario

Siobain Duffy
Rutgers University (SUNJ)
New Brunswick, New Jersey

Wassim El-Khoury
McGill University
Quebec, Ontario

Kenneth L. Gall
Cornell University Cooperative Extension
State University of New York
Stony Brook, New York

Thomas R. Hankinson
Produce Safety Solutions, Inc.
Toughkenamon, Pennsylvania

Adam D. Hoffman
Cornell University
Ithaca, New York

Wojciech Janisiewicz
USDA, ARS, Appalachian Fruit Research Station
Kearneysville, West Virginia

Vijay K. Juneja
USDA, ARS, Eastern Regional Research Center
Wyndmoor, Pennsylvania

Britta Leverentz
USDA, ARS, Produce Quality and Safety Lab
Beltsville, Maryland

Karl R. Matthews
Rutgers University (SUNJ)
New Brunswick, New Jersey

Brendan A. Niemira
USDA, ARS, Eastern Regional Research Center
Wyndmoor, Pennsylvania

John S. Novak
USDA, ARS, Eastern Regional
Research Center
Wyndmoor, Pennsylvania

Kathleen T. Rajkowski
USDA, ARS, Eastern Regional
Research Center
Wyndmoor, Pennsylvania

Gerald M. Sapers
USDA, ARS, Eastern Regional
Research Center
Wyndmoor, Pennsylvania

Donald W. Schaffner
Rutgers University (SUNJ)
New Brunswick, New Jersey

James P. Smith
McGill University
Quebec, Ontario

O. Peter Snyder, Jr.
Hospitality Institute of Technology and
Management
St. Paul, Minnesota

Christopher H. Sommers
USDA, ARS, Eastern Regional
Research Center
Wyndmoor, Pennsylvania

Gaurav Tewari
Tewari De-Ox Systems, Inc.
San Antonio, Texas

Martin Wiedmann
Cornell University
Ithaca, New York

James T. C. Yuan
American Air Liquide
Countryside, Illinois

Hong Zhuang
Redi-Cut Foods, Inc.
Franklin Park, Illinois

Table of Contents

SECTION I Variable Food Environments

SECTION II Pathogen Detection and Assessment

Section I

Variable Food Environments

1 Microbial Safety of Bakery Products

James P. Smith, Daphne Phillips Daifas, Wassim El-Khoury, and John W. Austin

CONTENTS

1-58716-041-2/03/$0.00+$1.50

INTRODUCTION

Bakery products have been an important part of a balanced diet for thousands of years. Today, the production of bread and other bakery products has evolved from a cottage industry into a large-scale, modern manufacturing industry. In 1998, sales of bakery products in the U.S. exceeded 10 million metric tons — a 14.5% increase over the previous 4 years (Kohn, 2000). In Canada, the bread and bakery industry shipped $2.3 billion of products in 2000, an increase of 36.3% from 1988 levels, and accounted for 4.2% of total food and beverage processing sector shipments (Agriculture and Agri-Food Canada, 2000). This sustained growth has been driven by consumer demands for convenient, premium baked goods that are fresh, nutritious, conveniently packaged, and shelf-stable. The increased demand is being met by various new processing and packaging technologies, including modified atmosphere packaging, a technology that has increased the availability and extended the shelf-life and market area of a wide variety of bakery products. At the same time, in-store bakeries have increased, as well as a renewed interest in "organic," ethnic, and artisan-type bakery products.

Today, a wide variety of bakery products is on supermarket shelves. These products include unsweetened goods (breads, rolls, buns, crumpets, muffins, bagels), sweet goods (pancakes, doughnuts, waffles, and cookies), and filled goods (fruit and meat pies, sausage rolls, pastries, sandwiches, cream cakes, pizza, and quiche). Most bakery products are marketed fresh and are stored at ambient temperature. However, other products, such as cream-, fruit-, and meat-filled pies and cakes, are stored under refrigerated or frozen storage conditions to achieve a longer shelf-life. Bakery products, like most minimally processed foods, are subject to physical, chemical, and microbiological spoilage. Although physical and chemical spoilage problems limit the shelf-life of low- and intermediate-moisture bakery products, microbiological spoilage is the main concern in high-moisture products. Furthermore, high-moisture unfilled and filled bakery products have also been implicated in outbreaks of food-borne illness and therefore pose safety concerns.

SAFETY CONCERNS OF BAKERY PRODUCTS

Minimal Processing

In order to achieve desirable textural and quality attributes, most bakery products receive a minimal heat treatment. For example, although bread is baked at high temperature, during baking the temperature in the center of the crumb rarely exceeds 100°C for more than a few minutes. According to Bryan et al. (1997), vegetative pathogenic microorganisms should be readily destroyed during baking; however, spore forming bacteria will survive baking and may grow to levels of public health concern if packaging and storage conditions are conducive to their growth. While vegetative microorganisms should be destroyed during baking, products may be subject to post-baking contamination from the air, equipment, and handlers (Sugihara, 1977). Furthermore, many ingredients, such as fresh and synthetic cream, cold custard, icings, spices, nuts, and fruit toppings or fillings, are added after baking and may be a potential source of contamination. Cross-contamination may also occur if bakery products are prepared or stored in the same area as raw foods such as eggs, meat, or milk.

Hazardous Products and Ingredients

Potentially hazardous foods have a pH >4.5 and an a_w >0.84. Many bakery products and their ingredients fall within this hazardous category. Bakery products can be conveniently classified into three groups according to pH: (1) high-acid bakery products with pHs <4.6, (2) low-acid bakery products with pHs >4.6 but <7, and (3) non-acid or alkaline bakery products with pHs >7. Examples of various products within these pH categories are shown in Table 1.1.

The a_w of bakery products is also an important indicator of their ability to support microbial growth. Smith and Simpson (1995) classified bakery products on the basis of their a_w as (1) low-moisture bakery products with a_w <0.6, (2) intermediate-moisture bakery products with a_w between 0.6 and 0.85, and (3) high-moisture bakery products with a_w >0.85 and generally between 0.95 and 0.99. Examples of products within each a_w category are shown in Table 1.2.

Many bakery products and ingredients have pH and a_w levels that restrict microbial growth, while others have levels conducive to microbial growth. For example, the pH of custard, used in many filled baked products, is 5.8 to 6.6 and is ideal for the growth of *Salmonella* (Bryan, 1976). It is also important to note that both pH and a_w may change during storage. For example, icing, which has a low a_w, is not usually a microbiological problem; however, the interface between the cake and icing may have a much higher a_w, which encourages microbial growth. Silliker and McHugh (1967) reported an incident in which *Staphylococcus aureus* grew at the interface of cake and icing.

TABLE 1.1
pH Range of Selected Bakery Products

Product	pH Range	Ref.
High Acid		
Sourdough bread	4.2–4.6	Martinez-Anaya et al. (1990)
Apple pie	4.2	Smith and Simpson (1995)
Low Acid		
White bread	5.7	Rosenkvist and Hansen (1995)
Whole wheat bread	5.6	Rosenkvist and Hansen (1995)
Chocolate nut bread	6.2–6.6	Denny et al. (1969)
Date nut bread	6.1–6.7	Denny et al. (1969)
Non-Acid		
Crumpets	6–8	Jenson et al. (1994)
Banana nut bread	7.2–7.9	Aramouni et al. (1994)
Carrot muffin	8.7	Smith and Simpson (1995)

TABLE 1.2
Water Activity (a_w) Range of Selected Bakery Products

Product	a_w
Low Moisture Content	
Cookies	0.2–0.3
Crackers	0.2–0.3
Intermediate Moisture Content	
Cake type doughnuts	0.85–0.87
Chocolate-coated doughnuts	0.82–0.83
Danish pastries	0.82–0.83
Cream-filled cake	0.78–0.81
Soft cookies	0.5–0.78
High Moisture Content	
Bread	0.96–0.98
Pita bread	0.9
Yeast-raised doughnuts	0.96–0.98
Fruit pies	0.95–0.98
Carrot cake	0.94–0.96
Custard cake	0.92–0.94
Cheesecake	0.91–0.95
Butter cake	0.9
Pizza crust	0.94–0.95
Pizza	0.99

Storage Conditions

Most bakery products, with the exception of cream-, custard-, and meat-filled products, are held at ambient temperature for maximum storage quality; however, such storage conditions may be conducive to microbial growth and may compromise safety. Furthermore, since most products are "cook and hold" and are not heated prior to consumption, there is no safety margin for destroying bacteria that may survive the baking process or may have been introduced during handling or storage. English style crumpets, a high-moisture snack food product held at ambient temperature, have been implicated in several food poisoning outbreaks involving *Bacillus cereus* (Jenson et al., 1994). For in-store bakeries, products are often displayed in bins or are loosely wrapped in paper. Although customers like this form of product presentation, a potential for contamination of these products from self-serve bins exists if they are handled without the use of tongs or glassine paper.

Products such as cream-, meat-, and cheese-filled cakes have an established history as vehicles of food-borne illness. Holding at refrigeration temperatures will delay microbial growth in these filled products, but it may not be sufficient to prevent the growth of psychrotrophic pathogens such as *Listeria monocytogenes*. Furthermore, there is always the potential of temperature abuse at all stages of the processing, distribution, and storage chain and in the home. Pathogens such as *Salmonella* spp. and *Staphylococcus aureus* can grow at temperatures as low as 8°C, i.e., mild temperature-abuse storage conditions. If products are frozen, bacterial growth will be slowed, but once the product is thawed, growth may resume, as has been shown in outbreaks involving *Salmonella* spp. (Schmidt and Ridley, 1985).

Modified Atmosphere Packaging

Modified atmosphere packaging (MAP) using CO_2-enriched gas atmospheres can extend the mold-free shelf-life and keeping quality of a wide variety of bakery products stored at ambient temperature. Examples of gas-packaged products on the marketplace include bread, pita bread, crumpets, sandwiches, pizza, and muffins. However, there are concerns about the safety of this technology, particularly with respect to the growth of facultative microorganisms such as *L. monocytogenes*, *Salmonella* spp., and *B. cereus*. Concern is also increasing about the potential growth of proteolytic strains of *Clostridium botulinum* in MAP high-moisture bakery products. Although this pathogen has not been implicated in any outbreaks involving bakery products, it has been shown to grow to hazardous levels in gas-packaged food stored at ambient temperature while products remained organoleptically acceptable to the consumer (Hintlian and Hotchkiss, 1986; Farber, 1991). Although gas packaging is widely used in Europe and is gaining acceptance in North America to extend the shelf-life of high-moisture, minimally processed bakery products, little data exist on the safety of such goods stored at ambient temperature.

Recent Market Trends

Recent consumer trends have resulted in novel products, such as preservative-free, low-fat, and reduced-calorie baked goods. However, modification of a product's formulation may also influence its a_w or pH to levels conducive to the growth of food-borne pathogens. Such novel products may be safe, but their safety must be assessed on an individual basis. This is even more critical if such products are packaged under modified atmospheres and stored at ambient temperature. The preceding safety concerns would appear justified because many high-moisture bakery products have been implicated in food-borne disease outbreaks.

BAKERY PRODUCTS ASSOCIATED WITH FOOD-BORNE DISEASE OUTBREAKS

Each year, thousands of North American consumers suffer from some form of food-borne illness, with symptoms ranging from mild to fatal. Mead et al. (1999) estimated that there are approximately 76 million food-borne illnesses, 325,000 hospitalizations, and 5,000 deaths each year in the U.S. Foods such as meat, fish, poultry, eggs, and dairy products are the most common vehicles of food-borne illnesses worldwide; however, bakery products have also been implicated in food-borne disease outbreaks (Todd, 1996). In the U.S. between 1988 and 1992, baked foods accounted for 29 outbreaks involving 820 cases out of a total of 2423 reported outbreaks of food-borne illnesses (Bean et al., 1996). In Canada, pizza, cheesecake, pies and tarts, bread, and muffins have all been implicated in outbreaks of food-borne illnesses (Todd, 1996). Seiler (1978) estimate that, between 1969 and 1972, 30% of food-borne illnesses in the U.K. were attributed to bakery products, with *S. aureus* being involved in most of these outbreaks. Pizza and noodles were responsible for 2 of the 72 outbreaks reported in Australia between 1980 and 1995 (Communicable Disease Network Australia and New Zealand [CDNANZ], 1997). More recently, *B. cereus* has been implicated in outbreaks of food-borne illness involving high-moisture English style crumpets (Jenson et al., 1994).

The rest of the world is not immune to food-borne disease outbreaks caused by bakery products. Todd (1996) reported that 35 to 47% of all food-borne illness outbreaks in Poland, Portugal, Bulgaria, and Switzerland were caused through bakery products. Cuba reported 186 outbreaks involving 8813 patients in the first 6 months of 1990. The major food vehicles were beef, pork, chicken, and cake. Between 1988 and 1990 in Brazil, several outbreaks were traced to white cheese and cream-filled cakes with *S. aureus* being the main organism involved (Potter et al., 1997).

This chapter will review the microorganisms of concern in minimally processed bakery products and the strategies used to enhance their safety

SPECIFIC MICROORGANISMS OF CONCERN

SALMONELLA SPECIES

Sources of Contamination

Salmonellosis is a common gastrointestinal food-borne illness that, although generally self-limiting, can result in chronic complications in the very young, the old, or the immunocompromised (D'Aoust, 1994). *Salmonella* species causing food-borne disease are commonly isolated from animals, their food products, and their processing environments (Wells et al., 2001; Swanenburg et al., 2001; Ebel et al., 1992). Although eggs are the most obvious source of *Salmonella* in bakery products, *Salmonella* species may also be introduced into these products through other ingredients, including flour, milk, cheese, butter, fruits, nuts, and spices. *Salmonella* can also be easily spread by cross-contamination when minimally processed or finished bakery products are in contact with other animal foods or contaminated surfaces during production, storage, and transportation.

However, the major source of *Salmonella* spp. in bakery products is eggs, which, although a potentially hazardous bakery ingredient, are invaluable for their foaming, emulsifying, and binding properties. *Salmonella* spp. may be found on the eggshell (Board, 1969). However, *Salmonella enteritidis* has also been found inside eggs as a result of transovarial transmission during formation. The U.S. Department of Agriculture (USDA) estimates that 1 in 100,000 eggs is contaminated with *S. enteritidis*. Higher numbers are found more often in whites than in yolks, but contaminated eggs occasionally contain extremely large numbers of *S. enteritidis*. Furthermore, handling contaminated eggs easily results in widespread contamination of work surfaces, equipment, and hands (Humphrey et al., 1991, 1994). Pasteurization of egg products effectively destroys *S. enteritidis* while maintaining their functional properties; however, the use of unpasteurized shell eggs in homemade raw cookie dough increases the risk of contamination of bakery products by *Salmonella* spp. (Food Safety and Inspection Service [FSIS], 1998).

Although eggs are one of the most common vehicles of transmission of *Salmonella* spp. in baked products, other ingredients may pose safety concerns. Pasteurization of milk destroys *Salmonella* spp.; however, dairy products, including fresh or dried milk, butter, cream, and cheese, may contain *Salmonella* spp. through inadequate pasteurization or through postpasteurization contamination in the dairy environment (Ahmed et al., 2000; Altekruse et al., 1998; Johnson et al., 1990; El-Gazzar and Marth, 1992). *Salmonella* spp. have also been found in flour. Richter et al. (1993) reported that 1.3% of 4000 samples of wheat flour contained *Salmonella* spp. Although flour is too dry for growth, cells can remain viable for several months (Dack, 1961).

Other bakery ingredients from which *Salmonella* spp. have been isolated, many of which have been implicated in illness, include cocoa and chocolate, particularly milk chocolate (D'Aoust 1977; Torres-Vitela et al., 1995), coconut, (Geopfert, 1980), peanuts and peanut butter (Scheil et al., 1998), fruit (Public Health Laboratory Services [PHLS], 1993; Golden et al., 1993), spices, and yeast flavorings (Lehmacher et al., 1995; Joseph et al., 1991). *Salmonella* spp. are resistant to desiccation and can survive for long periods of time on surfaces and in foods of low water activity, particularly those with a high fat content. Once hydrated, *Salmonella* spp. may grow rapidly in such products held at ambient temperature.

Salmonella spp. are heat labile and, consequently, they should be inactivated during baking or cooking. However, for minimally processed products such as cheesecake or custard- and meringue-type pies, puddings, or fillings, the mild heat treatment necessary to produce an acceptable product may be inadequate for complete destruction of this pathogen.

In unbaked products such as cold custard mixes, puddings, icings, and toppings, *Salmonella* spp., if present, may grow to hazardous levels. Although the a_w of some icings may provide a barrier to growth, the interface between the icing and the baked product may be more favorable for growth.

Associated Outbreaks

Because the symptoms of salmonellosis in people who are not at high risk may be mild, it is estimated that the disease is significantly under-reported (Todd, 1989). Outbreaks from contaminated chocolate and cheese suggest that the infective dose may be fewer than 10 *Salmonella* cells per 100 g of food (Hockin et al., 1989; Altekruse et al., 1998). Most reported outbreaks of salmonellosis caused by eating contaminated bakery products have involved *S. enteritidis* PT4, *S. enteritidis* PT7, and *S. typhimurium*. In most outbreaks, eggs were confirmed as the suspected vehicle of transmission.

The use of raw shell eggs in unbaked products has frequently resulted in outbreaks involving large numbers of people who consumed preprepared bakery products. Unbaked products prepared with raw shell eggs involved in outbreaks of salmonellosis include tiramisu, mousse, and charlotte russe (PHLS, 1993, 1999). Marshmallow cones, which typically have an a_w <0.7, have been associated with one food-borne illness outbreak (Lewis et al., 1996). The marshmallow was made by cold setting raw egg white combined with a commercial mix containing sugar and dried egg whites. The wrapped, filled cones were kept at room temperature and *Salmonella* spp. may have grown during storage in the bakery at the interface between the cone and the marshmallow or may have remained viable in the uncooked product. Consumers eating the cones became ill and *S. enteritidis* PT4 was isolated from the marshmallow but not from the mix or cones.

Many egg-based bakery products rely on minimal heat treatment to provide a light and tender product. Such products include sweet and savory custard fillings, pies, mousse, meringues, and cheesecake. Such minimal processing may be insufficient to destroy *Salmonella* spp. (Hao et al., 1999). Outbreaks are common in such foods, especially when prepared with unpasteurized shell eggs. The practice of cooking and holding at room temperature prior to consumption of products facilitates growth of surviving organisms. Custard pies prepared commercially and held for 21 h prior to consumption resulted in a large outbreak of salmonellosis in which one person died (Centers for Disease Control [CDC], 1990). Similar products implicated in outbreaks include bread pudding (CDC, 1990; Goodman et al., 1993), custard-filled cakes and pastries (Barnes and Edwards, 1992; PHLS, 1993), quiche (Holtby et al., 1995), meringue pies (Wright and Patterson, 1994), mousse, banana puddings (CDC, 1990; Holtby, 1992), and cheesecake (CDC, 2000; Wright et al., 1996).

Contaminated eggs or other ingredients may cross-contaminate equipment resulting in multiple outbreaks. Evans et al. (1996) reported consecutive salmonellosis outbreaks traced to the same bakery. The first outbreak was attributed to cross-contamination during preparation of a cold-custard mix in a bowl previously used for shell eggs; inadequate cleaning and disinfecting of nozzles used for piping cream was the source of cross-contamination in the second outbreak (Evans et al., 1996). Similarly, four outbreaks were caused through cross-contamination by raw eggs used in the production of gateaux and cheesecake (Wright et al., 1996).

Acidic products, such as fruit pies, have also been implicated in food-borne disease outbreaks. Apple pie was implicated in an unusual outbreak of salmonellosis involving *S. enteritidis* PT4. The commercially prepared pie was prepared from canned filling and the crust was glazed with a raw shell egg–milk glaze and baked. The 20-cm pies were cooled and held for 5 h at room temperature prior to consumption. The baking conditions were adequate for the destruction of *Salmonella* spp., and the low pH (3.6) of the baked pie filling should have been an additional barrier to control the growth of this pathogen (Bonner and Schweiger, 1994). However, cross-contamination of pies may have occurred during packaging or the egg glaze may have seeped through the high-fat pastry, resulting in growth of *S. enteritidis* at the product's interface.

Salmonella spp. resist desiccation and can survive for long time periods in foods of very low a_w. Outbreaks involving *Salmonella* spp. have been reported in cake mixes made with dry eggs (Skoll and Dillenberg, 1963), snack foods (Lehmacher et al., 1995; Killalea et al., 1996), and toasted oat cereal (CDC, 1998a).

Control Measures

The use of pasteurized eggs is critical to the control of *Salmonella* food-borne outbreaks. Pasteurized eggs should always be used in unbaked or lightly cooked products or in products that require the pooling or holding of shell eggs. Similarly, avoiding the use of cracked eggs significantly reduces the risk of illness (Todd, 1996).

Processing conditions also play a role in the destruction of *Salmonella* spp. Baking or cooking may destroy any *Salmonella* spp. Whether the organisms are completely destroyed will depend on the heat process, the type of bakery product, its ingredients, pH, and a_w. During baking of retail pumpkin pie with a pH of 5.6 and an a_w of 0.97, an internal temperature of 108°C for 1 min was sufficient to destroy 10 CFU/g of inoculated *S. typhimurium* (Wyatt and Guy, 1981a). A similar inoculum of *S. enteritidis* was destroyed during baking of cheesecake (pH 4.9, a_w 0.98), but when higher numbers (10^6 CFU/g) were inoculated into cheesecake, some *Salmonella* survived baking. In custard, a temperature of 60°C for 19 min was required to kill 10^7 CFU/g of *Salmonella*. However, if the temperature was increased to 65.7°C, similar levels of *Salmonella* could be inactivated in only 3.5 min (Angelotti et al., 1961). Sumner et al. (1991) studied the heat resistance of *S. typhimurium* in sugar syrups having a_w of 0.98 to 0.83 and found that at 65.6°C, the decimal reduction time ($D_{65.6°C}$) varied from less than 30 seconds to more than 40 min. Simply reboiling custard fillings after thickening ingredients were added, or rebaking pies after filling, was an effective method of inactivating *Salmonella* spp. (Cathcart et al., 1942). Regardless of the heat treatment, it is important that minimally processed bakery products are cooled rapidly to 2 to 4°C within 30 min of preparation.

Strict temperature control is also critical to ensure the safety of minimally processed bakery ingredients containing eggs or fresh cream. Refrigeration (4°C) of cream-filled pastries has been shown to prevent the growth of *Salmonella* spp. (Bryan, 1976); however, studies have shown that *Salmonella* spp. remain viable during refrigerated or frozen storage (Schmidt and Ridley, 1985). Therefore, temperature abuse at any stage during processing, distribution, and storage may compromise the safety of minimally processed bakery products or their ingredients. The importance of strict temperature control cannot be overemphasized because *Salmonella* spp. can grow at temperatures as low as 6°C. Additional barriers, such as potassium sorbate (2500 ppm), may be used to limit the growth of *Salmonella* spp. (Wyatt and Guy, 1981a).

Strict personal hygiene and good manufacturing practices are also critical to reduce cross-contamination of products by *Salmonella* spp. and limit the spread of this food-borne pathogen.

STAPHYLOCOCCUS AUREUS

Sources of Contamination

Staphylococcus aureus is a Gram-positive, facultatively anaerobic bacterium of which 50 to 70% of strains are estimated to be enterotoxigenic (Tranter, 1990). Staphylococcal food poisoning results in susceptible individuals following consumption of food in which *S. aureus* has grown to sufficient numbers (~10^6/g) to produce enterotoxin. Enterotoxin A is implicated in 75% of *S. aureus* outbreaks; however, outbreaks are occasionally attributed to enterotoxin B (Bergdall, 1989). The illness is characterized by nausea, vomiting, abdominal pain, and diarrhea, 2 to 6 h after eating the food containing staphylococcal enterotoxin. Recovery takes 1 to 3 days and death is rare (Tranter, 1990). Because symptoms of the illness are usually mild, medical attention is not always sought. Therefore, in many countries staphylococcal food poisoning is under-reported. It has been estimated that only 1 to 5% of cases are reported, and only when a large number of individuals is involved in an outbreak (Jablonski and Bohach, 1997).

The major reservoirs of *S. aureus* are humans and animals. *S. aureus* is carried by 30 to 50% of humans in nasal passages and throats, and on skin (Bergdoll, 1989). Humans harboring *S. aureus* therefore, are a major source of contamination of products during preparation or postpreparation handling. *S. aureus* is also ubiquitous in air, water, milk, sewage, and on food contact surfaces (Jablonski and Bohach, 1997). Postpreparation contamination is also possible from air, surfaces, and cross-contamination. Ingredients may also be a source of high numbers of *S. aureus*. Milk from mastitic cows can contain high levels of *S. aureus*. Although milk in North America is generally screened for mastitis, milk in some temperate countries is a source of *S. aureus* (Bergdall, 1989). In Brazil, surveys have reported an average of 10^3 CFU/g of *S. aureus* in milk, while 50% of pasteurized milk and cream have been shown to be contaminated with this pathogen (Santos et al., 1981). Other ingredients, such as reconstituted dried milk, whey and cream, have also been found to contain *S. aureus* enterotoxin (Bergdall, 1989).

Associated Outbreaks

In the U.S., cream-filled bakery products made from fresh or synthetic creams and custard fillings were once the main cause of all food poisoning outbreaks involving *S. aureus*. Due to good manufacturing practices, these products are now seldom implicated in outbreaks of *S. aureus* food poisoning; however, they continue to be a major source of illness in many temperate countries where refrigeration is still a problem (do Carmo and Bergdoll, 1990). Outbreaks in Brazil involving 122 cases were caused by eating tainted cream puffs; other outbreaks have been attributed to chocolate éclairs served on a flight from Rio De Janero to New York City. An outbreak of staphylococcal food poisoning, involving 215 cases, also occurred on a Caribbean cruise ship sailing from the U.S., with cream pastries being the implicated food (CDC, 1983; Waterman et al., 1987). One survey found that 55% of cream pies stored at room temperature from commercial establishments in Brazil were contaminated with *S. aureus* and 19.5% of the products contain levels >10^5 CFU/g (Anunciaçao et al., 1995). Desai and Kamat (1998) reported that 85.7% of seven pastry and biscuit creams from commercial establishments in India were positive for *S. aureus*, and had average counts of ~10^4 CFU/g. Leela et al. (1981) found that enterotoxigenic staphylococci in bakery products in India were always associated with cream and coconut fillings. *S. aureus* (3×10^1 to 6.5×10^4 CFU/g) producing enterotoxin A, B, and E were found in cakes, sweet puffs, vegetable puffs, and cream buns from five Indian bakeries. Breads and buns from the same bakeries were negative for *S. aureus* (Sankaran and Leela, 1983).

These outbreaks show that cream-type fillings are an excellent medium for growth of *S. aureus*. In cream pie fillings inoculated with 10^2 CFU/g of *S. aureus* FR-100, enterotoxin (3.9 μg/g) was first detected after 35 h at 20°C. In fillings stored at 37°C, 50 μg/g of enterotoxin was present after 14 h (Hirooka et al., 1987). McKinley and Clarke (1964) demonstrated that imitation cream fillings used in bakery products were also capable of supporting growth of enterotoxigenic strains of *S. aureus*. Although imitation cream on its own does not contain sufficient nutrients to support the growth of this pathogen, growth can occur at the interface of the cream and the baked product (McKinley and Clarke, 1964). Surkiewicz (1966) demonstrated that imitation cream pies spoiled within 48 h at ambient temperature and contained counts of *S. aureus* up to 10^6 CFU/g.

Other products have been shown to be contaminated with this pathogen. Sumner et al. (1993) isolated *S. aureus* from 9.8% of 214 bakery products, including oatmeal raisin cookies, apple muffins, cream puffs, and long johns. The a_w of the cookies was too low for growth of *S. aureus*; however, 3.3% of apple muffins, 30% of cream puffs, and 11% of long johns were positive for *S. aureus* and enterotoxin was produced by seven isolates. Concern was raised over enterotoxigenic strains of *S. aureus* isolated from low-moisture products, i.e., muffins, which are generally considered to be low risk products. This concern would seem justified since dry, flat bread was responsible for an outbreak of staphylococcal food poisoning in Norway (Aalvik and Haave, 1980).

Pizza is also frequently involved in food-borne outbreaks involving *S. aureus* (Todd, 1996). Contamination usually occurs through poor manufacturing practices and *S. aureus* can grow in high salt concentrations found in many pizza ingredients.

Although *S. aureus* is destroyed by heating, the enterotoxin is heat resistant and is not inactivated by pasteurization (Bergdoll, 1989). Hand-made tortillas had counts of approximately 10^8 CFU/g after preparation. Reheating of products resulted in only a 1- to 2-log reduction in counts and *S. aureus* enterotoxin survived the heating process (Capparelli and Mata, 1975). Therefore, food poisoning outbreaks caused by *S. aureus* may still occur, even in the absence of viable cells, if preformed enterotoxin is present in the product as a result of temperature abuse of ingredients or fillings prior to baking. Potential problems also exist if contaminated uncooked fillings and toppings, e.g., whipped cream, are used. *S. aureus* is a poor competitor, however; if contamination occurs after heating during which most other vegetative organisms have been destroyed, conditions will favor rapid growth.

An increasing variety of goods held at ambient temperature is available from in-store bakeries. These products are often handled manually in self-serve bins by employees and customers. This increases the potential for contamination by *S. aureus* and subsequent growth or enterotoxin production at ambient storage temperature.

Control Measures

The number of staphylococcal food poisoning outbreaks attributed to the consumption of cream- and custard-filled pastries in the U.S. has dramatically decreased in recent years due to improved sanitation, temperature control, modification of product formulations, and use of preservatives (Elliot, 1980). Good manufacturing practices (GMPs) have effectively reduced the level of contamination by *S. aureus* of frozen cream pies in North America. In a survey of the frozen cream pie industry, the Food and Drug Administration concluded that operating under good sanitary conditions resulted in products with no *S. aureus* in 0.1-g samples (Surkiewicz, 1966). A survey of all plants manufacturing frozen pies in the U.S. reported that levels of *S. aureus* were consistently <10 CFU/g. Guidelines have also been developed in Canada (Todd et al., 1983). Warburton and Weiss (1986) found that aerosol cream filling substitutes and powdered whipped toppings contained <5 CFU/g of *S. aureus* due to good manufacturing practices.

Methyl parabens, sodium benzoate, potassium sorbate, and calcium propionate have also been shown to inhibit *S. aureus* at pH of 5.2 to 5.6 and a_w of 0.86 to 0.90 (Boylan et al., 1976). Schmidt et al. (1969) found that potassium sorbate and sodium benzoate (1000 ppm) inhibited the growth of *S. aureus* in synthetic cream pies at 22 and 37°C at pH of 4.5 to 5.0, but not at pH of 5.2. Growth of *S. aureus* was found to be inhibited in fillings made from chocolate or cocoa. Although inhibition was initially attributed to a reduction in pH, substances in the non-fat part of the chocolate were later found to be responsible for inhibition of *S. aureus* (Cathcart and Merz, 1942). DL-serine has also been shown to inhibit the growth of this pathogen in cream pie fillings (Castellani, 1953).

Heating immediately after preparation can destroy *S. aureus* in fillings and filled products. Stritar et al. (1936) reported that heating custard-filled pastries previously inoculated with 10^5 CFU/g of *S. aureus* to 190.6°C for 25 min resulted in complete inactivation. Other temperature–time processes have been recommended (Gilcrease and Colman, 1941; Husseman and Tanner, 1947; Angelotti et al., 1961; Keller et al., 1978). Cathcart et al. (1942) found that simply reboiling custards after the thickening mix had been added also destroyed the inoculated *S. aureus*. However, care must be taken to avoid subsequent recontamination.

Strict temperature control (4°C) of finished products and their ingredients will inhibit growth and enterotoxin production by *S. aureus* (Bryan, 1976). Bergdoll (1989) reported that cakes held at room temperature resulted in food poisoning, while those held at 4°C did not. Cream-filled cakes inoculated with 10^6 CFU/g of an enterotoxin-producing strain of *S. aureus* did not support growth, nor was enterotoxin detected in cakes held under refrigeration. At room temperature (27 to 29°C), enterotoxin was detectable in cakes inoculated with 10^2 and 10^3 CFU/g after only 24 and 18 h, respectively. Such time–temperature abuse is not uncommon for cream-filled bakery products (Anunciaçao et al., 1995).

Because *S. aureus* can grow facultatively, MAP will not inhibit its growth or enterotoxin production (Hintlian and Hotchkiss, 1986). Park et al. (1988) demonstrated the importance of refrigerating fresh pasta stored under modified atmospheres. In pasta stored at 5°C, counts of *S. aureus* declined and no enterotoxin was detected. For fresh MAP pasta products with a_w <0.95, reductions in the product's a_w and pH is recommended as additional safety barriers against the growth of *S. aureus* (Castelvetri, 1991). Some products, such as meringue and butter creams, need to be formulated to an a_w of <0.86 because *S. aureus* can tolerate and even grow in products with such a high sugar content (Preonas et al., 1969). However, pockets of high moisture, and hence high a_w, may still occur within such products and encourage the growth of this pathogen.

Bacillus Species

Sources of Contamination

Bacillus cereus has been implicated in several outbreaks of food-borne illnesses involving bakery products. There is also evidence that *B. subtilis* and *B. licheniformis,* well known as rope-forming spoilage bacteria of bread, can also cause food-borne illness (Todd, 1982; Kramer and Gilbert, 1989; te Giffel et al., 1996). *B. cereus* causes two distinct forms of toxin-mediated gastroenteritis: the emetic type is generally associated with cereal-based foods, while the diarrheal type is most frequently associated with proteinaceous foods (Lund, 1990).

Bacillus species form spores, which are ubiquitously found in soils, dust, and water, and are commonly isolated from plants and animal products (Granum, 1997). *Bacillus* spp. attach to wheat milled into flour (Kirschner and Von Holy, 1989). Graves et al. (1967) examined flour from 11 U.S. mills and found that 18.2% of samples contained *B. cereus*, 9.1% contained *B. licheniformis*, and 45.3% contained *B. subtilis*. The level of contamination of *B. cereus* in wheat flour was generally less than 10^3 spores/g (Kaur, 1986; Kim and Goepfert, 1971; Rizk and Ebeid, 1989). Therefore, *Bacillus* spores are commonly found in flour and flour-based products as well as in the bakery environment. These spores are heat resistant and will survive baking and, under favorable conditions, may grow to levels associated with toxin production. Survival of spores during baking depends on the type of product and the internal temperature reached during baking, as well as the thermal resistance of the spore.

Bacillus spores are commonly found in milk and can survive pasteurization. Hence, they are a source of concern in dairy products such as cream, dried milk (Slaghuis et al., 1997; Larsen and Jorgensen, 1997), and whey concentrates (Pirttijarvi et al., 1998). Although *B. licheniformis* was initially found in higher numbers than *B. cereus* when milk products were held at room temperature, *B. cereus* rapidly dominated and reached levels associated with enterotoxin production (Crielly et al., 1994). Harmon and Kautter (1991) reported a six- to sevenfold increase in *B. cereus* in reconstituted non-fat dried milk held at room temperature for 10.5 h. Cold-adapted strains can even produce toxin at refrigerated temperature (Foegeding and Berry, 1997), especially in aerated products such as whipped cream (Christiansson et al., 1989).

Spices such as ginger, mace, allspice, cinnamon, garlic, and pizza spice usually contain low levels of *Bacillus* spores; however, higher levels of spores associated with toxin production have also been found in these bakery ingredients (Powers et al., 1976; Pafumi, 1986; Kneifel and Berger, 1994).

Other bakery ingredients that may be sources of *Bacillus* spp. include dried eggs (Shafi et al., 1970), soy protein (Becker et al., 1994), rice (Chung and Sun, 1986), yeast and improvers (Collins et al., 1991; Bailey and von Holy, 1993; te Giffel et al., 1996), dried fruits (Moreno et al., 1985; Aidoo et al., 1996), and cocoa (Gabis et al., 1970; te Giffel et al., 1996).

Associated Outbreaks

Approximately 10^5 spores of *B. cereus*/g of food are required to produce toxin; higher spore levels (10^6 to 10^9 spores/g) are required for *B. licheniformis* and *B. subtilis* (Lund, 1990). Food-borne illness caused by *Bacillus* spp. is under-reported because symptoms are generally mild and self-limiting (Terranova and Blake, 1978).

Although low numbers of spores may be present initially in flour (Kaur, 1986), spores that survive baking can grow rapidly in products held under suitable conditions (Rosenkvist and Hansen, 1995). *B. cereus* does not easily survive the baking of bread, but *B. subtilis,* which has a $D_{100°C}$ of 14 min, has been found in baked bread (Leuschner et al., 1998). Kaur (1986) reported that *B. cereus,* inoculated at ~10^4 spores/g, did not survive baking in 400-g loaves but survived baking in 800-g loaves. Although the actual number of *B. cereus* that survives baking is dependent on the loaf size and oven temperature–time combinations, the risk of food-borne illness from bread is minimal (Kaur, 1986; Collins et al., 1991; Rosenkvist and Hansen, 1995; Thompson et al., 1993). Kaur (1986) estimated that it would take 3 days at 27.5°C for *B. cereus* that had survived the baking process to reach levels of 10^5 CFU/g in bread.

Some strains of *B. subtilis* and *B. licheniformis* cause "ropiness," resulting in the sensory rejection of a product. However, this is not always the case and bread may have high numbers of *B. subtilis* yet still be organoleptically acceptable to the consumer. Illness attributed to consumption of bread with high numbers of *B. subtilis* or *B. licheniformis* has been reported (Todd, 1982; Sockett, 1991; Thompson et al., 1993). Therefore, control of *B. cereus*, *B. subtilis*, and *B. licheniformis* is critical to the safety of bread products.

Although the potential health hazard of contamination of bread by *B. cereus* is minimal, growth of this pathogen is of greater concern in bakery products that receive a minimal surface heat treatment. Examples of such griddle-baked bakery products include ethnic flat breads, crumpets, and waffles. Outbreaks of *B. cereus* gastroenteritis have been attributed to Naan bread (Cowden et al., 1995), crumpets (Lee, 1988), and pikelets (Murrell, 1978). Growth of *B. cereus* in these products is difficult to control because the heat treatment is insufficient to destroy spores and may actually heat shock spores, thus enhancing their growth at ambient storage temperature. There is also concern for growth in dry, reconstituted rice cereals particularly when reconstituted with milk (Jaquette and Beuchat, 1998).

B. cereus has also caused illness in bakery products containing dairy-based custards or creams. Pinegar and Buxton (1977) showed that custard used in the production of vanilla slices contained low numbers of spores (<500/g). However, immediately after cooking, levels of *B. cereus* increased to >10^6 CFU/g. *B. cereus* has also been isolated from prepared bakery products, including meat-filled bakery products (Khot and Sherikar, 1986), bread (Sadek et al., 1985), pies, and pastry (Pinegar and Buxton, 1977; Wyatt and Guy, 1981b).

Control Measures

Conventional methods of control include proper sanitation and testing of raw materials to reduce initial spore counts; however, these measures do not prevent germination and growth of *Bacillus* spp. in finished products. Growth of *Bacillus* species in baked products can be controlled with preservatives. Propionic acid, calcium or potassium propionate, and calcium acetate can delay germination and growth of some *Bacillus* spp., particularly rope producers (Kaur, 1986; Kirschner and Von Holy, 1989; Rosenquist and Hansen, 1998). Thompson et al. (1998) reported that white bread containing vinegar was more effective at preventing rope production than calcium propionate. Potassium sorbate, at a level of 2500 ppm, prevented growth of *B. cereus* in pumpkin pie (Wyatt and Guy, 1981b), while 2000 ppm of sorbic acid or 4000 ppm of potassium sorbate inhibited *B. subtilis* and *B. cereus* in the rice filling of Karelian pastry (Raevuori, 1976).

Lactic acid bacteria may also inhibit *Bacillus spp.* Corsetti et al. (1996) reported that 33% of 232 lactobacilli isolated from Italian sourdough inhibited strains of *B. subtilis*. Control of *B. subtilis* and *B. licheniformis* in bread was achieved by the addition of 10 to 15% sourdough fermented with *Lactobacillus plantarum* or *Lactobacillus sanfrancisco* L99. Bread had an increased acidity due to production of lactic and acetic acids and resulted in an increased shelf-life without the addition of preservatives (Rosenquist and Hansen, 1998). Some strains of lactic acid bacteria also produce bacteriocins. Nisin is a commercially available bacteriocin produced by *Lactococcus lactis* and is the only bacteriocin shown to have an antimicrobial effect against *Bacillus* spp. Growth of *B. cereus* in high-moisture English style crumpets was controlled in crumpets for up to 5 days at ambient storage temperature by the addition of 1 to 5 μg/g of nisin (Jenson et al., 1994). However, nisin at levels up to 100 μg/g had no effect on the growth of *B. subtilis* or *B. licheniformis* in wheat bread (Rosenquist and Hansen, 1998).

Spores of *Bacillus* are known for their heat resistance. Typical $D_{100°C}$ values for *B. cereus, B. subtilis*, and *B. licheniformis* have been reported as 40, 14, and 56 min, respectively (Wyatt and Guy, 1981b; Leuschner et al., 1998). Therefore, spores can readily survive the minimal heat processing conditions in baking. Thermal resistance may be decreased by decreasing the pH of the product. The D value of *B. cereus* spores decreased from 3.7 to 3.1 min at 90°C when the pH of custard was decreased from 7.2 to 6.2 (Bassen et al., 1989). A high pH (>9) may also provide some measure of control; however, very few bakery products are formulated to such pH levels (Leuschner et al., 1998; El-Khoury, 2000).

Cold storage may be a suitable control measure for some products, such as those containing cream and custard. However, up to 14% of *B. cereus* strains may be psychrotrophic (Granum, 1997) and so temperature alone is not a practical control measure. Custards and fillings should be prepared in small batches, cooled rapidly, and stored at 4°C.

Modified atmosphere packaging with CO_2 cannot control the growth of *Bacillus* spp. alone. Smith et al. (1983) reported that *B. subtilis* and *B. licheniformis* grew to 5×10^8 CFU/g after 3 days at 37°C or after 2 weeks at 20°C in crumpets packaged under modified atmospheres. However, combination treatments have proven effective in controlling this pathogen. El-Khoury (2000) controlled the growth of *B. cereus* in crumpets stored at 30°C for >28 days using a combination of 100% CO_2 and 600 S Negamold® an oxygen absorbent-ethanol vapor generator. However, products were organoleptically unacceptable due to the high levels of ethanol absorbed from the package headspace (El-Khoury, 2000).

Sutherland et al. (1996) examined the combined effects of ambient temperature, pH, salt, and CO_2 on the growth of *B. cereus* in carbohydrate-based foods. Quintavalla and Parolari (1993) also modeled the effects of pH, a_w, and temperature on the growth of *Bacillus* spp. isolated from bakery products. They reported a 12- to 15-day shelf-life for products stored at 20°C and reformulated to pH 5.2 and a_w 0.93. Reformulation to pH 4.3 and a_w 0.92 extended product shelf-life to 30 days at a similar storage temperature. However, such models are limited because the data cannot always be extrapolated to the food due to the complex nature of the food matrix.

CLOSTRIDIUM BOTULINUM

Sources of Contamination

Food-borne botulism is a rare, potentially fatal neuromuscular illness that can result from the consumption of food in which *C. botulinum* has grown and produced toxin (Hatheway, 1992; Dodds, 1992). Infant botulism results when germinating spores of neurotoxigenic clostridia colonize the infant intestine and produce neurotoxin *in situ* (CDC, 1998b). Botulism has also resulted from similar colonization in people who are immunocompromised.

The most likely contaminants of bakery products are *C. botulinum* type A and B spores since these are ubiquitously found in soil and in agricultural and animal products. Few surveys have been carried out to determine the levels of *C. botulinum* in finished bakery products; however, the levels of spores in bakery ingredients have been documented. *C. botulinum* spores have never been reported in flour (Elliot, 1980) and are rarely found in raw milk (Collins-Thompson and Woods, 1983). A recent survey of dairy foods in Italy found no spores of *C. botulinum* in raw, pasteurized, or clotted milk, butter, pasteurized cream, or ricotta cheese. However, higher levels (<10 spores/g) were found in mozzarella, soft, and processed cheeses commonly used as pizza ingredients. Furthermore, 32% of 1017 samples of mascarpone cream cheese were contaminated with *C. botulinum* spores (Franciosa et al., 1999). Cheese previously involved in an outbreak of food-borne botulism (Aureli et al., 1996) also contained neurotoxin type A (Franciosa et al., 1999). Fruits, vegetables, and spices may also be contaminated with *C. botulinum* spores. Outbreaks of food-borne botulism have been caused by a variety of fruits and vegetables (peppers, tomatoes, potatoes, mushrooms, onions, garlic, olives, peanuts, and hazelnut puree) used in the production of sweet and savory bakery foods (Notermans, 1992; Hauschild, 1989). Surveys of honey have shown levels of contamination of <1 to 10^1 spores/kg (Dodds, 1992), although higher levels (10^3 to 10^4 spores/kg) have been reported in honey associated with infant botulism (Hauschild, 1988).

Associated Outbreaks

Spores of *C. botulinum* are heat resistant, with proteolytic spores having a reported $D_{100°C}$ value in the range of 25 min (Dodds and Austin, 1997). Since the internal temperature of bread during baking rarely reaches >100°C, proteolytic spores of *C. botulinum* will survive baking (Soloski and Cryns, 1950; Marston and Wannan, 1976) and have the potential to germinate, grow, and produce toxin in products stored under favorable conditions. Daifas et al. (1999a) showed that the heat treatment required to bake crumpets of acceptable quality represented only 1% of the treatment necessary for complete spore destruction. Whether spores are present in ingredients prior to baking or in fillings or toppings, or result from environmental contamination, they have the potential to germinate, outgrow, and produce neurotoxin if conditions are suitable.

Bakery products with a pH <4.5 and an a_w >0.95 can support growth of *C. botulinum*. Other conditions that may allow for growth and toxin in bakery products include gas packaging, which excludes oxygen, vacuum packaging in cans or jars, or packaging in airtight conditions in which spoilage organisms may consume residual headspace oxygen (Aramouni et al., 1994; Austin et al., 1998; Daifas et al., 1999a, b). However, the inclusion of oxygen in the package atmosphere does not ensure the safety of MAP products. Studies have shown that toxin may be formed in foods containing high levels of oxygen in the package headspace (20 to 100% v/v) and even in food stored aerobically (Snyder, 1996; Clavero et al., 2000; Dufresne et al., 2000).

The potential of bakery products to support the growth of proteolytic strains of *C. botulinum* has been well established in inoculation experiments (Edmondson et al., 1923). Bever and Halvorson (1947) also demonstrated that neurotoxin could be produced by *C. botulinum* in a medium of sterile bread crumbs containing calcium or sodium propionate at levels of 0.2 to 1.4% when the pH of the medium was between 4.5 and 9.6. Concern was also expressed about the safety of canned bread intended for military rations (Soloski and Cryns, 1950). Subsequent studies confirmed that canned bread could support growth and neurotoxin production and determined that the conditions necessary for the safe production of this product would be a pH <5.4 and a moisture content of 35% corresponding to an a_w of ~0.95 (Soloski and Cryns, 1950; Kadavy and Dack, 1951; Ulrich and Halvorson, 1949). However, increasing the pH of the product resulted in toxigenesis.

Growth and toxin production was observed in inoculated canned bread of pH 5.8 and a moisture content >36% and in canned steamed chocolate nut bread of pH 6.8 and 36% moisture (Wagenaar and Dack, 1954, 1960). Bread of pH 4.8 was determined to be safe because no growth occurred and the number of inoculated spores decreased during ambient storage in an equilibrium relative humidity (e.r.h.) of 97% (Ingram and Handford, 1957). However, the safety of the bread depended on maintaining its pH throughout storage because viable spores still existed after 6 months. Temperature differences within the canned bread that occurred during freezing, thawing, or heating resulted in moisture migration and uneven moisture distribution in the bread (Weckel et al., 1964). The pH of thawed canned bread was unchanged, but the pH of the moisture that developed within the can was high enough (pH 5.5 to 6.7) to support the growth of *C. botulinum*. These early studies demonstrated the need for multiple barriers in a food product to ensure its safety, particularly for products stored at ambient temperature. Bread is no longer canned for military rations; however, canned breads are commercially available in Japan and South Africa (Lombard et al., 2000).

Home-style canned quick breads containing fruits and vegetables are hermetically sealed in Mason jars and then baked. Such breads are available by mail order and are promoted in newspapers and on the Internet; a shelf-life at room temperature of 6 months or longer is claimed. Aramouni et al. (1994) demonstrated the survival of inoculated spores of *C. sporogenes* in canned home-style banana bread and concluded that further work was needed to determine safe processing procedures for this type of product. More recently, studies in the authors' laboratory have shown that quick breads, containing zucchini, apple, and banana and packaged under modified atmospheres, may be a potential hazard. When challenged with proteolytic spores of *C. botulinum,* packaged with an oxygen absorbent, and stored at room temperature, products became toxic within 28 days (unpublished results).

More recently, English style crumpets, a high-moisture snack food that has been implicated in several outbreaks of *B. cereus* in Australia, were investigated for their potential to support the growth of *C. botulinum.* Crumpets were inoculated with proteolytic strains of *C. botulinum,* packaged in various gas atmospheres (60% CO_2/40% N_2, in air, or in air with an oxygen absorbent), and stored at ambient temperature. *C. botulinum* increased from 2.6 log CFU/g to 7 log CFU/g and produced toxin in less than 1 week in all crumpets, regardless of the packaging atmosphere (Daifas et al.,1999b). Sensory analysis indicated that crumpets were sensorially acceptable at the time of toxin detection (Daifas et al., 1999b). When crumpets were packaged in 100% CO_2, toxin production was slightly delayed, regardless of the crumpet pH (6.5 and 8.3); however, toxin production again preceded spoilage (Daifas et al., 1999c).

These studies confirm earlier studies that bakery products are a suitable substrate for the growth of *C. botulinum.* Furthermore, they demonstrate that if high bakery moisture products are contaminated with *C. botulinum* spores, they could present a serious health hazard because they would be toxic yet still acceptable to the consumer. Fortunately, no outbreaks of botulism caused by bakery products have been reported in North America. However, in India, a suspected outbreak of food-borne botulism affecting 34 children, including three fatalities, was attributed to the growth and production of neurotoxin type E by *C. butryicum* in sevu, a crisp flat bread prepared from gram (pulse) flour. Growth and neurotoxin production was attributed to improper storage of the implicated crisp bread (Chaudhry et al., 1998). In Italy, an outbreak of botulism resulted from consumption of tiramisu, a bakery confectionery product made with contaminated mascarpone cheese (Aureli et al., 1996).

Control Measures

Although cases of food-borne botulism are rare, the severity of the intoxication means that the potential for growth and toxin production in a bakery product should be considered and additional barriers to growth and toxin production by this pathogen need to be incorporated into the product and/or packaging. Based on previous empirical data (Castelvetri, 1991), reformulation of bakery products to an a_w of ~0.95 and a pH of ~5.7 should prove effective against growth of *C. botulinum* and should ensure the safety of most bakery products stored at ambient temperature.

The a_w of bakery products is one of the most important factors influencing the growth of, and toxin production by, *C. botulinum.* Denny et al. (1969) demonstrated the importance of a_w on the growth of proteolytic *C. botulinum.* In seven varieties of inoculated canned, low-acid fruit and vegetable bread stored for up to 6 years, toxin was produced at a_w >0.95 but not an a_w of <0.95. *C. botulinum* did not grow or produce toxin in low-acid, apple coffee cake or spice cake with an a_w of 0.93 when inoculated with 10^5 spores/g and incubated at 30°C (Powers et al., 1988). More recently, high-moisture, low-acid crumpets, pizza, and bagels were inoculated with 10^4 spores/g of *C. botulinum* and stored at ambient temperature. Crumpets (a_w 0.990) and pizza (a_w 0.960) became toxic while bagels (a_w 0.944) did not (Daifas et al., 1999a).

Another barrier that can be used to enhance the safety of bakery products is pH. Growth and toxin production was inhibited in cooked shelf-stable noodles acidified to pH <4.6. However, toxin production occurred in one sample of noodles in which the pH had increased as a result of microbial growth demonstrating the importance of incorporating multiple barriers on the growth of *C. botulinum* (Ikawa, 1991). Daifas et al. (1999c) also showed that growth of, and toxin production by, *C. botulinum* could be inhibited in high-moisture crumpets stored at ambient temperature if products were reformulated to pH 4.6, regardless of the packaging atmosphere.

Well-known for its antimycotic effect, ethanol has also been investigated as a potential barrier to inhibit the growth of *C. botulinum*. Recent studies have confirmed the antibotulinal action of ethanol vapor in crumpets challenged with proteolytic strains of *C. botulinum* and packaged with ethanol vapor generators (Ethicap®) sizes 2, 4, and 6G (Daifas et al., 2000). Toxin was detected after 5 days in all English style crumpets inoculated with 500 spores/g of *C. botulinum* and packaged in air. Packaging with Ethicap (2G) delayed toxin for 10 days at ambient temperature while complete inhibition for 21 days was observed in all crumpets packaged with Ethicap 4 or 6G. However, all crumpets were organoleptically unacceptable due to their absorption of ethanol (>2% w/w) from the package headspace. Nevertheless, these initial studies have shown that, in the vapor phase, ethanol has the potential to enhance the safety of high-moisture English-style crumpets stored at ambient temperature (Daifas et al., 2000).

OTHER MICROORGANISMS OF CONCERN

LISTERIA MONOCYTOGENES

Listeria monocytogenes is widespread in nature, occurring in soil, vegetation, and water as well as in many animal and plant products (Lovett and Twetd, 1988). Furthermore, it can grow over wide pH, a_w, and temperature ranges. Outbreaks of listeriosis have resulted from consumption of soft cheese, cream, and butter (Linnan et al., 1988; Lyytikäinen et al., 2000). *L. monocytogenes* is therefore of concern in bakery products containing dairy ingredients. *L. monocytogenes* is also classified as a psychrotrophic pathogen and can readily grow at refrigerated storage temperatures at which cream-, cheese-, or butter-based iced or filled pastries and cakes are stored and consumed without heating. Furthermore, since *L. monocytogenes* is ubiquitous in the bakery environment, postprocessing contamination of finished products is possible.

Gifford et al. (1991) analyzed 28 bakery products for *L. monocytogenes*; however, no products were contaminated with this pathogen. In another survey of 300 pastries from 100 bakeries in France, Ferron and Michard (1993) found that 14% of pastries were contaminated with *L. monocytogenes*. One sample contained 7×10^5 CFU/g — levels that can cause listeriosis. The authors concluded that the risk of listeriosis from pastry was at least equivalent to that from meat and delicatessen products.

Richter et al. (1991) demonstrated that heating flour inoculated with *L. monocytogenes* for 5 min at 80°C was sufficient to destroy this pathogen. However, when inoculated, nonthermally treated flour was stored at room temperature, *L. monocytogenes* survived for at least 2 weeks. *L. monocytogenes* is also unlikely to survive boiling of custards used in the preparation of certain filled products.

Strict hygiene and temperature control is essential to prevent contamination and growth of this pathogen. However, because temperature abuse is commonplace in the food industry, temperature alone is not an adequate barrier to ensure the safety of products contaminated with *L. monocytogenes*. Furthermore, it can grow under aerobic, microaerophilic, and anaerobic conditions, as well as in the presence of elevated levels of CO_2. *L. monocytogenes* can grow at low pH/a_w levels (~5.2, 0.92, respectively), so product reformulation is not commercially viable because it may result in sensory and textural changes in the finished product.

Mycotoxigenic Molds

Although bacteria are most commonly implicated in outbreaks of food-borne illness, molds, which often limit the shelf-life of high and intermediate bakery products, can also be of public health concern. Although moldy products will be rejected by consumers, molds may secrete mycotoxins into bakery products without visible signs of spoilage. Some molds, including *Alternaria, Aspergillus*, *Fusarium*, and *Penicillium* spp., may secrete one or more toxins that may be teratogenic and carcinogenic (Visconti and Bottalico, 1983; Tabibi and Salehian, 1974).

Mycotoxins have been found in many foods, including cereals and grain products, seeds, nuts, fruits, vegetables, and dairy products (Malloy and Marr, 1997). In a survey of flour, Weidenborner et al. (2000) found that *Aspergillus* species were the predominant isolates, of which 93.3% of isolates were toxigenic. Increased mycotoxin levels in grains are influenced by climatic conditions (drought, rain) and improper storage. Abouzied et al. (1991) surveyed 92 samples of grains from retail stores for aflatoxin B1, zearalenone, and deoxynivalenol (DON [vomitoxin]). Only one sample of buckwheat flour was positive for aflatoxin B1, but zearalenone was found in 26% of samples. DON was found in 50% of samples at levels in excess of the FDA-recommended level of 1 ppm. Furthermore, 88% of corn cereals, wheat flour, muffin mixes, and rice cereals tested positive for DON (Abouzied et al., 1991).

Control measures include proper storage of wheat to avoid moisture pick-up and moisture migration thereby preventing mold growth. However, outbreaks of mycotoxicosis have occurred from flour made from rain-damaged wheat (Bhat et al., 1989). Antimycotic agents, such as propionates and sorbates, greatly reduce the risk of mold growth and mycotoxin production in bread (Lennox and McElroy, 1984). Genetic engineering is also playing a role in the production of grains with increased resistance to DON-producing molds (Campbell et al., 2000).

Viruses

Norwalk-like viruses (NLVs), also known as small, round-structured viruses (SRSV), are ubiquitous in the environment and are highly resistant. They can cause viral gastroenteritis illness from foods prepared or handled under unsanitary conditions (Hedberg and Osterholm, 1993). An active carrier of NLVs can contaminate large quantities of food in a short time period; therefore, outbreaks can involve a large number of cases. One outbreak in Minnesota, which involved 3000 cases, was attributed to the consumption of frosted bakery products (MacDonald and Griffin, 1986; Kuritsky et al., 1984). Norwalk-like viral gastroenteritis, involving U.S. Army trainees in Texas, was associated with the consumption of contaminated crumb cake, pie, and rolls (Arness et al., 1999). Another outbreak, involving 250 cases of school children, was associated with hamburger buns and cookies contaminated by symptomless excreters of the virus (Bean et al., 1996). Custard slices prepared from dried custard mix and, inadvertently, with contaminated water were associated with a large-scale community outbreak of Norwalk virus (Brugha et al., 1999).

Another virus that can be transmitted from infected handlers to bakery products is Hepatitis A. Outbreaks of Hepatitis A have been attributed to unbaked sherry trifle (Chaudhuri et al., 1975), breads, rolls and sandwiches (Sockett et al., 1993), and pastries covered with glaze or icing applied after baking (Schoenbaum et al., 1976). More recently, a community outbreak in 1994 in New York was traced to an infected bakery worker who contaminated cooked doughnuts while applying a sugar glaze (Weltman et al., 1996).

Reduction of viral transmission from infected bakery workers to food depends on education, good personal hygiene, and proper sanitation. Furthermore, all ill workers should be excluded from production areas and, if possible, food handlers should be immunized.

CONCLUSION

Despite improvements in technology and manufacturing practices and stricter food hygiene and safety regulations, the number of food-borne illnesses remains fairly constant each year. Minimally processed bakery products have contributed to this trend. The implication of bakery products in food-borne illnesses outbreaks can be attributed to a number of causes (Farber, 1989; CDNANZ, 1997), including:

1. Bakery products containing lower levels of traditional hurdles to microbial growth, i.e., preservatives, salt, sugar in response to consumer demands for low-calorie, preservative-free, "fresh" food
2. Novel methods of packaging, such as modified-atmosphere packaging, that may promote the growth of certain pathogenic bacteria
3. Increased time between preparation and consumption of bakery products, particularly cream- and meat-filled products, which increases the potential of temperature abuse
4. Increased globalization in sources of raw materials, food production and distribution, and centralization of food operations

5. Emerging pathogens and changes in mechanisms of transmission, infective doses, and microbial resistance to temperature, acid, and antimicrobial agents
6. Increased surveillance and awareness of food-borne illnesses

Although poultry, eggs, meat, fish, and dairy products are the main vehicles of transmission of most pathogenic microorganisms in food, constant vigilance is still required to ensure the safety of minimally processed bakery products, particularly those containing meat or cream. This is even more critical in view of the above-mentioned reasons and globalization of the food industry. Recommended guidelines proposed by regulatory agencies on the use of multiple barriers should also be incorporated into high-risk minimally processed bakery products to ensure their shelf-life and safety. Continued education of bakery personnel in food hygiene and good manufacturing practices are also imperative to minimize contamination and spread of food-borne illnesses through high-moisture, minimally processed bakery products.

REFERENCES

Aalvik, B. and Haave, I.J.J., 1980, Dry, flat bread as the cause of staphylococcal food poisoning, *Norsk Veterinaertidsskrift*, 92:746–747.

Abouzied, M.M. et al., 1991, Immunochemical assessment of mycotoxins in 1989 grain foods: evidence for deoxyivalenol (vomitoxin) contamination, *Appl. Environ. Microbiol.*, 57(3):672–677.

Agriculture and Agri-Food Canada, 2000, The Canadian bread and bakery industry, sub-sector profile, Food Bureau, Agriculture and Agri-Food Canada, 1–10.

Ahmed, R. et al., 2000, Epidemiologic typing of *Salmonella enterica* serotype *enteritidis* in a Canada-wide outbreak of gastroenteritis due to contaminated cheese, *J. Clin. Microbiol.*, 38(6):2403–2406.

Aidoo, K.E. et al., 1996, The composition and microbial quality of pre-packed dates purchased in greater Glasgow, *Int. J. Food Sci. Technol.*, 31(5):433–438.

Altekruse, S.F. et al., 1998, Cheese-associated outbreaks of human illness in the United States, 1973–1992: sanitary manufacturing practices protect consumers, *J. Food Prot.*, 61(10):1405–1407.

Angelotti, R., Foter, M.J., and Lewis, K.H., 1961, Time-temperature effects on Salmonellae and Staphylococci in foods. III. Thermal death time studies, *Appl. Microbiol.*, 9:308–315.

Anunciaçao, L.L.C. et al., 1995, Production of staphylococcal enterotoxin A in cream-filled cake, *Int. J. Food Microbiol.*, 26:259–263.

Aramouni, F. et al., 1994, Growth of *Clostridium sporogenes* PA 3679 in home-style canned quick breads, J. *Food Prot.*, 57(10):882–86.

Arness, M. et al., 1999, Norwalk-like viral gastroenteritis in U.S. Army trainees — Texas, 1998, *Morbidity Mortality Wkly. Rep.*, 48(11):225–227.

Aureli, P., Franciosa, G., and Pourshaban, M., 1996, Foodborne botulism in Italy, *Lancet*, 348:1594.

Austin, J.W. et al., 1998, Growth and toxin production by *Clostridium botulinum* on inoculated fresh-cut vegetables, *J. Food Prot.*, 61(3):324–328.

Bailey, C.P. and von Holy, A., 1993, *Bacillus* spore contamination associated with commercial bread manufacture, *Food Microbiol.*, 10:287–294.

Barnes, G.H. and Edwards, A.T., 1992, An investigation into an outbreak of *Salmonella enteritidis* phase-type 4 infection and the consumption of custard slices and trifles, *Epidemiol. Infect.*, 109:397–403.

Bassen, M.K. et al., 1989, Thermal resistance of *Bacillus cereus* spores in custard preparations, *J. Appl. Bacteriol.*, 5:511–516.

Bean, N.H. et al., 1996, Surveillance for foodborne disease outbreaks — United States, 1988–1992, *Morbidity Mortality Wkly. Rep.*, 45(SS-5):1–55.

Becker, H. et al., 1994, *Bacillus cereus* in infant foods and dried milk products, *Int. J. Food Microbiol.*, 23(1):1–15.

Bergdoll, M.S., 1989, *Staphylococcus aureus,* in *Foodborne Bacterial Pathogens*, Doyle, M.P., Ed., New York: Marcel Dekker, Inc., 463–524.

Bever, J.S. and Halvorson, H.O., 1947, The possibilities of the growth of *Clostridium botulinum* in canned bread, *Hormel Instit. Annu. Rep.*, 43:53–58.

Bhat, R.V. et al., 1989, Outbreak of trichothecene mycotoxicosis associated with consumption of mould-damaged wheat products in Kashmir Valley, India, *Lancet*, 1(8628):35–37.

Board, R.G., 1969. Microbiology of the hen's egg, *Adv. Appl. Microbiol.*, 11:245–281.

Bonner, D. and Schweiger, M., 1994, Apple pie: an unusual vehicle for food poisoning, *Commun. Dis. Rep. Rev.*, 4(5):R60–R61.

Boylan, S.L., Acott, K.A., and Labuza, T.P., 1976, *Staphylococcus aureus* challenge study in an intermediate moisture food, *J. Food Sci.*, 41:918–921.

Brugha, R. et al., 1999, A community outbreak of food-borne small round-structured virus gastroenteritis caused by a contaminated water supply, *Epidemiol. Infect.*, 122(1):145–154.

Bryan, F.F., Guzewich, J.J., and Todd, E.C.D., 1997, Surveillance of foodborne disease. III. Summary and presentation of data on vehicles and contributory factors; their value and limitations, *J. Food Prot.*, 60 (6):701–714.

Bryan, F.L., 1976, Public health aspects of cream-filled pastries. a review, *J. Milk Food Technol.*, 39:289–296.

Campbell, H. et al., 2000, Mycotoxins in barley and oat samples from Eastern Canada, *Can. J. Plant Sci.*, 80:977–980.

Capparelli, E. and Mata, L., 1975, Microflora of maize prepared as tortillas, *Appl. Microbiol.*, 29(6):802–806.

Castellani, A.G., 1953, Inhibiting effects of amino acids and related compounds upon the growth of enterotoxigenic micrococci in cream pastry, *Appl. Microbiol.*, 1:195–199.

Castelvetri, F., 1991, Microbiological implications of modified atmosphere packaging of fresh pasta, CAP'91 Schotland Business Research Inc., Princeton, NJ.

Cathcart, W.H. and Merz, A., 1942,*Staphylococcus* and *Salmonella* control in foods. III. Effect of chocolate and cocoa fillings on inhibiting growth of staphylococci, *Food Res.*, 7:96–99.

Cathcart, W.H., Merz, A., and Ryberg, R.E., 1942, Staphylococci and *Salmonella* control in foods. IV. Effect of cooking bakery custards, *Food Res.*, 7:100–103.

Centers for Disease Control, 1983, Staphylococcal food poisoning on a cruise ship, *Morbidity Mortality Wkly. Rep.*, 32(22):294–295.

Centers for Disease Control, 1990, Epidemiologic notes and reports update: *Salmonella enteritidis* infections and grade A shell eggs — United States, 1989, *Morbidity Mortality Wkly. Rep.*, 38(51–52):877–880.

Centers for Disease Control, 1998a, Multistate outbreak of *Salmonella* serotype *agona* infections linked to toasted oats cereal — United States, April–May, 1998, *Morbidity Mortality Wkly. Rep.*, 47(22):462–464.

Centers for Disease Control, 1998b, Botulism in the United States, 1899–1966, *Handbook for Epidemiologists, Clinicians, and Laboratory Workers*, U.S. Dept. Health and Human Services.

Centers for Disease Control, 2000, Outbreaks of *Salmonella* serotype *enteritidis* infection associated with eating raw or undercooked shell eggs — United States, 1996–1998, *Morbidity Mortality Wkly. Rep.*, 40(4):73–79.

Chaudhuri, A.K., Cassie, G., and Silver, M., 1975, Outbreak of food-borne type-A hepatitis in greater Glasgow, *Lancet*, 2(7927):223–225.

Chaudhry, R. et al., 1998, Outbreak of suspected *Clostridium butyricum* botulism in India, *Emerging Infect. Dis.*, 4:506–507.

Christiansson, A. et al., 1989, Toxin production by *Bacillus cereus* dairy isolates in milk at low temperatures, *Appl. Environ. Microbiol.*, 55(10):2595–2600.

Chung, K.T. and Sun, H.L., 1986, Distribution and characteristics of *Bacillus cereus* isolated from rice in Taiwan, *J. Food Sci.*, 51(5):1208–1212.

Clavero, M.R.S. et al., 2000, Influence of water activity and storage conditions on survival and growth of proteolytic *Clostridium botulinum* in peanut spread, *Food Microbiol.*, 17:53–61.

Collins, N.E., Kirschner, L.M., and von Holy, A., 1991, Characterization of *Bacillus* isolates from ropey bread, bakery equipment and raw materials, *S. Afr. J. Sci.*, 87:62–66.

Collins-Thompson, D.L. and Wood, D.S., 1993, Control in dairy products, in *Clostridium botulinum: Ecology and Control in Foods*, Hauschild, A.H. W. and Dodds, K.L., Eds., New York: Marcel Dekker, Inc., 261–277.

Communicable Disease Network Australia and New Zealand, 1997, Foodborne disease. Towards reducing foodborne illness in Australia, Technical Report Series #2, Foodborne Disease Working Party for Communicable Disease Network Australia and New Zealand.

Corsetti, A., Gobbetti, M., and Smacchi, E., 1996, Antibacterial activity of sourdough lactic acid bacteria: isolation of a bacteriocin-like inhibitory substance from *Lactobacillus sanfrancisco C57*, *Food Microbiol.*, 13:447–456.

Cowden, J.M. et al., 1995, Outbreaks of foodborne infectious intestinal disease in England and Wales 1992 and 1993, *Commun. Dis. Rep.*, 5(8):R109–R117.

Crielly, E.M., Logan, N.A., and Anderton, A., 1994, Studies on the *Bacillus* flora of milk and milk products, *J. Appl. Bacteriol.*, 77(3):256–273.

D'Aoust, J.Y., 1977, *Salmonella* and the chocolate industry. A review, *J. Food Prot.*, 40(10):718–727.

D'Aoust, J.Y., 1994, *Salmonella* and the international food trade, *Int. J. Food Microbiol.*, 24(1/2):11–31.

Dack, G.M., 1961, Public health significance of flour bacteriology, *Cereal Sci. Today*, 6:9–10.

Daifas, D.P. et al., 1999a, Challenge studies with *Clostridium botulinum* in high moisture bakery products packaged under modified atmospheres, *Food Austr.*, 51:507–511.

Daifas, D.P. et al., 1999b, Growth and toxin production by *Clostridium botulinum* in English-style crumpets packaged under modified atmospheres, *J. Food Prot.*, 62(4):349–355.

Daifas, D.P. et al., 1999c, Effect of pH and CO_2 on growth and toxin production by *Clostridium botulinum* in English-style crumpets packaged under modified atmospheres, *J. Food Prot.*, 62(10):1157–1161.

Daifas, D.P. et al., 2000, Effect of ethanol vapor on growth and toxin production by *Clostridium botulinum* in a high moisture bakery product, *J. Food Safety*, 20:111–125.

Denny, C.B., Goeke, D.J., Jr., and Steinberg, R., 1969, Inoculation tests of *Clostridium botulinum* in canned breads with special reference to water activity, Res. Rep. No. 4–69, Washington Research Laboratory National Canner's Association, Washington, D.C.

Desai, B. and Kamat, M.Y., 1998, Recovery and characterization of enterotoxigenic strains of staphylococci and microbiological quality of processed Indian foods, *J. Food Sci. Technol.*, 35:461–464.

do Carmo, L.S. and Bergdall, M.S., 1990, Staphylococcal food poisoning in Belo Horizonte (Brazil), *Rev. Microbiol.*, 21:320–323.

Dodds, K.L., 1992, Worldwide incidence and ecology of infant botulism, in Clostridium botulinum: *Ecology and Control in Foods*, Hauschild, A.H.W. and Dodds, K.L., Eds., New York: Marcel Dekker, Inc., 105–120.

Dodds, K.L. and Austin, J.W., 1997, *Clostridium botulinum,* in *Food Microbiology — Fundamentals and Frontiers*, Doyle, M.P., Beuchat, L.R., and Montville, T.J., Eds., Washington, D.C.: ASM Press, 288–304.

Dufresne, I. et al., 2000, Effect of headspace oxygen and films of different oxygen transmission rate on toxin production by *Clostridium botulinum* type E in rainbow trout fillets stored under modified atmospheres, *J. Food Safety*, 20(3):157–176.

Ebel, E.D., David, M.J., and Mason, J., 1992, Occurrence of *Salmonella enteritidis* in the U.S. commercial egg industry: report on a national spent hen survey, *Avian Dis.*, 36(3):646–654.

Edmonson, R.B., Thom, C., and Giltner, L.T., 1923, Experiments with *Bacillus (Clostridium) botulinum* under household conditions, *Food Technol.*, 23:143–145.

El-Gazzar, F.E. and Marth, E.H., 1992, *Salmonellae*, salmonellosis, and dairy foods: a review, *J. Dairy Sci.*, 75(9):2327–2343.

El-Khoury, W., 2000, Control of *Bacillus cereus* in English style crumpets, M.S. Thesis, McGill University, Montreal, Quebec, Canada.

Elliot, R.P., 1980, Cereals and cereal products, in *Microbial Ecology of Foods*, vol. 2: *Food Commodities*, International Commission on Microbiological Specifications for Foods, New York: Academic Press, Inc., 669–730.

Evans, M.R. et al., 1996, Consecutive *Salmonella* outbreaks traced to the same bakery, *Epidemiol. Infect.*, 116:161–167.

Farber, J.M., 1989, Foodborne pathogenic microorganisms: characteristics of the organisms and their associated diseases. I. Bacteria, *Can. Inst. Food Sci. Technol. J.*, 22:311–321.

Farber, J.M., 1991, Microbiological aspects of modified atmosphere packaging technology — a review, *J. Food Prot.*, 54(1):58–70.

Ferron, P. and Michard, J., 1993, Distribution of *Listeria* spp. in confectioners' pastries from Western France: comparison of enrichment methods, *Int. J. Food Microbiol.*, 18(4): 289–303.

Foegeding, P.M. and Berry, E.D., 1997, Cold temperature growth of clinical and food isolates of *Bacillus cereus, J. Food Prot.*, 60(10):1256–1258.

Food Safety and Inspection Service, 1998, *Consumer Information.* Volume 3 (4). *The Food Safety Educator New Consumer Research Driving Food Safety Education*, United States Department of Agriculture, Washington, D.C.

Franciosa, G. et al., 1999, *Clostridium botulinum* spores and toxin in mascarpone cheese and other milk products, *J. Food Prot.*, 62(8):867–871.

Gabis, D.A., Langlois, B.E., and Rudnick, A.W., 1970, Microbiological examination of cocoa powder, *Appl. Microbiol.*, 20:644–645.

Gifford, H., Lorenz, K., and Sofos, G., 1991, Analysis of bakery products for the presence of *Listeria monocytogenes*, *Lebensmittel-wissenschaft Technologie*, 2(5):476–477.

Gilcrease, F.W. and Colman, M.B., 1941, Studies of rebaking cream filled pastry, *Am. J. Public Health*, 31:956–958.

Goepfert, J.M., 1980, Vegetable, fruits, nuts and their products, in *Microbial Ecology of Foods*, vol. 2: *Food Commodities*, International Commission on Microbiological Specifications for Foods, New York: Academic Press, Inc., 606–642.

Golden, D.A., Rhodehamel, E.J., and Kautter, D.A., 1993, Growth of *Salmonella* spp. in cantaloupe, watermelon, and honeydew melons, *J. Food Prot.*, 56(3):194–196.

Goodman, L.J. et al., 1993, Evaluation of an outbreak of foodborne ilness initiated in the emergency department, *Ann. Emergency Med.*, 22(8):1291–1294.

Granum, P.E., 1997, *Bacillus cereus,* in *Food Microbiology — Fundamentals and Frontiers*, Doyle, M.P., Beuchat, L.R., and Montville, T.J., Eds., Washington, D.C.: ASM Press, Inc., 327–336.

Graves, R.R. et al., 1967, Bacterial and actinomycete flora of Kansas-Nebraska and Pacific Northwest wheat and wheat flour, *Cereal Chem.*, 44:288–299.

Hao, Y-Y., Scouten, A.J., and Brackett, R.E., 1999, Cheesecake: a potential vehicle for salmonellsis?, *J. Food Prot.*, 62(1):26–29.

Harmon, S.M. and Kautter, D.A., 1991, Incidence and growth potential of *Bacillus cereus* in ready-to-serve foods, *J. Food Prot.*, 54(5):372–374.

Hatheway, C.L., 1992, *Clostridium botulinum* and other clostridia that produce botulinum neurotoxin, in *Clostridium botulinum — Ecology and Control in Foods*, Hauschild, A.H.W. and Dodds, K.L., Eds., New York: Marcel Dekker, Inc., 3–20.

Hauschild, A.H.W., 1988, *Clostridium botulinum* in honey, syrups and dry infant cereals, *J. Food Prot.*, 51(11):892–894.

Hauschild, A.H.W., 1989, *Clostridium botulinum,* in *Foodborne Bacterial Pathogens*, Doyle, M.P., Ed., New York: Marcel Dekker, Inc., 111–189.

Hedberg C.W. and Osterholm, M.T., 1993, Outbreaks of food-borne and waterborne viral gastroenteritis, *Clin. Microbiol. Rev.*, 6:199–210.

Hintlian, C.B. and Hotchkiss, J.H., 1986, The safety of modified atmosphere packaging: a review — do modified atmospheres enhance pathogenesis but delay signs of spoilage?, *Food Technol.*, 40(12):70–6.

Hirooka, E.Y., De Salzberg, S.P.C., and Bergdoll, M.S., 1987, Production of staphylococcal enterotoxin a and thermonuclease in cream pies, *J. Food Prot.*, 50(11):952–955.

Hockin, J.C. et al., 1989, An international outbreak of *Salmonella nima* from imported chocolate, *J. Food Prot.*, 52(1):51–54.

Holtby, I., 1992, Food poisoning in two homes for the elderly, *Commun. Dis. Rep. Rev.*, 2(11):R125.

Holtby, I. et al., 1995, Outbreak of *Salmonella enteritidis* phage type 4 infection associated with cheese and onion quiche, *Commun. Dis. Rep. Rev.*, 5(8):R118-R119.

Humphrey, T.J., Martin, K.W., and Whitehead, A., 1994, Contamination of hands and work surfaces with *Salmonella enteriditis* PT 4 during the preparation of egg dishes, *Epidemiol. Infect.*, 113:403–409.

Humphrey, T.J. et al., 1991, Numbers of *Salmonella enteritidis* in the contents of naturally contaminated hens' eggs, *Epidemiol. Infect.*, 106:489–496.

Husseman, D.L. and Tanner, F.W., 1947, Relation of certain cooking procedures to *Staphylococcus* food poisoning, *Am. J. Public Health*, 37:1407–1414.

Ikawa, J.Y., 1991, *Clostridium botulinum* growth and toxigenesis in shelf-stable noodles, *J. Food Sci.*, 56:264–265.

Ingram, M. and Handford, P.M., 1957, The influence of moisture and temperature on the destruction of *Clostridium botulinum* in acid bread, *J. Appl. Bacteriol.*, 20:442–453.

Jablonski, L.M. and Bohach, G.A., 1997, *Staphylococcus aureus,* in *Food Microbiology — Fundamentals and Frontiers*, Doyle, M.P., Beuchat, L.R., and Montville, T.J., Eds., Washington, D.C.: ASM Press, Inc., 353–375.

Jaquette, C.B. and Beuchat, L.R., 1998. Survival and growth of psychrotrophic *Bacillus cereus* in dry and reconstituted infant rice cereal, *J. Food Prot.*, 61(12):1629–1635.

Jenson, L., Baird, L., and Delves-Broughton, J., 1994, The use of nisin as a preservative in crumpets, *J. Food Prot.*, 57(10):847–877.

Johnson, E.A., Nelson, J.H., and Johnson, M., 1990, Microbiological safety of cheese made from heat-treated milk. III. Technology, discussion, recommendations, bibliography, *J. Food Prot.*, 53(7):610–623.

Joseph, C.A. et al., 1991, A national outbreak of salmonellosis from yeast flavoured products, *Commun. Dis. Rep. Rev.*, 2(1):R16-R18.

Kadavy, J.L. and Dack, G.M., 1951, The effect of experimentally inoculating canned bread with spores of *Clostridium botulinum* and *Bacillus mesentericus, Food Res.*, 16:328–337.

Kaur, P., 1986, Survival and growth of *Bacillus cereus* in bread, *J. Appl. Bacteriol.*, 60:513–516.

Keller, G.M., Hanson, R.S., and Bergdoll, M.S., 1978, Molar growth yields and enterotoxin B production of *Staphylococcus aureus* S-6 with amino acids as energy sources, *Infect. Immunol.*, 20:151–157.

Khot, J.B. and Sherikar, A.T., 1986, Microbiology of meat food bakery products sold in the city of Bombay, *Indian Vet. Med. J.*, 9:125–130.

Killalea D. et al., 1996, International epidemiological and microbiological study of an outbreak of *Salmonella agona* infection from a ready-to-eat savoury snack — I: England and Wales and the United States, *Br. Med. J.,* 313(7065):1105–1107.

Kim, H.U. and Goepfert, J.M., 1971, Occurrence of *Bacillus cereus* in selected dry food products, *J. Appl. Bacteriol.*, 60:513–516.

Kirschner, L.M. and Von Holy, A., 1989, Rope spoilage of bread, *S. Afr. J. Sci.*, 85:425–427.

Kneifel, W. and Berger, E., 1994, Microbiological criteria of random samples of spices and herbs retailed on the Austrian market, *J. Food Prot.*, 57(10):893–901.

Kohn, S., 2000, An update of the U.S. baking industry, *Cereal Foods World*, 45(3):94–97.

Kramer, J.M. and Gilbert, R.J., 1989, *Bacillus cereus* and other *Bacillus* species, in *Foodborne Bacterial Pathogens*, Doyle, M.P., Ed., New York: Marcel Dekker, Inc., 22–70.

Kuritsky, J.N. et al., 1984, Norwalk gastroenteritis: a community outbreak associated with bakery product consumption, *Ann. Int. Med.*, 100:519–521.

Larsen, H.D. and Jorgensen, K., 1997, The occurrence of *Bacillus cereus* in Danish pasteurized milk, *Int. J. Food Microbiol.*, 34(2):179–186.

Lee, H.W., 1988, *Bacillus cereus* food poisoning associated with crumpets, *Abstr., 6th Austr. Food Microbiol. Conf.*, September 4–7, Surfers Paradise, Queensland, Australia.

Leela, R.K., Sankaran, R., and Vijayaraghavan, P.K., 1981, Occurrence of enterotoxigenic staphylococci in raw, processed, and ready-to-eat foods — I. Isolation and biochemical characteristics, *Indian J. Nutr. Diet.*, 18:197–203.

Lehmacher, A., Bockemuhl, J., and Aleksic, S., 1995, nationwide outbreak of human salmonellosis in Germany due to contaminated paprika and paprika-powdered potato chips, *Epidemiol. Infect.*, 115(3):501–511.

Lennox, J.E. and McElroy, L.J., 1984, Inhibition of growth and patulin synthesis in *Penicillium expansum* by potassium sorbate and sodium propionate in culture, *Appl. Environ. Microbiol.*, 48(2):1031–1033.

Leuschner, R.G.K., O'Callaghan, M.J.A., and Arendt, E.K., 1998, *Bacillus* spoilage in parbaked and rebaked brown sodabread, *J. Food Sci.*, 63:915–918.

Lewis, D.A. et al., 1996, Marshamallows cause an outbreak of infection with *Salmonella enteritidis* phage type, *Commun. Dis. Rep. Rev.*, 6(13):R183-R186.

Linnan, M.J. et al., 1988, Epidemic listeriosis associated with Mexican-style cheese, *New Engl. J. Med.*, 319:823–828.

Lombard, G.E. et al., 2000, Preservation of South African steamed bread using hurdle technology, *Lebensm.-Wiss.u-Technology*, 33:138–143.

Lovett, J. and Twetd, R.M., 1988, *Listeria*, *J. Food Technol.*, 42(4):188–191.

Lund, B.M., 1990, Foodborne disease due to *Bacillus* and *Clostridium* species, *Lancet*, 336:982–986.

Lyytikäinen, O. et al., 2000, An outbreak of *Listeria monocytogenes* serotype 3A infections from butter in Finland, *J. Infect. Dis.*, 181:1838–1841.

MacDonald, K.L. and Griffin, P.M., 1986, Foodborne disease outbreaks, annual summary, 1982, *Morbidity Mortality Wkly. Rep.*, 35(1SS):7–10.

Malloy, C.D. and Marr, J.S., 1997, Mycotoxins and public health: a review, *J. Public Health Manage. Pract.*, 3:61–69.

Marston, P.E. and Wannan, T.L., 1976, Bread baking: the transformation from dough to bread, *Bakers Dig.*, 50(8):24–8, 49.

Martinez-Anaya, M.A. et al., 1990, Microflora of the sourdoughs of wheat flour bread x interactions between yeasts and lactic acid bacteria in wheat doughs and their effects on bread quality, *Cereal Chem.*, 67(1):85–91.

McKinley, T.W. and Clarke, E.J., Jr., 1964, Imitation cream filling as a vehicle of staphylococcal food poisoning, *J. Milk Food Technol.*, 27:302–304.

Mead, P.S. et al., 1999, Food-related illness and death in the United States, *Emerging Infect. Dis.*, 5(5):607–625.

Moreno, A.J. et al., 1985, *Bacillus cereus*, diarrhea, and home-dried apples, *Ann. Int. Med.*, 102(6):868–869.

Murrell, W.G., 1978, *Bacillus cereus* food poisoning from pikelets, Communicable Disease Intelligence, Australian Environmental Health Branch, Department of Health.

Notermans, S.H.W., 1992, Control in fruits and vegetables, in *Clostridium botulinum*: *Ecology and Control in Foods*, Hauschild, A.H. W. and Dodds, K.L., Eds., New York: Marcel Dekker, Inc., 233–260.

Pafumi, J., 1986, Assessment of the microbiological quality of spices and herbs, *J. Food Prot.*, 49(12):958–963.

Park, C.E., Szabo, R., and Jean, A., 1988, A survey of wet pasta packaged under a CO_2:N_2 (20:80) mixture for staphylococci and their enterotoxins, *Can. Inst. Food Sci. Technol. J.*, 21: 109–111.

Pinegar, J.A. and Buxton, J.D., 1977, An investigation of the bacteriological quality of retail vanilla slices, *J. Hyg.*, 78:387–394.

Pirttijarvi, T.S. et al., 1998, *Bacillus cereus* in a whey process, *Int. J. Food Microbiol.*, 44(1/2):31–41.

Potter, M.E., Ayala, S.G., and Silarug, N., 1997, Epidemiology of foodborne diseases, in *Food Microbiology — Fundamentals and Frontiers*, Doyle, M.P., Beuchat, L.R., and Montville, T.J., Eds., Washington, D.C.: ASM Press, Inc., 376–390.

Powers, E.M., Latt, T.G., and Brown, T., 1976, Incidence and levels of *Bacillus cereus* in processed spices, *J. Milk Food Technol.*, 39:668–670.

Powers, E.M. et al., 1988. Studies of tray pack coffee cakes and spice cakes adjusted to water activities 0.86 to 0.93 and inoculated with *Clostridium botulinum,* Technical Report, Natick/TR-88/056, United States Army Natick RD&E Center, Natick, MA.

Preonas, D.L. et al., 1969, Growth of *Staphylococcus aureus* MF31 on the top and cut surfaces of Southern custard pies, *Appl. Microbiol.*, 18:68–75.

Public Health Laboratory Services, 1993, Recent outbreaks of Salmonellosis, *Commun. Dis. Rep.*, 3(3):1.

Public Health Laboratory Services, 1999, General outbreaks of foodborne illness in humans, England and Wales: quarterly reports, *Commun. Dis. Rep.*, 9(11):103–104.

Quintavalla, S. and Parolari, G., 1993, Effects of temperature, a_w and pH on the growth of *Bacillus* cells and spores: a response surface methodology study, *Int. J. Food Microbiol.*, 19(3):207–216.

Raevuori, M., 1976, Effect of sorbic acid and potassium sorbate on growth of *Bacillus cereus* and *Bacillus subtilis* in rice filling of Karelain pastry, *Appl. Microbiol.*, 31:205–213.

Richter, K.S. et al., 1993, Microbiological quality of flours, *Cereal Foods World*, 38(5):367–369.

Richter, K.S. et al., 1991, Survival of *Listeria monocytogenes* in cake and straight grade flour heat and non-heat processed, 76th Annu. Meet. Am. Assoc. Cereal Chem., Seattle, Washington, October 13–16, 1991, *Cereal Foods World*, 36(8):729.

Rizk, I.R. and Ebeid, H.M., 1989, Survival and growth of *Bacillus cereus* in Egyptian bread and its effect on bread staling, *Nahrung*, 33(9):839–844.

Rosenkvist, H. and Hansen, Å., 1995, Contamination profiles and characterisation of *Bacillus* species in wheat bread and raw materials for bread production, *Int. J. Food Microbiol.*, 26:353–363.

Rosenkvist, H. and Hansen, Å., 1998, The antimicrobial effect of organic acids, sour dough and nisin against *Bacillus subtilis* and *Bacillus licheniformis* isolated from wheat bread, *J. Appl. Microbiol.*, 85:621–631.

Sadek, M.A. et al., 1985, Isolation and enumeration of *Bacillus cereus* and some other microorganisms from Balady bread, *Zeitschrift fur die Gesamte Hygiene und Ihre Grenzgebiete*, 31(11):623–624.

Sankaran, R. and Leela, R.K., 1983, Prevalence of enterotoxigenic staphylococci in bakery products, *J. Food Prot.*, 46(2):95–97.

Santos, D.E.C., Genigeorgis, C., and Farver, T.B., 1981, Prevalence of *Staphylococcus aureus* in raw and pasteurized milk used for commercial manufacturing of Brazilian minas cheese, *J. Food Prot.*, 44(3):172–176.

Scheil, W. et al., 1998, A South Australian *Salmonella mbandaka* outbreak, *Aust. N. Zealand J. Public Health*, 22(5):536–539.

Schmidt, D. and Ridley, S.C., 1985, Microbiological and sensory characteristics of a home style confectionary product prepared with cream cheese, *J. Food Qual.*, 7(4):283–288.

Schmidt, E.W., Jr., Gould, W.A., and Weiser, H.H., 1969, Chemical preservatives to inhibit the growth of *Staphylococcus aureus* in synthetic cream pies acidified to pH 4.5 to 5.0, *Food Technol.*, 23: 1197–1220.

Schoenbaum, S.C., Baker, O., and Jezek, Z., 1976, Common-source epidemic of hepatitis due to glazed and iced pastries, *Am. J. Epidemiol.*, 104(1):74–80.

Seiler, D.A.L., 1978, The microbiology of cake and its ingredients, *Food Trade Rev.*, 48:339–344.

Shafi, R., Cotterill, O.J., and Nichols, M.L., 1970, Microbial flora of commercially pasteurized egg products, *Poultry Sci.*, 49:578–585.

Silliker, J.H. and McHugh, S.A., 1967, Factors influencing microbial stability of butter-cream-type filling, *Cereal Sci. Today*, 12(1):63–65, 73–74.

Skoll, S.L. and Dillenberg, H.O., 1963, *Salmonella thompson* in cake mix, *Can. J. Public Health*, 54:325–329.

Slaghuis, B.A. et al., 1997, Effect of pasteurization on the incidence of *Bacillus cereus* spores in raw milk, *Int. Dairy J.*, 7(4):201–205.

Smith, J.P. and Simpson, B.K., 1995, Modified atmosphere packaging of bakery and pasta products, in *Principles of Modified-Atmosphere and Sous Vide Product Packaging*, Farber, J.M. and Dodds, K.L., Eds., Lancaster, PA: Technomic Publishing Company, Inc., 207–242.

Smith, J.P., Jackson, E.D., and Ooraikul, B., 1983, Storage study of a gas-packaged bakery product, *J. Food Sci.*, 48:1370–1371.

Snyder, O.P., Jr., 1996, Redox potential in deli foods: botulism risk?, *Dairy, Food Environ. Sanit.*, 16:546–548.

Sockett, P.N., 1991, Food poisoning outbreaks associated with manufactured foods in England and Wales 1980–1989, *Commun. Dis. Rep.*, 1(R10):R105-R109.

Sockett, P.N. et al., 1993, Foodborne disease survey in England and Wales 1989–1991, *Commun. Dis. Rep.*, 3(12):R164-R173.

Soloski, T. and Cryns, J., 1950, Practical and theoretical considerations in the manufacture of canned bread, *Am. Assoc. Cereal Chem.*, 8:107–12.

Stritar, J., Dack, G.M., and Jungewaelter, F.G., 1936, The control of staphylococci in custard-filled puffs and éclairs, *Food Res.*, 1:237–246.

Sugihara, T.F., 1977, Non-traditional fermentations in the production of baked goods, *Bakers Dig.*, 51(5):76, 78, 80, 142.

Sumner, S.S., Albrecht, J.A., and Peters, D.L., 1993, Occurrence of enterotoxigenic strains of *Staphylococcus aureus* and enterotoxin production in bakery products, *J. Food Prot.*, 56:722–724.

Sumner, S.S. et al., 1991, Heat resistance of *Salmonella typhimurium* and *Listeria monocytogenes* in sucrose solutions of various water activities, *J. Food Sci.*, 56(6):1741–1743.

Surkiewicz, B.F., 1966, Bacteriological survey of the frozen prepared foods industry. I. Frozen cream-type pies, *Appl. Microbiol.*, 14:21–26.

Sutherland, J.P., Aherne, A., and Beaumone, A.L., 1996, Preparation and validation of a growth model for *Bacillus cereus*: the effects of temperature, pH, sodium chloride and carbon dioxide, *Int. J. Food Microbiol.*, 30:359–372.

Swanenburg, M. et al, 2001, *Salmonella* in the lairage of pig slaughterhouses, *J. Food Prot.*, 64(1):12–16.

Tabibi, M. and Salehian, A.A., 1974, Food poisoning in bread caused by *Aspergillus flavatoxin* (L'intoxication alimentaire du pain par *Aspergillus flavatoine* hiver 1967 à Teheran), *Acta Medica Iranica*, 17(1–2):63–69.

te Giffel, M.C. et al., 1996, Incidence of *Bacillus cereus* and *Bacillus subtilis* in foods in the Netherlands, *Food Microbiol.*, 13:53–58.

Terranova, W. and Blake, P.A., 1978, *Bacillus cereus* food poisoning, *New Engl. J. Med.*, 298:143–144.

Thompson, J.M., Dodd, C.E.R., and Waites, W.M., 1993, Spoilage of bread by *Bacillus, Int. J. Biodeteriorat. Biodegradat.*, 32:55–66.

Thompson, J.M., Waites, W.M., and Dodd, C.E.R., 1998, Detection of rope spoilage in bread caused by *Bacillus* species, *J. Appl. Microbiol.*, 84:481–486.

Todd, E.C.D., 1989, Preliminary estimates of costs of foodborne disease in Canada and costs to reduce Salmonellosis, *J. Food Prot.*, 52(8):586–594.

Todd, E.C.D., 1982, Foodborne and waterborne disease in Canada — 1977. Annual summary, *J. Food Prot.*, 45(9):865–873.

Todd, E.C.D., 1996, Worldwide surveillance of foodborne disease: the need to improve, *J. Food Prot.*, 59(1): 82–92.

Todd, E.C.D. et al., 1983, Microbiological quality of frozen cream–type pies sold in Canada, *J. Food Prot.*, 46(1):34–40.

Torres-Vitela, M.R., Escartin, E.F., and Castillo, A., 1995, Risk of Salmonellosis associated with consumption of chocolate in Mexico, *J. Food Prot.*, 58(5):478–481.

Tranter, H.S., 1990, Foodborne illness. Foodborne staphylococcal illness, *Lancet*, 336:1044–1046.

Ulrich, J.A. and Halvorson, H.O., 1949, The possibilities of the growth of *Clostridium botulinum* in canned bread. Annual report, 1948–1949, The Hormel Institute of the University of Minnesota, 45–50.

Visconti, A. and Bottalico, A., 1983, High levels of ochratoxins A and B in moldy bread responsible for mycotoxicosis in farm animals, *J. Agric. Food Chem.*, 31(5):1122–1123.

Wagenaar, R.O. and Dack, G.M., 1954, Further studies on the effect of experimentally inoculating canned bread with spores of *Clostridium botulinum, Food Res.*, 19:521–529.

Wagenaar, R.O. and Dack, G.M., 1960, Studies on canned chocolate nut roll experimentally inoculated with spores of *Clostridium botulinum* types A and B, *Food Res.*, 25:646–653.

Warburton, D.W. and Weiss, K.F., 1986, Microbiological quality of non-dairy creamers, fillings, and toppings, *J. Food Prot.*, 49(8):621–622.

Waterman, S.H. et al., 1987, Staphylococcal food poisoning on a cruise ship, *Epidemiol. Infect.*, 99:349–353.

Weckel, K.G., Hawley, R., and McCoy, E., 1964, Translocation and equilibration of moisture in canned frozen bread, *Food Technol.*, 18:234–236.

Weidenborner, M. et al., 2000, Whole wheat and white wheat flour — the mycobiota and potential mycotoxins, *Food Microbiology,* 17(1):103–107.

Wells, S.J. et al., 2001, Fecal shedding of *Salmonella* species by dairy cows on farm and at cull cow markets, *J. Food Prot.*, 64(1):3–11.

Weltman, A.C. et al., 1996, An outbreak of Hepatitis A associated with a bakery, New York, 1994: The 1968 West Branch, Michigan outbreak repeated, *Epidemiol. Infect.*, 117:333–341.

Wright, J.P. and Patterson, W.J., 1994, Outbreak forum — VIII. Food poisoning at a Masonic lodge, *Commun. Dis. Rep. Rev.*, 4(5):R58-R60.

Wright, J.P. et al., 1996, Four outbreaks of *Salmonella enteritidis* phage type 4 food poisoning linked to a single baker, *Commun. Dis. Rep. Rev.*, 6(8):R112-R115.

Wyatt, C.J. and Guy, V.H., 1981(a), Growth of *Salmonella typhimurium* and *Staphylococcus aureus* in retail pumpkin pies, *J. Food Prot.*, 44(6):418–421.

Wyatt, C.J. and Guy, V.H., 1981(b), Incidence and growth of *Bacillus cereus* in retail pumpkin pies, *J. Food Prot.*, 44(6):422–424.

2 Concerns with Minimal Processing in Apple, Citrus, and Vegetable Products

Kathleen T. Rajkowski
and Elizabeth A. Baldwin

CONTENTS

INTRODUCTION

Although the incidence of food-borne illnesses linked to fresh produce is low, there is increased awareness that fruits and vegetables can be contaminated with microbiological pathogens. For its microbiological sampling program of certain fresh fruits and vegetables, the U.S. Food and Drug Administration (FDA) conducted surveys of both imported and domestic produce. A 4% (44 of 1003 sampled) contamination rate was reported in published results for imported product (http://www.cfsan.fda.gov/~dms/prodsur6.html). In the interim report on domestic product, there was a 1.6% violation rate (http://www.cfsan.fda.gov/~dms/prodsur9.html). Two microbiological pathogens that can cause food-borne illnesses were present on the produce: *Salmonella* and *Shigella.*

With the shift in diet toward the consumption of more fresh fruits and vegetables and greater distribution distances from new geographic sources, there are more reported illnesses involve fresh produce (Tauxe et al., 1997). In the U.S. from 1988

1-58716-041-2/03/$0.00+$1.50

to 1992, 64 outbreaks of disease were related to the consumption of fruits and vegetables resulting in 9 deaths (Bean et al., 1996). From 1993 to 1997, in 66 outbreaks with 2 deaths, fruits and vegetables were the vehicles of transmission (Olson et al., 2000). Because food-borne infections are sporadic and many go unreported, the exact number of cases related to produce is unknown (Tauxe et al., 1997). Due to changes in the food supply and food consumption patterns in the U.S., deaths caused by food-borne diseases are even more difficult to estimate (Mead et al., 1999). The food-borne outbreaks cause gastroenteritis and in severe cases, hospitalization is required. Bean et al. (1996) present guidelines listing syndrome, incubation time, and clinical identification procedures for the confirmation of gastrointestinal food-borne-disease outbreaks by etiological agent.

Beuchat (1996) and Sumner and Peters (1997) discussed and listed the microflora isolated from raw fruits and vegetables, including those involved in food-borne outbreaks. These authors listed the pathogens identified and the produce involved. The microbiology of minimally processed fresh fruits and vegetables was reviewed by Nguyen-The and Carlin (1994); these reviews discuss the human pathogens and spoilage microorganisms recovered and identified from fresh produce. Table 2.1 lists bacterial pathogens associated with fruits and vegetables.

In January of 1997, President Clinton announced a Food Safety Initiative, and a report by the U.S. Department of Health and Human Services, the U.S. Department of Agriculture, and the U.S. Environmental Protection Agency identified domestic fresh produce as an area of concern. In October of that same year, President Clinton initiated a plan entitled *Produce and Imported Food Safety Initiative* to assure safety of imported fruits and vegetables in the American diet. An alarming survey of imported produce in 1999, including broccoli, cantaloupe, celery, cilantro, loose-leaf lettuce products (radicchio, escarole, endive, chicory), parsley, scallions, and strawberries, found 40 out of 1000 samples tested positive for bacterial pathogens. Of this contaminated produce, 35 samples were contaminated with *Salmonella* and 9 with *Shigella*. No *Escherichia coli* O157:H7 was found (FDA, 2001).

Fruit and vegetable contamination problems can occur in the growing environment. During growth the fruit or vegetable can become contaminated from sources such as soil, animals, birds, and insects. Following production, the processes of harvesting, washing, cutting, slicing, packaging, and shipping can create additional conditions where contamination can occur. When produce is consumed in the raw, as is the case with fresh-cuts, harmful microorganisms may be present and ingested. In October of 1998, the Food and Drug Administration issued *Guidance for Industry — Guide to Minimize Microbial Food Safety Hazards for Fresh Fruits and Vegetables,* which describes good agricultural and manufacturing practices (GAPs and GMPs, respectively). These guides cover water quality, manure management, worker training, field and facility sanitation, and transportation (FDA, 2001).

Traditionally, fresh fruits and vegetables were considered safe to eat raw — straight from the field, but now bacterial pathogens are being found on or in the fruit or vegetable. Today the consumer is advised through the news media to wash fresh fruits and vegetables before eating. Commercial washing kits are available to clean the raw produce. Consumers, particularly the young, elderly, or immunocompromised

TABLE 2.1
Bacterial Pathogens and Associated Fruit, Vegetable, and Juice Vehicles

Pathogen	Fruit/Vegetable/Juice	Ref.
Aeromonas spp.	Vegetables, sprouts	Escuder et al., 1999; NACMCF, 1999; Velázquez et al., 1998
Listeria monocytogenes	Cabbage, lettuce, salad	Beuchat, 1996; Kakiomenou et al., 1998; NACMCF, 1999; Nguyen-The and Carlin, 1994; Odumeru et al., 1997; Thompson and Powell, 2000
E. coli	Lettuce, melons, cantaloupe, cabbage, unpast. apple juice	Ackers et al., 1998; Beauchat, 1996; Castro-Rosas and Escartin, 2000; Cody et al., 1999; NACMCF, 1999; Nguyen-The and Carlin, 1994; Park and Beuchat, 1999; Tauxe et al., 1997
Salmonella spp.	Tomato, alfalfa sprout, salad, melon	Castro-Rosas and Escartin, 2000; Kakiomenou et al., 1998; Maxcy, 1978; NACMCF, 1999; Park and Beuchat, 1999; Thompson and Powell, 2000; Wei et al., 1995
Staphylococcus aureus	Salads, sprouts	Fowler and Foster, 1976; Maxcy, 1978; Thompson and Powell, 2000
Bacillus cereus	Sprouts	NACMCF, 1999; Thompson and Powell, 2000
Shigella	Onion, lettuce, cabbage	Beuchat, 1996; Escudero et al., 1999; Satchell et al., 1990
Vibrio cholerae	Alfalfa sprouts	Castro-Rosas and Escartin, 2000
Klebsiella	Dried bush okra, sprouts	Mpuchane and Gasha, 1996; NACMCF, 1999; Thompson and Powell, 2000
Campylobacter	Vegetables	Escudero et al., 1999; Park and Sanders, 1992
Pseudomonas	Vegetables	Nguyen-The and Carlin, 1994
Clostridium botulinum	Vegetables	Olsen et al., 2000
Yersinia enterliticus	Lettuce	NACMCF, 1999

for any reason, is now warned by government advisories that eating raw sprouts or drinking unpasteurized fruit drinks can make them ill (Tauxe et al., 1997).

Because of high water activity (a_w) and nutrient content, fresh produce can support the growth of a variety of disease-causing microorganisms (Sumner and Peters, 1997). Table 2.1 lists produce types from which the bacterial pathogens were isolated and identified. Some produce types (sprouts, lettuce, salad, and melons) were identified as vehicles of more than one pathogen species.

Today fresh fruits and vegetables can be purchased in a variety of forms. The produce can be whole, minimally processed (peeled) whole, cut into pieces, or in the form of unpasteurized juices. When whole, the produce is sold individually and usually not packaged. Wells and Butterfield (1999) have shown that pathogenic microorganisms such as *Salmonella* can grow in the wound area of a damaged, intact fruit or vegetable. When minimally processed, the produce can be packaged in pouches with and without modified atmosphere, in plastic clam packs, or in trays covered with a polymeric wrap. The packaging is necessary because the fruit has lost all or part of its protective covering (peel). The cut and exposed surfaces of the minimally processed fruits or vegetables are a main concern because microorganisms can grow there (Nguyen-The and Carlin, 1994).

Another form is unpasteurized fresh juice processed in a food plant, in the store, or in the home. If a bacterial pathogen is present on or in the produce or on any processing surface, it will have an almost unlimited food source for growth in the unpasteurized juice, resulting in a food-borne outbreak (Cook et al., 1998, Steele et al., 1982). Pathogens can grow in apple and orange juices if the produce is not pasteurized before packaging, is temperature abused, or is kept for a long time (Cook et al., 1998; Doyle and Mazzota, 2000). Because food-borne illness involving young children has been traced to drinking untreated juices, the FDA has issued a consumer alert about the health risk of drinking untreated juices (http://www.cfsan.fda.gov/~dms/juicsaf2.html).

It was thought that the high acidity of some juices would inhibit the growth of pathogenic bacteria. In apple cider, which has a naturally low pH, *E. coli* O157:H7 was reported to survive and was acid tolerant in trypticase soy broth >pH 4 (Miller and Kasper, 1994). In 1996 an outbreak of *E. coli* O157:H7 infection from drinking unpasteurized commercial apple juice resulted in one death (Cody et al., 1999). *Salmonellae* are reported to survive in orange juice of pH 3.5 at refrigerated temperatures (Parish et al., 1997). *Salmonella* Hartford was associated with a food-borne outbreak linked to unpasteurized orange juice (Cook et al., 1998).

FRESH-CUT FRUITS AND VEGETABLES

High consumer demand for healthy foods has led to the exponential growth of the fresh-cut produce industry. The International Fresh-cut Produce Association (IFPA) defines fresh-cut products as fruits or vegetables that have been trimmed, peeled, or cut into 100% usable product that is bagged or prepackaged to offer consumers high nutrition, convenience, and flavor while still maintaining freshness (IFPA, 2001). Fresh-cuts are one category of minimally, or lightly, processed fruits and vegetables. These products should be in a raw state, not frozen or thermally processed, and ready to eat or cook (Anonymous, 1998a, b). Minimally processed products can also include unpasteurized, fresh juices.

The rapid growth of fresh-cuts in the U.S. over the past 10 years can be attributed to an increased consumer awareness of the nutritional benefits of fruits and vegetables, busy lifestyles, and spending power. Fresh-cut retail sales in the U.S. in 1994 were $5.8 billion (Hodge, 1995) and have reached more than 10% of the U.S. fresh vegetable and fruit market ($8.8 billion in 1998); sales are projected to increase to $19 billion by 2003 (Greenleaf, 1999). This value-added product has great potential for the fresh produce industry; however, several problems have limited commercial development, especially for fresh-cut fruits. Among the most serious are browning, softening, flavor deterioration, microbial decay, and safety. More specifically, browning, softening, and microbial decay were found to be limiting factors for the shelf-life of fresh-cut apples (Lakakul et al., 1999), pears (Gorny et al., 2000; Dong et al., 2000), peaches (Gorny et al., 1998), and kiwi fruit (Agar et al., 2000).

Maintenance of quality and control of microbial populations are problems for the fresh-cut produce industry. A combination of treatments is often necessary to maintain physiological quality and limit microbial growth. Solutions have included various chemical dips (Gorny et al., 1999; McHugh and Senesi, 2000; Dong et al., 2000), the use of certain cultivars and maturity levels that have superior stability (Gorny et al., 2000), controlled or modified atmospheres (Lakakul et al., 1999; Gorny et al., 1999), high-pressure processing (Boynton, 1999), and irradiation (Prakash et al., 2000).

New technologies, such as modified atmosphere packaging (MAP), have extended shelf-life and reduced decay and spoilage organisms, resulting in a shift in microbial population dynamics that may favor growth of human pathogens. In addition, when spoilage organisms are controlled, the food may look and smell edible even when it is not safe to eat. The change in atmosphere created by MAP, and fruit respiration (relative increase in CO_2 and decrease in O_2), may allow *Clostridium botulinum* to grow and form toxin, especially at elevated storage temperatures. Lower O_2 levels created by MAP reduce growth of aerobic spoilage organisms (Farber, 1991; Hotchkiss and Banco, 1992). Germination of *C. botulinum* spores is stimulated by elevated CO_2 (Enfors and Molin, 1978; Foegeding and Busta, 1983), as has been demonstrated with packaged, cut honeydew melons. Melons were inoculated with *C. botulinum*, then treated with UV light to inactivate vegetative organisms and packaged using passive MAP. This resulted in marginal spoilage and botulinal toxin formation (Larson and Johnson, 1999).

Fresh-cut produce deteriorates faster than its intact counterpart (Cantwell, 1995) due to the wounding associated with processing. This affects the stability of the produce as a result of biochemical and physiological changes (Brecht, 1995; Saltveit, 1997). Problems include changes in color (browning), flaccidity (loss of water), and microbial contamination at the cut surface (Brecht, 1995; King and Bolin, 1989; Varoquaux and Wiley, 1994). The surface structure of lettuce has been shown to protect *E. coli* O157:H7 cells or other pathogens from chlorine inactivation (Liao and Sapers, 2000; Takeuchi and Frank, 2001; Ukuku and Sapers, 2001). Wounding induces signals that elicit physiological and biochemical responses throughout the tissue (Ke and Saltveit, 1989; Saltveit, 1997). Wounding plant tissues makes them more susceptible to attack by plant pathogenic microorganisms and contamination with human pathogens.

The cut surface of any processed vegetable can support microbial growth. In addition, the cut vegetable continues to metabolize, producing more nutrients that become available for microbes. The natural microbiological counts of minimally processed produce including leaf and cut lettuce were reported by Beuchat (1996), Nguyen-The and Carlin (1994), and Sumner and Peters (1997). Aerobic mesophilic counts can range from 10^3 to 10^8/g, depending on the produce variety and geographic location (Beuchat, 1996). Recent studies (Escudero et al., 1999; Park and Beuchat, 1999; Velázquez et al., 1998; Wei et al., 1995) have documented the growth of inoculated pathogens (*Salmonella, Aeromonas, Yersinia, Vibrio. cholerae*) on produce; *E. coli* O157:H7 was confirmed as the agent in an outbreak involving leaf lettuce (Ackers et al., 1998). It has been shown (Kakiomenou et al., 1998; Jacxsens et al., 1999) that when minimally processed products are contaminated and stored at 4 to 7°C, psychotropic bacterial pathogens can grow on the produce. Kakiomenou et al. (1998) demonstrated that when fresh-cut salad vegetables were packaged under modified atmosphere and stored at 4°C, *S. Enteritidis* and *Listeria monocytogenes* survived, and Jacxsens et al. (1999) showed that *L. monocytogenes* and *Aeromonas* spp. survived in fresh-cut salad vegetables stored at 7°C.

Bacteria account for the major portion of microflora on vegetable salads, but exact numbers vary according to time of year and geographic location (Garcia-Gimeno and Zurera-Cosano, 1997; Hagenmaier and Baker, 1998; King et al., 1991; Manvell and Ackland, 1986). Pathogens such as *L. monocytogenes* can also grow in contaminated produce at refrigerated temperatures (Odumeru et al., 1997). When the storage temperature of packaged salads was increased (abuse), lactic acid bacteria were the dominant species. Garcia-Gimeno and Zurera-Cosano (1997) suggested modeling the growth of lactic acid bacteria to predict the shelf-life of packaged salads; as a temperature and spoilage indicator, this model can be used to set the pull date of the product before the pathogens outgrow the nonpathogens.

Whole fresh fruits and vegetables with bacterial soft rot and fungal rot were shown to have a high incidence of contamination with *Salmonella* spp. (Wells and Butterfield, 1997, 1999). These areas of contamination become the inoculum for cross-contamination if the cutting utensil is not cleaned properly. Wells and Butterfield (1997) reported that contamination of wash water during rinsing of soft rotted fruit and vegetable could cause problems. Such contamination during washing can explain the increase of *Salmonella* and *E. coli* O157:H7 outbreaks associated with unpasteurized fresh juices (Cody et al., 1999; Cook et al., 1998).

If washing equipment becomes contaminated with the pathogen, a biofilm may develop. The biofilm structure can provide a protective environment for pathogens and reduce the effectiveness of any sanitizer or inhibitory agents used. The results of different sanitizers (free chlorine, acidified sodium chlorite, hydrogen peroxide, lactic acid, lactic acid plus chlorine and Tsunami [peroxyacetic acid]) as effective agents against *Salmonella*, *E. coli* O157:H7, *A. hydrophila*, and *Y. enterocolitica* on fresh produce were reviewed by Escudero et al. (1999), Park and Beuchat (1999), and Velázquez et al. (1998). Not one single agent was shown to be effective against all pathogens. Rajkowski and Thayer (2000) recently reported that the use of irradiation at a level of 2 kGy was effective in achieving a 5-log kill of *Salmonella* and *E. coli* O157:H7 on radish sprouts. Further investigation of the use and keeping quality of irradiated produce is necessary.

One developing technology is the use of antimicrobial edible packaging based on cellulosic ethers, fatty acids, and nisin. Nisin is an antimicrobial peptide that is effective against gram-positive bacteria (Coma et al., 2001).

MINIMALLY PROCESSED APPLES AND CITRUS

Cut apples are a nutritious and popular snack, ideal for airlines and schools and as a packaged product. Commercial products have appeared on the East and West Coasts (Anonymous, 1998c). New methods for washing whole apples prior to cutting include hot-water immersion (Fleischman et al., 2001), flatbed brush washing (Annous et al., 2001), and the use of sanitizers for reducing populations of *E. coli*

(Sapers et al., 1999). These methods were effective only if the *E. coli* was not internalized. Hot-water immersion is effective in reducing or eliminating the microorganism if the inoculum is dropped onto the surface of the apple (Fleischman et al., 2001), whereas, if inoculated by submersion, brush washing and sanitizers can reduce only the surface population (Buchanan et al., 1999; Sapers et al., 1999; Annous et al., 2001).

However, browning and product safety are issues for the cut fruit. Browning can be inhibited by erythorbic or ascorbic treatments (Sapers et al., 1990; Sapers and Ziolkowski, 1987). Human pathogens and spoilage organisms are not so easily controlled. Fresh-cut apples were recalled due to possible *L. monocytogenes* contamination (SafetyAlerts.com, 2001); *L. monocytogenes* survived and increased on "Delicious" apple slices stored at 10 or 20°C in 0.5% O_2 and 15% CO_2 (Conway et al., 2000). When *L. monocytogenes* was inoculated into decayed apple tissue with the spoilage organism *Glomerella cingulata*, it increased in population, but it did not survive after 5 days in fruit infected with the spoilage organism *Penicillium expansum* (Conway et al., 2000). This may have been a pH effect because the pH increased in decaying tissue with *Glomerella cingulata* while, with *Penicillium expansum*, the pH decreased (Conway et al., 2000).

Total CFU/ml blended apple tissue on plate count agar was 4×10^3 for cut apple stored 2 weeks at 4°C without surface treatment and for 8 in. slices treated with a cellulose coating containing ascorbic acid, citric acid, and potassium sorbate (Baldwin et al., 1996). After 4 weeks of storage, the microbial populations on the fruit without surface treatment were 10^4, while treated fruit populations were 78 CFU/ml (Baldwin et al., 1996). Included in the storage atmospheres of natural apple volatiles, hexanal and trans-2-hexenal were shown to prolong the lag phases of inoculated yeast (*Pichia Subpelliculosa*) and reduced growth potential of naturally occurring bacteria, thus extending product shelf-life (Corbo et al., 2000). Hexanal vapor inhibited hyphae growth of fungi, *Pencillium expansum* and *Botrytis cinerea*, on apple slices (Song et al., 1996). The effect of these vapors on human pathogens is not known.

Like apple, peeled citrus segments, particularly orange and grapefruit, would be a popular and nutritious snack. The peel of citrus consists of a thin layer of colored flavedo and an inner layer of spongy white albedo tissue, which is difficult to remove from the fruit. Subsequent section production is labor intensive in that each section must be cut by hand from the peeled fruit. Traditionally, the fruits were steamed to loosen the peel, peeled by machine, and treated with lye. Cut sections were small and friable and necessitated liquid packing (Bruemmer et al., 1978).

The USDA Citrus and Subtropical Products Laboratory has developed a novel technology consisting of vacuum infusion of pectinases to facilitate removal of the peel from oranges and grapefruit (Bruemmer and Griffin, 1978; Bruemmer, 1981). This has resulted in fruit devoid of the adhering albedo, and sections that were easily separated. The mode of action of the process is the de-polymerization of middle lamellar pectins by the food-grade pectin-degrading enzymes, resulting in digested albedo and diminished segment membrane integrity. The resulting unpasteurized, dry-pack segments could provide a product with the quality of fresh fruit if flavor, texture, and microbial deterioration could be controlled (Baker and Bruemmer,

1989). This procedure and similar technologies have been commercialized in Florida (Stanley, 1996) and South Africa.

The vacuum infusion method requires that freshly washed orange or grapefruit be warmed to 30°C core temperature by immersion in warm water or holding overnight at 32°C. Fruit is then immersed in boiling water for 90 seconds to minimize contamination of segments by surface microflora, followed by scoring radically three times from stem to blossom end, which penetrates the flavedo and facilitates infusion of the enzyme solution. The technology is temporarily licensed to Pre-peeled Fruit Inc. in Groveland, Florida, and the pre-peeled product is marketed in Florida grocery stores (Stanley, 1996).

Other methods of peeling citrus have been developed. One patented by Adams and Kirk (1991) involves pressure rather than vacuum infusion of enzymes. Vacuum or pressure can be used to infuse water (without enzymes) into the pectin rich inner portion of the peel albedo (Pao et al., 1996). Infusion of water alone facilitates peeling and results in firm fruit with less juice leakage than with enzyme peeling (Pao et al., 1996), but a thin layer of albedo tissue and threads remains attached to the peel.

While these methods are promising, they each have inherent problems. One is microbial instability. Because the segments are not pasteurized, they contain indigenous microflora. To monitor microbial problems, grapefruit segments were stored in polyethylene bags, with or without 0.2% (w/v) potassium sorbate, at 2°C and examined periodically for yeast and bacteria. While bacterial contamination of segments was found to be low and declined during storage, initially low yeast contamination increased rapidly after 3 weeks of storage. This was controlled by 0.2% potassium sorbate dips (Baker and Bruemmer, 1989). Pao et al. (1998) challenged peeled sections with human pathogens including *E. coli* 0157:H7, *L. monocytogenes*, and *Staphylococcus aureus*. Refrigeration of inoculated sections reduced growth of all pathogens and caused population reduction of *Salmonella* spp. and *S. aureus*, but growth was observed with all pathogens at 24°C. Juice leakage may reduce the pH of contacted surface portions of the cut or peeled fruit. The reduction of surface pH inhibits some competitive microflora but favors the growth of spoilage yeasts and molds (Pao et al., 1997). Irradiation of fruit salad containing cantaloupe, pineapples, and orange sections using 0.5 kGy reduced the microbial populations by more than 90%, but populations subsequently increased during storage (Hagenmaier and Baker, 1997).

Unlike the fresh-cut or peeled products, unpasteurized juices from apple and citrus have been commercially available for many years and have a record of safety problems (Parish, 1997) (Table 2.2). Inoculated *Salmonellae* survived 27 days at pH 3.5, 46 days at pH 3.8, 60 days at pH 4.1, and 73 days at pH 4.4 in inoculated pasteurized orange juice (Parish et al., 1997). In 1996, nonpasteurized apple juice and juice blends from a processing facility in California were determined to be contaminated with *E. coli* 0157:H7 (CFDC, 1996). Freshly pressed, unpreserved apple cider can support *E. coli* O157:H7 organisms; the risk for this can be reduced by washing and brushing apples before pressing and by preserving the cider with sodium benzoate (Besser et al., 1993).

In 1995 a salmonellosis outbreak from orange juice implicated a citrus-processing facility and was linked to amphibians (Parish 1997; CFDC, 1995). *Salmonella* cells in the orange juice were associated with population levels of fecal coliforms and *E.*

TABLE 2.2
Disease Outbreaks from Consumption of Apple and Orange Juices

Year	Disease Vehicle	Causative Microorganism	Ref.
1923	Sweet cider	*Salmonella typhi*	Paquet, 1923
1994	Orange juice	*S. typhi*	Duncan et al., 1946
1962	Orange juice	Hepatitis A	Eisenstein et al., 1963
1966	Orange juice	Gastroenteritis agent	Tabershaw et al., 1967
1974	Apple cider	*S. Typhimurium*	Centers for Disease Control, 1967
1980	Apple cider	Enterotoxigenic *E. coli*	Steele et al., 1982
1989	Orange juice	*S. Typhi*	Birkhead et al., 1993
1991	Apple cider	*E. coli* 0157:H7	Besser et al., 1993
1992	Orange juice	Enterotoxigenic *E. coli*	Singh et al., 1995
1993	Apple cider	*Cryptosporidium*	Millard et al., 1994
1994	Orange juice	Gastroenteritis agent	FDA, 1994
1995	Orange juice	*S. hartford, S. gaminara, S. rubislaw*	Centers for Disease Control, 1995
1995	Orange juice	*E. coli*	Singh et al., 1995
1996	Apple juice	*E. coli* 0157:H7	Centers for Disease Control, 1996
1996	Apple juice	*E. coli* 0157:H7	Centers for Disease Control, 1997
1996	Apple juice	*Cryptosporidium parvum*	Centers for Disease Control, 1997

Source: Modified from Parish, M.E., *Crit. Rev. Microb.*, 23:109–119, 1997.

coli (Parish, 1998). Inoculation of *S. hartford* into refrigerated orange juice resulted in a maintained population level for 5, 10, 15, and 20 days at pH 3.5, 3.8, 4.1, and 4.4, respectively (Parish et al., 1997). Inactivation of *Lactobacillus plantarum* in orange–carrot juice was accomplished by means of high-intensity pulsed electric fields as a nonthermal preservation method (Rodrigo et al., 2001). Although non-thermal pasteurization of fruit juices has been demonstrated, it is not yet commercially practical. Isostatic high pressure, pulsed light, pulsed electric field, and filtration are several methods reported in the literature (Parish, 1997). A combined low-temperature, high-pressure treatment reduced counts of *E. coli* O157:H7 and various serovars of *Salmonella* in fruit juices. The pathogens were found to be most sensitive in grapefruit juice and least sensitive in apple juice at 615 MPa for 2 min at 15°C (Teo et al., 2001). UV light pasteurization is being tested on apple juice under simulated commercial conditions at a pilot plant at Illinois Institute of Technology's National Center for Food Safety and Technology in Sumit Argo, Illinois (Hollingsworth, 2001).

SPROUTS AND SPROUT SEEDS

Sprouts from a large variety of seeds (alfalfa, clover, broccoli, radish, and sunflower) are usually consumed raw, whereas mung and soy sprouts are usually cooked. The seeds are germinated and grown hydroponically in trays or drums; growth temperature will vary depending on the time of year and geographic location of the sprouting facility (Beuchat, 1996). The ambient temperature and moist environment are ideal for raising sprouts, but they are also ideal for microbial growth. In the past decade, more than 12 reported outbreaks of sprout-related illness worldwide have been reviewed extensively (Beuchat, 1996; NACMCF, 1999a, b; Taormina et al., 1999; Tauxe et al., 1997; Thompson and Powell, 2000). Of the pathogens isolated, *Salmonella* spp. is the most common, followed by *E. coli* O157:H7 (27) and, more recently, *Klebsiella pneumoniae* (Thompson and Powell, 2000). As a result of these outbreaks, two recent reports were issued on the microbiological safety (NACMCF, 1999a, b) and the risk factors associated with consumption of fresh sprouts (Thompson and Powell, 2000). Of the different varieties of sprouts, alfalfa sprouts were found to have the largest number of outbreaks associated with their consumption. In 1999, the U.S. Department of Health and Human Services advised consumers in a press release of the risks associated with eating raw sprouts, especially alfalfa sprouts (U.S. Dept. Health and Human Services, 1999).

The pathogens found on or in sprouts (*Salmonella* spp. and *E. coli* O157:H7) originate from the seeds used for sprouting (NACMCF, 1999a, b; Thompson and Powell, 2000). In addition to these pathogens, Castro-Rosas and Estartín (2000) reported that alfalfa sprouts grown from *V. cholerae* O1-contaminated seeds were positive for the microorganism, even after treating the seeds with 100 mg/L of free chlorine from sodium hypochlorite.

Rice seeds are one of the other seeds being considered for sprouting because of the increased nutrient benefits. Piernas and Guiraud (1997) found that a treatment of 5 min at 60°C in a 1000-ppm solution of sodium hypochlorite resulted in up to a 5-log reduction in aerobic plate counts without affecting germination. There was no information whether this treatment would be effective against pathogens. To date, in the U.S. the only approved FDA guideline for the sprout industry is the use of 20,000 ppm of calcium hypochlorite solution, which still cannot guarantee a safe product. Dr. William Fett of the Agriculture Research Service (Wyndmoor, PA) stated through personal communication that the 20,000-ppm treatment with calcium hypochlorite solution does negatively affect the germination of some sprout seed varieties such as wheat. The other limiting factors are the disposal problems and safety of the workers.

EMERGING PATHOGENS

Three pathogens isolated from fruits and vegetables currently being studied are *Campylobacter, K. pneumoniae,* and *Shigella* spp. *C. jejuni* is the most common reported cause of food-borne infection in the U.S. (Altekruse et al., 1999) and is now more common than salmonellosis in England, Canada, and Australia (Park and Sanders, 1992). Two produce-related *C. jejuni* outbreaks occurred between 1973 and 1989 (Smith and Tamplin, 2000). In their sampling of fresh produce, Park and Sanders (1992) reported that *Campylobacter* isolates were found at rates of 1.6 to 3.3% in unwashed vegetables.

The other pathogen, *K. pneumoniae*, was believed to have contaminated alfalfa sprouts recalled in Canada in 1999 (Wargurton, 1999). Since this recall, Health Canada has developed a method for the isolation and identification of *K. pneumoniae* (Warburton, 1999).

Shigella spp. was isolated and identified from the domestic and imported fresh fruits and vegetables surveyed by the FDA (http://www.cfsan.fda.gov/~dms/prodsur9/htm and ~dms/prodsur6.htm).

CAUSES OF PRODUCE CONTAMINATION

The means by which fresh produce becomes contaminated with bacteria are outlined in reviews by Beuchat (1996) and Sumner and Peters (1997). Cross-contamination occurs during processing in the cutting or shredding operation (Garg et al., 1990). During this operation, proper cleaning is recommended (NACMCF, 1999b). There are concerns about the adequacy of some sanitation practices for fruit destined for unpasteurized juice or fresh-cut products because the raw fruit quality and degree of contamination affects the fresh-cut or juice product (Eleftheriadou et al., 1998). Sometimes the microflora on fruit from the field are not sufficiently removed in the packing and processing houses (Pao and Brown, 1998).

Microorganisms present on the peel contaminate the cut surface of fruits and vegetables after cutting. This can be surveyed, as was citrus in Florida (Pao and Brown, 1998). Citrus fruit surface microbial populations were monitored during washing and waxing in several commercial Florida packinghouses. The average total aerobic plate counts and yeast and fungal counts were 4.0 log CFU/cm^2 and 3.3 log CFU/cm^2, respectively, for incoming fruits and 2.1 log CFU/cm^2 and 1.3 log CFU/cm^2, respectively, after processing (washing and waxing). In this survey, no *Salmonellae* were found at any point and no *E. coli* was found after processing. When *E. coli* was applied to the fruit prior to processing at 4.8 log CFU/cm^2, the levels were reduced to 1.4 log CFU/cm^2 after washing and waxing (Pao and Brown, 1998).

Sapers et al. (1999) investigated better ways to sanitize apples contaminated with *E. coli*. They compared commercial washing formulations with 200 ppm Cl_2 and 5% H_2O_2 or combinations of H_2O_2 and commercial detergent formulations, both heated at 50°C or unheated and applied to apples inoculated with nonpathogenic *E. coli*. Heated commercial formulations resulted in a 2.5-log reduction in *E. coli* populations, 200-ppm Cl_2 resulted in a 2-log reduction, and heated H_2O_2 combined with surfactants resulted in a 3- to 4-log reduction in *E. coli* load.

Soil and water present on the surface of produce can support the growth and survival of pathogens. *L. monocytogenes* was detected in soil and vegetation and isolated from sewage sludge cake, commonly used as agricultural fertilizer in Iraq (Al-Gahazali and Al-Azawi, 1990). Tauxe et al. (1997) suggested that the increased use of manure instead of chemical fertilizers, especially in less developed countries, could introduce food-borne pathogens into the soil if the manure is not treated properly. Bryan (1977) reviewed those illnesses associated with contaminated foods by wastewater. The microbes listed in Table 2.1 are also listed in Bryan's review (1977). Rajkowski and Rice (1999) reported that when the coliform growth response (a microbial assay of water quality) of wastewater was greater than 2, *E. coli* O157:H7, *Salmonella* spp., and *V. cholerae* spp. could survive and grow.

There is also concern that farm workers harvesting produce from fields irrigated with wastewater can be a source of cross-contamination. A study by Ait Melloul and Hassani (1999) observed farm workers irrigating many vegetables, some of which were consumed raw. The children of these farm workers had a higher incidence of salmonellosis than those children in other geographic areas. The affected children were observed playing in the field where the untreated wastewater was used as irrigation. The families of these farm workers also ate raw vegetables harvested from these fields.

With increased transportation and demand for fresh fruits and vegetables, the produce of such irrigated fields may be the source of future food-borne outbreaks. In addition to wastewater, river water is reported to support the growth of pathogens (Rajkowski and Rice, 1999).

Robinson and Adams (1978) studied the effect of ultraviolet treatment on contaminated irrigation water used to irrigate a celery crop. Their results showed that the u-v treatment was effective in cleaning up the polluted water supply. If polluted waters were used at any point in the fruit or vegetable operation, the produce, too, became contaminated (Ait Melloul and Hassani, 1999; Beuchat, 1996; Castro-Rosas and Escartin, 2000). Proper disinfecting of water used in produce operations is essential and, when economically feasible, sanitizing agents such as chlorine should be used. As with any water used on fresh produce, the microbiological quality of the irrigation water should be monitored.

SOLUTION

Each step in processing affects the microflora of fresh-cut vegetables (Garg et al., 1990), as does postprocessing handling and packing (Zagory, 1999). Identifying each point in the process and the possible ways contamination can occur is essential. The citrus packing line study is such an example (Pao and Brown, 1998). The National Advisory Committee on Microbiological Criteria for Foods (NACMCF, 1999b) has recommended several steps to assure microbiological safety of fresh produce and has developed seven specific recommendations. They are (1) good agricultural practices, (2) good manufacturing practices, (3) hazard analysis critical control point programs (HACCP), (4) training, (5) risk assessment, (6) research, and (7) better produce identification–tracing systems to investigate produce-related outbreaks. HACCP was considered the best guideline to assure produce safety. Kvenberg et al. (2000) developed a generic HACCP plan mandated for production of fruit and vegetable juices. In addition to HACCP, in the processing of fresh fruit and vegetables, guidance to ensure the microbiological safety of the produce for small catering operations worldwide was outlined by Mossel et al. (1999). Their guidelines parallel those recommended by NACMCF and their endeavor was to provide an unconditionally safer food supply.

FUTURE

With the proposed guidelines in place, there are still areas of research needed to assure a safe produce supply. Doyle and Mazzotta (2000) reviewed studies on the heat resistance of *Salmonellae* and suggested that research on the thermal resistance of *Salmonella* in fruit juices is needed. Research is also needed to determine the radiation resistance of the pathogens on seeds used for sprouts and other fresh-cut fruits and vegetables. As mentioned previously, not one sanitizer is effective at the same exposure time, temperature, and strength for all possible pathogens and products. Research is needed to find one universal sanitizer that would be effective for all edible fruit and vegetable products, resulting in safer produce supply.

REFERENCES

Ackers, M.L. et al., 1998, An outbreak of *Escherichia coli* O157:H7 infections associated with leaf lettuce consumption, *J. Infect. Dis.*, 177:1588–1593.

Adams, B. and Kirk, B., 1991, Process for enzyme peeling of fresh citrus fruit, U.S. Patent 5,000,967, March 19, 1991.

Agar, I.T. et al., 2000, Postharvest CO_2 and ethylene production and quality maintenance of fresh-cut kiwifruit slices, *J. Food Sci.*, 64:433–440.

Ait Melloul, A. and Hassani, L., 1999, *Salmonella* infection in children from wastewater-spreading zone of Marrakesh city (Morocco), *J. Appl. Microbiol.*, 87:536–539.

Al-Ghazali, M.R. and Al-Azawi, K., 1990, *Listeria monocytogenes* contamination of crops grown on soil treated with sewage sludge cake, *J. Appl. Bacteriol.*, 69:642–647.

Altekruse, S.F. et al., 1999, *Campylobacter jejuni* — an emerging foodborne pathogen, *Emerging Infect. Dis.*, 5:28–35.

Annous, B.A. et al., 2001, Efficacy of washing with a commercial flatbed brush washer, using conventional and experimental washing agents, in reducing populations of *Escherichia coli* on artificially inoculated apples, *J. Food Prot.*, 64:159–163.

Anonymous, 1998a, Agricultural Marketing Service, Quality through verification program for the fresh-cut produce industry, The Federal Register, 63(172) (September 4, 1998): 47220–47224.

Anonymous, 1998b, Food and Drugs. Food labeling. Code of Federal Regulations. 21. Section 101.95151.

Anonymous, 1998c, Convenience driving fresh apple slices, *Fresh Cut*, December.

Baker, R.A. and Bruemmer, J.H., 1989, Quality and stability of enzymically peeled and sectioned citrus fruit, in *Quality Factors of Fruits and Vegetables: Chemistry and Technology*, Jen, J.J., Ed., ACS Symposium Series 405, Washington, D.C.: American Chemical Society, 140–148.

Baldwin, E.A. et al., 1996, Improving storage life of cut apple and potato with edible coating, *Postharvest Biol. Technol.*, 9:151–163.

Bean, N.H. et al., 1996, Surveillance for foodborne disease outbreaks United States, 1988–1992, MMWR 45, #SS-5.

Besser, R.E. et al., 1993, An outbreak of kearrhea and hemolytic uremic syndrome from *Eshcerichia coli* O157:H7 in fresh pressed apple cider, *J. Am. Med. Assoc.*, 269:2217–2220.

Beuchat, L.R., 1996, Pathogenic microorganisms associated with fresh produce, *J. Food Prot.*, 59:204–216.

Birkhead, G.S. et al., 1993, Typhoid fever at a resort hotel in New York: a large outbreak with an unusual vehicle, *J. Infect. Dis.*, 167:1228–1232.

Boynton, B.B., 1999, Quality and stability of pre-cut mangoes and carambolas subjected to high pressure processing, M.S. thesis, University of Florida, Gainesville.

Brecht, J.K., 1995, Physiology of lightly processed fruits and vegetables, *Hortic. Sci.*, 30:18–22.

Bruemmer, J.H., 1981, Method of preparing citrus fruit sections with fresh fruit flavor and appearance, U.S. Patent #4284651.

Bruemmer, J.H. and Griffin, A.W., 1978, Sectionizing grapefruit by enzyme digestion, *Proc. Fla. State. Hortic. Soc.*, 91:112–114.

Bruemmer, J.H., Griffin, A.W., and Onayemi, O., 1978, Sectionizing grapefruit by enzyme digestion, *Proc. Fla. State Hort. Soc.*, 91:112–114.

Bryan, F., 1977, Diseases transmitted by foods contaminated by wastewater, *J. Food Prot.*, 40:45–56.

Buchanan, R.L. et al., 1999, Contamination of intact apples after immersion in an aqueous environment containing *Escherichia coli* O157:H7, *J. Food Prot.*, 62:444–450.

Cantwell, M.C., 1995, Food safety: microbiological concerns, *Perishables Handling Newsl.*, 81:15–16.

Castro-Rosas, J. and Escartín, E.F., 2000, Survival and growth of *Vibrio cholerae* O1, *Salmonella typhi*, and *Escherichia coli* O157: H7 in alfalfa sprouts, *J. Food Sci.*, 65:162–165.

CFDC: Centers for Disease Control, 1995, Outbreak of *Salmonella Hartford* infections among travelers to Orlando, Florida, EPI-AID Trip Rep., 95–92.

CFDC: Centers for Disease Control, 1996, Outbreak of *Escherichia coli* O157:H7 infections associated with drinking unpasteurized commercial apple juice — British Columbia, California, Colorado and Washington, October, *Morbidity Mortality Wkly. Rep.*, 45:975.

CFDC: Centers for Disease Control, 1997, Outbreaks of *Eshcerichia coli* O157:H7 infection and cryptosporidiosis associated with drinking unpasteurized apple juice — Connecticut and New York, October 1996, *Morbidity Mortality Wkly. Rep.*, 46:4.

Cody, S.H. et al., 1999, An outbreak of *Escherichia coli* O157:H7 infection from unpasteurized commercial apple juice, *Ann. Intern. Med.*, 130:202–209.

Coma, V. et al., 2001, Antimicrobial edible packaging based on cellulosic ethers, fatty acids, and nisin incorporation to inhibit *Listeria innocua* and Staphylococcus aureus, *J. Food Prot.*, 64:470–475.

Conway, W.S., Leverentz, B., and Saftner, R.A., 2000, Survival and growth of *Listeria monocytogenes* on fresh-cut apple slices and its interaction of *Glomerella cingulata* and *Penicillium expansum*, *Plant Dis.*, 84:177–181.

Cook, K.A. et al., 1998, Outbreak of *Salmonella* serotype Hartford infections associated with unpasteurized orange juice, *JAMA*, 280:1504–1509.

Corbo, M.R. et al., 2000, Effects of hexanal, *trans*-2-hexenal, and storage temperature on shelf-life of fresh sliced apples, *J. Agric. Food Chem.*, 48:2401–2408.

Dong, X., Wrolstad, R.E., and Sugar, D., 2000, Extending shelf-life of fresh-cut pears, *J. Food. Sci.*, 65:181–186.

Doyle, M.E. and Mazzotta, A.S., 2000, Review of studies on the thermal resistance of *Salmonellae*, *J. Food Prot.*, 63:779–795.

Duncan, T.G. et al., 1946, Outbreak of typhoid fever with orange juice as the vehicle, illustrating the value of immunization, *Am. J. Pub. Health*, 36:34.

Eisenstein, A.B. et al., 1963, An epidemic of infectious hepatitis in a general hospital, *JAMA*, 185:171.

Eleftheriadou, M. et al., 1998, Factors affecting quality and safety of freshly squeezed orange juice (FSOJ), *Dairy, Food Environ. Sanit.*, 18:14–23.

Enfors, S.O. and Molin, G., 1978, The influence of high concentrations of carbon dioxide on the germination of bacterial spores, *J. Appl. Bacteriol.*, 45:279–285.

Escudero, M.E. et al., 1999, Effectiveness of various disinfectants in the elimination of *Yersinia enterocolitica* on fresh lettuce, *J. Food Prot.*, 62:665–669.

Farber, J.M., 1991, Microbiological aspects of modified-atmosphere packaging — a review, *J. Food Prot.*, 54:58–70.

Fleischman, G.J. et al., 2001, Hot water immersion to eliminate *Escherichia coli* O157:H7 on the surface of whole apples: thermal effects and efficacy, *J. Food Prot.*, 64:451–455.

Foegeding, P.M. and Busta, F.F., 1983, Effect of carbon dioxide nitrogen and hydrogen gases on germination of *Clostridium botulinum* spores, *J. Food Prot.*, 46:987–989.

FDA: Food and Drug Administration, 2001, http://www.cfsan.fda.gov/~dms/prodsur6.html.

FDA: Food and Drug Administration, 2001, http://www.cfsan.fda.gov/~dms/prodsur9.html.

FDA: Food and Drug Administration, 1994, Recalls and field corrections: foods – class II, April 20.

Fowler, J.L. and Foster, J.F., 1976, A microbiological survey of three fresh green salads — can guidelines be recommended for these foods, *J. Milk Food Technol.*, 39:111–113.

García-Gimeno, R.M. and Zurera-Cosano, G., 1997, Determination of ready to eat vegetable salad shelf-life, *Int. J. Food Microbiol.*, 36:31–38.

Garg, N., Churey, J.J., and Splittstoesser, D.F., 1990, Effect of processing conditions on the microflora of fresh-cut vegetables, *J. Food Prot.*, 53:701–703.

Gorny, J.R., Hess-Pierce, B., and Kader, A.A., 1998, Effects of fruit ripeness and storage temperature on the deterioration rate of fresh-cut peach and nectarine slices, *Hortic. Sci.*, 33:110–113.

Gorny, J.R., Hess-Pierce, B., and Kader, A.A., 1999, Quality changes in fresh-cut peach and nectarine slices as affected by cultivar, storage atmosphere and chemical treatments, *J. Food Sci.*, 64:429–432.

Gorny, J.R. et al., 2000, Quality changes in fresh-cut pear slices as affected by cultivar, ripeness stage, fruit size, and storage regime, *J. Food Sci.*, 65:541–544.

Greenleaf, C., 1999, Making the cut, *Amer. Veg. Grower*, March:10–11.

Hagenmaier, R.D., and Baker, R.A., 1997, An evaluation of gamma irradiation for preservation of citrus salads in flexible packaging, *Proc. Fla. State Hortic. Soc.*, 110:243–245.

Hagenmaier, R.D., and Baker, R.A., 1998, A survey of the microbial population and ethanol content of bagged salad, *J. Food Prot.*, 61:357–359.

Hodge, K., 1995, Fresh-cut and the "perfect meal," *Fresh Cut*, 3(8):12.

Hotchkiss, J.H. and Banco, M.J., 1992, Influence of new packaging technologies on the growth of microorganism in produce, *J. Food Prot.*, 55:815–820.

IFPA: International Fresh Cut Association, 2001, http://www.fresh-cuts.org.

Jacxsens, L. et al., 1999, Behavior of *Listeria monocytogenes* and *Aeromonas* spp. on fresh-cut produce packaged under equilibrium-modified atmosphere, *J. Food Prot.*, 62:1128–1135.

Kakiomenou, K., Tassou, C., and Nychas, G.J., 1998, Survival of *Salmonella enteritidis* and *Listeria monocytogenes* on salad vegetables, *World J. Microbiol. Biotechnol.*, 14:383–387.

Ke, D. and Saltveit, M.E., 1989, Wound-induced ethylene production, phenolic metabolism and susceptibility to russet spotting in iceberg lettuce, *Physiol. Plant.*, 76:412–418.

King, A.D. and Bolin, H.R., 1989, Physiological and microbiological storage stability of minimally processed fruits and vegetables, *Food Technol.*, 43:132–135, 139.

King, A.D., Jr. et al., 1991, Microbial flora and storage quality of partially processed lettuce, *J. Food Sci.*, 56:459–461.

Kvenberg, J. et al., 2000, HACCP development and regulatory assessment in the United States of America, *Food Control*, 11:387–401.

Lakakul, R., Beaudry, R.M., and Hernandez, R.J., 1999, Modeling respiration of apple slices in modified-atmosphere packages, *J. Food Sci.*, 64:105–110.

Larson, A.E. and Johnson, E.A., 1999, Evaluation of botulinal toxin production in packaged fresh-cut cantaloupe and honeydew melons, *J. Food Prot.*, 62:948–952.

Liao, C.-H. and Sapers, G.M., 2000, Attachment and growth of *Salmonella Chester* on apple fruits and *in vivo* response to attached bacterial to sanitizer treatment, *J. Food Prot.*, 63:876–883.

Manvell, P.N. and Ackland, M.R., 1986, Rapid detection of microbial growth in vegetable salads at chill and abuse temperatures, *Food. Microbiol.*, 3:59–65.

Maxcy, R.B., 1978, Lettuce salad as a carrier of microorganisms of public health significance, *J. Food Prot.*, 41:435–438.

McHugh, T.H. and Senesi, E., 2000, Apple wraps: a novel method to improve the quality and extend the shelf-life of fresh-cut apples, *J. Food Sci.*, 65:480–485.

Mead, P.S. et al., 1999, Food-related illness and death in the U.S., *Emerging Infec. Dis.*, 5:607–625.

Millard, P.S., Gensheimer, K.F., and Addiss, D.G., 1994, An outbreak of cryptosporidiosis from fresh pressed apple cider, *JAMA*, 272:1592–1596.

Miller, L.G. and Kasper, C.W., 1994, *Escherichia coli* O157:H7 acid tolerance and survival in apple cider, *J. Food Prot.*, 57:460–464.

Mossel, D.A.A., Jansen, J.T., and Struijk, C.B., 1999, Microbiological safety assurance applied to smaller catering operations world-wide. From angst through ardour to assistance and achievement, *Food Control*, 10:195–211.

Mpuchane, S.F. and Gashe, B.A., 1996, Presence of *Escherichia coli*, *Klebsiella pneumoniae* and *Enterobacter* species in dried bush okra (*Corchorus olitorius*) and African spider herb (*Cleome gynandra*), *Food Control*, 7:169–172.

NACMCF: National Advisory Committee on Microbiological Criteria for Foods, 1999a, Microbiological safety evaluation and recommendations on sprouted seeds, *Int. J. Food Microbiol.*, 52:123–153.

NACMCF: National Advisory Committee on Microbiological Criteria for Foods, 1999b, Microbiological safety evaluation and recommendations on fresh produce, *Food Control*, 10:117–143.

Nguyen-The, C. and Carlin, F., 1994, The microbiology of minimally processed fresh fruits and vegetables, *Crit Rev. Food Sci. Nutr.*, 34:371–401.

Odumeru, J.A. et al., 1997, Assessments of the microbiological quality of ready-to-use vegetables for health-care food services, *J. Food Prot.*, 60:954–960.

Olsen S.J. et al., 2000, Surveillance for foodborne-disease outbreaks — U.S., 1993–1997, *Mortality Morbidity Wkly. Rep.*, 49:SS-1.

Pao, S. and Brown, G.E., 1998, Reduction of microorganisms on citrus fruit surfaces during packinghouse processing, *J. Food Prot.*, 61:903–906.

Pao, S., Davis, C.L., and Kelsey, D.F., 2000, Efficacy of alkaline washing for the decontamination of orange fruit surfaces inoculated with *Escherichia coli*, *J. Food Prot.*, 63:961–964.

Pao, S., Petracek, P.D., and Brown, G.E., 1996, Effect of infusion method on peel removal and storage quality of citrus, *Hortic. Technol.*, 6:409–413.

Pao, S., Widmer, W.W., and Petracik, P.D., 1997, Effects of cutting on juice leakage, microbiological stability and bitter substances of peeled citrus, *Libensm. Wiss. U. Technol.*, 30:670–675.

Pao, S., Brown, G., and Schneider, K.R., 1998, Challenge studies with selected pathogenic bacteria on freshly peeled hamlin orange, *J. Food Sci.*, 63:359–362.

Paquet, P., 1923, Epidemie de fievre typhoide determinee par la consommation de petit cidre, *Revue d'Hygiene*, 45:165.

Parish, M.E., 1997, Public health and nonpasteurized fruit juices, *Crit. Rev. Microbiol.*, 23:109–119.

Parish, M.E., 1998, Coliforms, *Escherichia coli* and *Samonella serovars* associated with a citrus processing facility implicated in salmonellosis outbreak, *J. Food Prot.*, 61:280–284.

Parish, M.E., Narciso, J.A., and Fredrich, L.M., 1997, Survival of *Salmonellae* in orange juice, *J. Food Safety*, 17:273–281.

Park, C.E. and Sanders, G.W., 1992, Occurrence of thermotolerant campylobacters in fresh vegetables sold at farmers' outdoor markets and supermarkets, *Can. J. Microbiol.*, 38:313–316.

Park, C.M. and Beuchat, L.R., 1999, Evaluation of sanitizers for killing *Escherichia coli* O157:H7, *Salmonella* and naturally occurring microorganisms on cantaloupes, honeydew melons, and asparagus, *Dairy Food Environ. Sanit.*, 19:842–847.

Piernas, V. and Guiruad, J.P., 1997, Disinfection of rice seeds prior to sprouting, *J. Food. Sci.*, 62:611–615.

Prakash, A. et al., 2000, Effects of low-dose gamma irradiation on the shelf-life and quality characteristics of cut romaine lettuce packaged under modified atmosphere, *J. Food Sci.*, 65:549–553.

Rajkowski, K.T. and Rice, E.W., 1999, Recovery and survival of *Escherichia coli* O157:H7 in reconditioned pork-processing wastewater, *J. Food. Prot.*, 62:731–734.

Rajkowski, K.T. and Thayer, D.W., 2000, Reduction of *Salmonella* spp. and strains of *Escherichia coli* O157:H7 by gamma radiation of inoculated sprouts, *J. Food. Prot.*, 63:871–875.

Robinson, I. and Adams, R.P., 1978, Ultra-violet treatment of contaminated irrigation water and its effect on the bacteriological quality of celery harvest, *J. Appl. Bacteriol.*, 45:83–90.

Rodrigo, D. et al., 2001, Study of inactivation of *Lactobacillus plantarum* in orange-carrot juice by means of pulsed electric fields: comparison of inactivation kinetics models, *J. Food Prot.*, 64:259–263.

Saltveit, M.E., 1997, Physical and physiological changes in minimally processed fruits and vegetables, in *Phytochemistry of Fruits and Vegetables*, Tomas-Barberan, F.A. and Robins, R.J., Eds., London: Oxford University Press, 205–220.

SafetyAlerts, 2001, Crunch park apple packages recalled in 17 states, http://www.safetyalerts.com/reca/f/01/crnckpak.htm.

Sapers, G.M., Garzarella, L., and Pilizota.V., 1990, Application of browning inhibitors to cut apple and potato by vacuum and pressure infiltration, *J. Food Sci.*, 55:1049–1053.

Sapers, G.M., Miller, R.L., and Mattrazzo, A.M., 1999, Effectiveness of sanitizing agents in inactivating *Escherichia coli* in Golden Delicious apples, *J. Food Sci.*, 64:734–737.

Sapers, G.M. and Ziolkowski, M.A., 1987, Comparison of erythorbic and ascorbic acids as inhibitors of enzymatic browning in apple, *J. Food Sci.*, 52:1732–1733,1747.

Satchell, F.B. et al., 1990, The survival of *Shigella sonni* in shredded cabbage, *J. Food Prot.*, 53:558–562.

Singh, B.R., Kulshreshtha, S.B., and Kapoor, K.N., 1995, An orange juice-borne diarrhoeal outbreak due to enterontoxigenic *Escherichia coli*, *J. Food Sci. Technol.*, 32:504–506.

Smith, J.L. and Tamplin, M., 2000, Produce as a vector for Campylobacteriosis, http://www.arserrc.gov/mfs/HotSeat.htm.

Song, J. et al., 1996, Hexanal vapor is a natural metabolizable fungicide: inhibition of fungal activity and enhancement of aroma biosynthesis in apple slices, *J. Am. Soc. Hortic. Sci.*, 121:937–942.

Stanley, D., 1996, Zip-skinned citrus, *Agric. Res.*, 44:8–9.

Steele, B.T., Murphy, N., and Rance, C.P., 1982, An outbreak of hemolytic uremic syndrome associated with ingestion of fresh apple juice, *J. Pediatr.*, 101:963–965.

Sumner, S. and Peters, D.L., 1997, Microbiology of vegetables, in *Processing Vegetable Science and Technology*, Smith, D.S. et al., Eds., Lancaster, PA: Technomic Publishing Co., Inc., 87–106.

Tabershaw, I.R., Schmelzer, L.L., and Bruyn, H.B., 1967, Gastroenteritis from an orange juice preparation, *Arch. Environ. Health*, 15:72–77.

Takeuchi, K. and Frank, J.F., 2001, Quantitative determination of the role of lettuce leaf structures in protecting *Escherichia coli* O157:H7 from chlorine disinfection, *J. Food Prot.*, 64:147–151.

Taormina, P.J., Beuchat, L.R., and Slutsker, L., 1999, Infections associated with eating seed sprouts: an international concern, *Emerging Infect. Dis.*, 5:626–634.

Tauxe, R. et al., 1997, Microbial hazards and emerging issues associated with produce: a preliminary report to the National Advisory Committee on Microbiological Criteria of Foods, *J. Food Prot.*, 60:1400–1408.

Thompson, S. and Powell, D.A., 2000, Risks associated with the consumption of fresh sprouts, Food Safety Network Technical Report #16, http://www.plant.uoguelph.ca/safefood/micro-haz/sprouts-risk-sylvanus-juloo.htm.

Ukuku, D.O. and Sapers, G.M., 2001, Effect of sanitizer treatments on *Salmonella* Stanley attached to the surface of cantaloupe and cell transfer to fresh-cut tissues during cutting practices, *J. Food Prot.*, 64:1286–1291.

U.S. Department of Health and Human Services, Consumer advised of risks associated with raw sprouts, Press Release No. P99–13, Washington, D.C., July 9, 1999.

Varoquaux, P. and Wiley, R.C., 1994, Biological and biochemical changes in minimally processed refrigerated fruits and vegetables, in *Minimally Processed Refrigerated Fruits and Vegetables*, Wiley, R.C., Ed., New York: Chapman and Hall, 226–268.

Velázquez, L. et al., 1998, Survival of *Aeromonas hydrophilia* in fresh tomatoes (*Lycopersicum esculentum* Mill) stored at different temperatures and treated with chlorine, *J. Food Prot.*, 61:414–418.

Warburton, D., 1999, Analysis of sprouts for coliforms, *Escherichia coli*, and *Klebsiella pneumoniae*, Government of Canada — Health Protection Branch, MFLP-64.

Wei, C.I. et al., 1995, Growth and survival of *Salmonella montevideo* on tomatoes and disinfection with chlorinated water, *J. Food Prot.*, 58:829–836.

Wells, J.M. and Butterfield, J.E., 1997, *Salmonella* contamination associated with bacterial soft rot of fresh fruits and vegetables in the marketplace, *Plant Dis.*, 81:867–872.

Wells, J.M. and Butterfield, J.E., 1999, Incidence of *Salmonella* on fresh fruit and vegetables affected by fungal rots or physical injury, *Plant Dis.*, 83:722–726.

Zagory, D., 1999, Effects of post-processing handling and packaging on microbial populations, *Postharvest Biol. Technol.*, 15:313–321.

3 The Microbial Safety of Minimally Processed Seafood with Respect to *Listeria monocytogenes*

Adam D. Hoffman, Kenneth L. Gall, and Martin Wiedmann

CONTENTS

1-58716-041-2/03/$0.00+$1.50

Listeria monocytogenes is a food-borne pathogen that is frequently isolated from ready-to-eat (RTE) foods and from food processing plant environments. It rarely causes human disease, but it is generally serious when it does occur. Recent estimates suggest that a total of 2500 human listeriosis cases occur annually in the U.S. (Mead et al., 1999). While only a few human listeriosis cases have been linked to the consumption of seafood worldwide, control of *L. monocytogenes* contamination represents a major concern for producers of RTE seafood products.

PRODUCTION OF MINIMALLY PROCESSED SEAFOOD

Seafood products can be separated into six major categories, each characterized by the type of thermal process or preservation technique used (Table 3.1). These categories range from products that are eaten raw, to lightly preserved or mildly heat-preserved smoked products, to dried, fermented, and fully heat-processed products. For the purpose of this chapter, two of these broad categories will be considered in the following discussion of minimally processed RTE seafood products. **Lightly preserved** seafood products are processed with little to no thermal processing and are preserved through the addition of salt, smoke, or acidic ingredients to decrease pH. **Mildly heat-preserved** seafood products include hot smoked fish and cooked crustaceans, which receive a thermal process but are not packed commercially sterile and may be handled prior to packaging.

The following seafood products do not fit our definition of minimally processed and are not discussed here:

- Raw RTE seafood, like raw fish and shellfish, that are harvested and eaten raw without any cooking or the addition of preservative ingredients
- Fresh and frozen seafood products cooked by the consumer before consumption
- Canned or other thermally processed products
- Dried fish with greater than 6% salt

For a thorough discussion of the incidence and behavior of *L. monocytogenes* in all fish and seafood products, other references such as Jinneman et al. (1999) may be useful.

Description of Lightly Preserved Seafood Products

Cold-smoked seafood probably represents the most important category of lightly preserved seafood products in terms of the amount consumed in the U.S. and the

TABLE 3.1
Seafood Categories and Relevant Characteristics

	Relevant Characteristics	Most Likely Sources of *L. monocytogenes* Contamination
Lightly Preserved		
Cold-smoked fish	<6% NaCl (w/w) water phase	Post-processing contamination
Gravlax	pH > 5.0	Raw material contamination
Salted fish		
Marinated fish		
Fermented fish		
Mildly Heat Preserved		
Hot-smoked fish	Thermally processed	Postprocessing contamination
Pre-cooked crustaceans		
Raw RTE Seafood		
Bi-valve molluscan shellfish	Harvested and consumed raw	Raw material contamination
Raw fish		Processing contamination
Heat-Processed		
Heat-processed and sealed	Commercially sterile	Consumer handling and preparation
Fresh Fish		
Fresh and frozen fish, shellfish, and crustaceans	Cooked by consumer	Consumer handling and preparation
Dried Fish		
Dried	>6% NaCl (w/w) water phase	Contamination unlikely
Dry-salted		
Smoked-dry fish		

potential for *Listeria* contamination. A variety of different types of fish and shellfish are cold-smoked. Salmon is the predominant cold-smoked fish product in the U.S.; other species that are cold-smoked include trout, tuna, marlin, sea bass, and haddock. The cold-smoking process is generally carried out at temperatures ranging from 18 to 24°C, which are insufficient to destroy microorganisms including *L. monocytogenes* (Gram and Huss, 2000; IFT, 2001). However, the smoking process does impart some preservative and antimicrobial characteristics to the fish as well as its distinctive flavor. Prior to the smoking, fish are brined using a solution of salt, sugar, and/or nitrite by soaking or injection. Some products are dry cured by applying salt and sugar directly to the surface of the product before smoking.

Current guidelines from the U.S. Association of Food and Drug Officials (AFDO) require air-packaged cold-smoked fish to have at least 2.5% water phase salt in the loin muscle of the finished product and vacuum- or modified atmosphere-packaged cold-smoked products to have 3.5% water phase salt or 3.0% water phase salt, and between 100 and 200 ppm sodium nitrite in the finished product (AFDO, 1991). The smoking process also dries the outer layer of the fish, lowering water activity and reducing opportunities for microbial growth. Current AFDO guidelines require that cold-smoked fish be produced by a controlled process that utilizes a temperature monitoring system to assure that the temperature in the chamber does not exceed 90°F during a drying and smoking period that does not exceed 20 hours, or that the temperature in the smoking chamber does not exceed 50°F during a drying and smoking period that does not exceed 24 hours.

Liquid smoke and natural smoke are commonly used in the smoked-fish processing industry. While the characteristics of smoke vary, a common characteristic is the presence of phenols and aldehydes that have anti-microbial properties. Reductions of *Listeria* during natural smoke application range between 1 and 3 logs, depending on the length and temperature of the smoking process (Sabanadesan et al., 2000). The reduction of *L. monocytogenes* by liquid smoke is highly dependent on the phenol concentration. At low levels of phenols (10 to 20 ppm), *L. monocytogenes* growth is inhibited, but numbers are not reduced. At relatively high levels of phenol (200 ppm), *L. monocytogenes* numbers decline over time (Thurette et al., 1998).

Gravlax (also known as **gravadlax**, gravlox, or dill salmon) is a marinated salmon product that originated in Scandinavia as a traditional method of preserving salmon. It is prepared by coating raw salmon fillets with a mixture of sugar, salt, pepper, and dill. Two coated fillets are pressed together under light pressure and stored under refrigeration for up to two days or more, depending on the product thickness, the temperature, and the amount of pressure applied (Dore, 1993). The finished marinated product is thinly sliced and used like traditional cold smoked "nova lox"-type products.

Semipreserved fish include a wide variety of fermented, marinated, spiced, pickled, and salted fish products. Many of these products have a history dating back to regional and cultural techniques and traditions that have been used for thousands of years. There are many variations in the ingredients and procedures used to produce these products. However, each product utilizes similar preservation techniques that create barriers or inhibit the growth of spoilage microorganisms and of pathogens like *L. monocytogenes*. Growth is inhibited by reducing water activity and pH, but *L. monocytogenes* and other microorganisms are not eliminated from the product. Several references provide additional information on the variety of different semi-preserved, specialty seafood products produced around the world (Jarvis, 1987; Gall et al., 2000).

Mildly Heat-Preserved Seafoods

Production of **hot-smoked seafood** is a process similar to the cold-smoking process described earlier, although it is carried out at higher temperatures. Hot-smoking temperatures may vary, but typical temperatures reported in the literature include 65°C for 20 minutes followed by 60° for 45 minutes (Heinitz and Johnson, 1998). Current AFDO guidelines require that the production of hot-smoked fish be air packaged and utilize a controlled process to heat fish to a continuous temperature of at least 145°F for a minimum of 30 minutes, and that brined fish contain no less than 2.5% water phase salt in the loin muscle of the finished product. For hot-smoked fish that will be vacuum or modified atmosphere packaged, the product must be heated to at least 145°F for at least 30 continuous minutes and contain not less than 3.5% water phase salt or 3.0% water phase salt and 100 to 200 ppm sodium nitrite in the loin muscle of the finished product. A variety of different fish and shellfish are hot smoked, including salmon and other salmonid species, bluefish, herring, whitefish, mussels, and other products. Jemmi and Keush (1992) found that when *L. monocytogenes* has been experimentally introduced to fish, it does not survive the hot-smoking process. Based on this work, they postulated that *Listeria* contamination of hot-smoked fish is due to postprocessing contamination.

A number of different types of **crustaceans** such as crabs, crayfish, and lobsters are cooked during processing. This step is utilized to facilitate the removal of the edible meat from the shell and to reduce the microbial load present in the raw product. Processing steps that occur after this heat treatment frequently involve contact with workers' hands or processing equipment. Precooked crustaceans that are ready to eat are susceptible to postprocessing contamination, even though the cooking step should adequately destroy any *Listeria* on the raw product. Of particular concern are products that are cooked and manually separated after cooking, because the risk of cross-contamination from employees or raw product to the finished product can be high.

RELEVANT CHARACTERISTICS OF *LISTERIA MONOCYTOGENES*

The genus *Listeria* contains five species, namely, *L. monocytogenes*, *L. ivanovii*, *L. innocua*, *L. seeligeri*, and *L. welshimeri*. All members of this genus are Gram-positive, non-sporeforming rods, closely related to the genera *Lactobacillus* and *Streptococcus*.

Virulence and Pathogenesis

Of the species classified in the genus *Listeria,* only *L. monocytogenes* and *L. ivanovii* are considered pathogens. While both are hemolytic, facultative intracellular pathogens, *L. ivanovii* is predominantly associated with disease (specifically abortions) in sheep. *L. monocytogenes*, on the other hand, causes human and animal disease and is the only *Listeria* species that represents a human public health concern. Human listeriosis can occur as sporadic cases and as epidemic outbreaks. Most epidemic outbreaks have been linked to consumption of contaminated foods and 99% of sporadic cases are also thought to be food-borne (Farber and Peterkin, 1991; Schuchat et al., 1991; Mead et al., 1999). The majority of human listeriosis cases occur in pregnant women, neonates, the immunosuppressed, and elderly individuals. Nevertheless, healthy people can also be infected by *L. monocytogenes.* Symptoms in these individuals include diarrhea, fever, and muscle aches; the disease is generally mild and self-limiting. More serious complications associated with human listeriosis include stillbirth, septicemia, and infections of the central nervous system (meningoencephalitis, encephalitis) (Rocourt et al., 2000). The Centers for Disease Control and Prevention (CDC) estimate that 2500 cases of clinical listeriosis occur annually in the U.S., resulting in a total of 500 deaths (Mead et al., 1999).

Data from the Foodborne Diseases Active Surveillance Network (FoodNet) found that, although human listeriosis is less common than many other food-borne diseases (e.g., those caused by *Escherichia coli* O157:H7, *Campylobacter*, or *Salmonella*), it is by far the most severe. *L. monocytogenes* has a 90% hospitalization rate as well as the highest case fatality rate, resulting in death for 20% of those infected (Mead et al., 1999). As an increasing segment of the population falls into high-risk groups for *L. monocytogenes* infection, e.g., the elderly, those with pre-existing diseases, and the immunocompromised (including AIDS patients, transplant recipients, and cancer patients), listeriosis will continue to be a major public health concern. Although high-risk groups have been identified epidemiologically, belonging to a high-risk group does not mean that exposure to *L. monocytogenes* will necessarily result in illness; the frequency of listeriosis in these groups is not high despite frequent exposure (Notermans and Hoornstra, 2000). While it is likely that exposure to fairly high doses is required for infection and disease, strain differences in virulence and differences in host susceptibility may also contribute to the fact that exposure to *L. monocytogenes* from contaminated foods rarely appears to cause disease (Notermans and Hoornstra, 2000; Norton et al., 2001b; Wiedmann et al., 1997).

Environmental Survival

All *Listeria* spp. are aerobic, microaerophilic, facultative anaerobic, catalase positive, oxidase negative, and esculin hydrolysis positive. *L. monocytogenes* is considered a ubiquitous organism, which can be isolated from many different environmental sources (surface water, soil, sewage, and vegetative material). This organism has the ability to grow and survive under a variety of different conditions. *L. monocytogenes* grows from close to 0 to 44°C and is thus considered a psychrotolerant organism.

Contamination of food with *L. monocytogenes* is compounded by growth characteristics that allow it to multiply under preservative conditions. Although the infectious dose for *L. monocytogenes* is unknown, the risk of infection increases as the organism reaches higher concentrations, making growth during refrigerated storage a concern (Anonymous, 2001; Notermans et al., 1998). *L. monocytogenes* not only multiplies at refrigeration temperatures but is also highly salt tolerant and able to survive and multiply at the 2 to 4% salt concentrations typical for many smoked and lightly preserved seafoods (Pelroy et al., 1994a; Sabanadesan et al., 2000; Thurette et al., 1998). Another barrier used to preserve smoked fish is vacuum packaging, which creates a low-oxygen environment inhibiting strict aerobes such as *Staphylococcus*. Under these conditions the organism of greatest concern is *Clostridium botulinum,* a Gram-positive spore former that grows only in anaerobic conditions and produces a deadly toxin (Heinitz and Johnson, 1998). While *C. botulinum* is inhibited from germinating and growing by the addition of nitrite (100 to 200 ppm) or by a minimum 3.5% salt concentration in absence of nitrites, *L. monocytogenes* is able to grow in these microaerobic conditions uninhibited by these levels of nitrite and salt (Pelroy et al., 1994a).

PRESENCE AND INCIDENCE OF *LISTERIA* SPP. AND *L. MONOCYTOGENES* IN MINIMALLY PROCESSED SEAFOOD, PROCESSING ENVIRONMENTS, AND RAW MATERIALS

FINISHED PRODUCTS

L. monocytogenes causes relatively few human disease cases, particularly compared to many other food-borne pathogens; however, it does appear to be commonly present in raw and RTE foods. For example, USDA data from 1998 indicate that 2.5% of 3547 samples of RTE products tested (including salads, jerky, sausage, cooked poultry, ham, and roast beef) were positive for *L. monocytogenes* (http://www.fsis.usda.gov/oa/topics/lm_action.htm). To further illustrate, in 1999 about 4.6% of sliced ham and luncheon meats and 2.7% of roast beef and corned beef samples tested by USDA were positive for *L. monocytogenes*. A variety of reports exists on the incidence of *L. monocytogenes* in foods (reviewed by Farber and Peterkin, 1991), most of which report a considerable incidence of *L. monocytogenes* in various foods tested (2 to 10%).

Some studies also report much higher incidence in certain foods. For example, raw chicken seems to be commonly contaminated with *L. monocytogenes*; surveys show contamination rates ranging from 12 to 60% (Farber and Peterkin, 1991). The incidence of *L. monocytogenes* in cold-smoked salmon and cooked fish products has been reported to range from 6 to 36%, and the recent Food and Drug Administration (FDA)/USDA risk assessment (see below) estimated that 15% of all smoked fish is contaminated with *L. monocytogenes* (Embarek, 1994; Anonymous, 2001; Heinitz and Johnson, 1998). Few data exist on the incidence of *L. monocytogenes* in minimally processed seafoods other than smoked fish. In one study, McCarthy (1997) reported that 17% of vacuum-packed crawfish tails were contaminated, while all boiled crabs tested were negative for *L. monocytogenes*. In a survey of lightly pickled fish in Switzerland, *L. monocytogenes* was isolated from 25.8% of samples (Jemmi, 1990).

RAW MATERIALS

Raw materials may be a potential source of *L. monocytogenes* contamination of finished, minimally processed seafoods. Raw material contamination for finfish (mainly salmon) designated for cold and hot smoking has been reported to range from 2 to 30% of incoming product (Eklund et al., 1995; Hoffman et al., 2002; Norton et al., 2000). Jeyasekaran et al. (1996) and Rawles et al. (1995) reported that 12% of raw shellfish and 8% of raw crabs tested positive for *L. monocytogenes*.

PROCESSING PLANT ENVIRONMENTS

L. monocytogenes has also been regularly isolated from the environment of RTE food processing facilities, including seafood processing facilities (Hoffman et al., 2002; Norton et al., 2001a; Autio et al., 1999; Rørvik et al., 1997). The observation that contamination rates reported for food environments vary widely between processing plants and between studies may often reflect real differences between plants, but it also may reflect differences in sample collection practices (e.g., timing of sampling relative to processing runs and sanitation, selection of sampling sites). Rørvik et al. (1997) isolated *L. monocytogenes* from 25 of 40 processing plants surveyed. Norton et al. (2001a) found that 27.7% of environmental samples collected from three smoked-fish processing plants were positive for *L. monocytogenes.* Studying one processing plant, Autio et al. (1999) found, not surprisingly, that sanitized and cleaned surfaces sampled before production were contaminated less frequently (13%) than surfaces sampled during production (30%). Hoffman et al. (2002) found that in one smoked-fish processing plant, 43.8% of all environmental samples collected were positive for *L. monocytogenes*, with drain samples the most frequently contaminated (62.5%) and sanitized food contact surfaces the least contaminated (3.1%). Interestingly, a second smoked-fish plant sampled at the same times with the same methodology showed significantly lower environmental *L. monocytogenes* contamination rates (1.2%).

SOURCES OF *L. MONOCYTOGENES* CONTAMINATION IN MINIMALLY PROCESSED SEAFOODS

It has been well established that *L. monocytogenes* is inactivated by commercial heat treatments used for the production of cooked RTE foods. While *L. monocytogenes* present in raw fish may survive process treatments typical for many minimally processed seafoods, such as cold-smoked products (Eklund et al., 1995), contamination from the processing plant environment during or after processing appears to be the major source of finished product contamination for these foods as well as for other RTE foods (Autio et al., 1999; Gravani, 1999; Norton et al., 2001a; Rørvik et al., 1997; Rørvik et al., 2000). Contamination of RTE products that support growth of *L. monocytogenes* even with small quantities of this organism is a particular concern to the food industry due to its ability to multiply at refrigeration temperature during storage.

Processing Plant Environment

L. monocytogenes survives extremely well in the non-host environment, including processing plants. *L. monocytogenes* may be introduced into processing plants through a variety of routes, including raw materials, employees' shoes or clothes, and equipment that moves between the inside and the outside of a plant. Additionally, food contact surfaces, non-food contact surfaces (boxes, crates, carts, tools), and physical structures (drains, refrigeration coils, etc.) can become sources once contaminated and provide opportunities to move *L. monocytogenes* throughout the plant.

Studies using molecular fingerprinting techniques have significantly contributed to an improved understanding of the ecology, sources, and spread of *L. monocytogenes* and *Listeria* spp. in processing plant environments. While a diversity of different *L. monocytogenes* strains is found in most processing plants (including seafood plants), individual processing facilities often harbor unique *L. monocyto-*

genes populations and strains that persist for months or years despite sanitation protocols designed to eliminate them (Autio et al., 1999; Hoffman et al., 2002; Norton et al., 2001a; Rørvik et al., 2000). Patterns of persistent processing plant contamination have been reported for a variety of food processing environments, including those for smoked fish, poultry, meat, and dairy foods (Arimi et al., 1997; Lawrence and Gilmour, 1995; Nesbakken et al., 1996; Rørvik et al., 1995). These studies indicate that although a variety of *L. monocytogenes* may be introduced (probably daily) into the plant environment from different sources, most are eliminated by cleaning and sanitation. Some subtypes appear to colonize specific niches in the plant environment, though, and persist over time. Thus, continuous monitoring for the presence and reintroduction of persistent *L. monocytogenes* contamination must be a component of every *L. monocytogenes* control strategy (Hoffman et al., 2002).

Persistent *L. monocytogenes* contamination in processing plants represents a major concern for the food industry and for public health professionals. Some studies using molecular subtyping of *L. monocytogenes* isolates specifically showed that the subtypes persisting in respective plants were responsible for the majority of finished product contamination (Norton et al., 2001a). Eradication of persistent strains from the plant will likely considerably reduce the risk of finished product contamination from the environmental sources (Autio et al., 1999).

Exposure of consumers to persistent *L. monocytogenes* subtypes through contaminated finished products over prolonged times may cause disease in a number of people and thus lead to a outbreak that can be traced back to a specific plant or food product. For example, environmental postprocessing contamination is thought to have been the source of a 1998–1999 multistate human listeriosis outbreak linked to the consumption of contaminated hot dogs and deli meats (Anonymous, 1998; 1999a). Specifically, an increased level of environmental *Listeria* contamination (possibly associated with a construction event in the implicated plant) appears to have coincided with the time when product contamination with the outbreak strain first occurred. Apparently, environmental contamination was responsible for finished product contamination over an extended time period (>4 months), thus leading to a large outbreak.

Raw Materials

Elimination of *L. monocytogenes* from finished products will require identification and control of all potential contamination sources. Many minimally processed RTE seafoods may not receive a sufficient heat treatment to inactivate the organism if it is present in the raw material. Since some data indicate that raw materials used for production of minimally processed seafoods may occasionally be contaminated with *L. monocytogenes* (see above), raw materials represent a potential source of finished product contamination (Autio et al., 1999; Eklund et al., 1995; Farber, 2000; Hoffman et al., 2002; Norton et al., 2001a; Rørvik et al., 2000). Cold-smoked seafoods probably represent the most important category in which raw materials are a concern as a source of finished product contamination (Eklund et al., 1995). However, some data indicate that processing steps and conditions involved in production of minimally processed seafoods often inhibit *L. monocytogenes* growth and may even reduce *Listeria* numbers present on the raw materials (Sabanadesan et al., 2000). Recent in-plant studies using molecular subtyping strategies also indicate that raw materials rarely seem to be responsible for finished product contamination in the production of cold-smoked fish. Instead, the processing plant environment seems to be responsible for most incidences of finished product contamination (Autio et al., 1999; Norton et al., 2001a).

Employees and Processing Personnel

Employees and processing personnel not only represent a potential source for the introduction of *L. monocytogenes* in the processing plant environment, but they may also serve as direct sources of contamination if they are involved in postprocessing handling of products. It has been shown that up to 1 to 10% of healthy adults may be fecal carriers of *L. monocytogenes* (Farber and Peterkin, 1991; Schuchat et al., 1991).

Retail and Consumer

Although very little data exist on sources of *L. monocytogenes* contamination after production and during distribution, it is likely that contamination sources at the retail and consumer level may also be responsible for contamination of minimally processed seafoods. Slicing and handling at the retail level may be responsible for cross-contamination of minimally processed seafoods (Dauphin et al., 2001; Hudson and Mott, 1993; Uyttendaele et al., 1999). Similarly, handling by the consumer may allow for *L. monocytogenes* cross-contamination from other sources. To illustrate, one study found that 21% of households surveyed had *L. monocytogenes* present on surfaces such as sinks, toothbrushes, washcloths, and the refrigerator interior (Beumer et al., 1996b).

CONTROL STRATEGIES FOR PREVENTION OF *L. MONOCYTOGENES* CONTAMINATION

Reduction and, ideally, elimination of pathogenic *L. monocytogenes* from RTE foods are crucial for efforts to reduce the number of food-borne listeriosis cases and to protect the food and seafood industries from costly recalls, law suits, negative publicity, and listeriosis cases linked to a product. As outlined previously, the most important sources for *L. monocytogenes* contamination of RTE minimally processed seafoods that are under the control of the processing industry include (1) processing plant environments, (2) raw materials, and (3) employees and processing personnel. Good manufacturing practices and sanitation standard operation procedures (SSOPs) are crucial underlying policies to prevent and minimize *Listeria* contamination (Elliot and Kvenberk, 2000). Monitoring for and tracking of *L. monocytogenes* contamination represent an integral part of *Listeria* control strategies.

Control Strategies

Although this chapter cannot attempt a comprehensive guideline on the development of environmental *Listeria* control strategies, an overview on this subject is provided here. For a more comprehensive coverage of *Listeria* control strategies in the food and seafood processing industry, see recent publications by the FAO (Anonymous, 1999b) and Tompkin et al. (1999).

Cleaning and Sanitation

As outlined earlier, *L. monocytogenes* is found commonly in many environments and can be introduced in the food chain and into food processing plants by many routes. Cleaning and sanitation in processing plants and in slaughter operations thus represent the most important strategy to control *L. monocytogenes* and the development and implementation of SSOPs are critical for processing plants. It is important to assure that cleaning and sanitation procedures eliminate transient (recently introduced) as well as persistent *Listeria* contaminations. The efficiency of cleaning and sanitation protocols needs to be monitored through an environmental *Listeria* testing plan (see below) and information gained from microbial testing should be integrated into the cleaning and sanitation regimen. While extensive efforts to control and monitor *L. monocytogenes* can reduce plant contamination and the risk of finished product contamination, it is impossible, given currently available technology, to eradicate this organism from processing plants or from the food chain (Tompkin et al., 1999).

Successful control of *L. monocytogenes* requires consistency and attention to detail (Tompkin et al., 1999). If at all possible, a designated crew that is not distracted with other processing duties and time constraints should perform cleaning and sanitation. Generally, food contact surfaces should be cleaned and sanitized daily to reduce cross-contamination between processing days and prevent microorganisms from establishing biofilms and persistent contamination patterns. The processing environment should be kept clean, sanitized, and, whenever possible, dry (Anonymous, 1999b).

Specifically, cleaning and sanitation should follow these steps (taken from Tompkin et al., 1999): (1) dry clean, (2) pre-rinse the equipment, (3) visually inspect the equipment, (4) foam and scrub the equipment, (5) rinse the equipment, (6) visually inspect the equipment, (7) clean the floors, (8) sanitize the equipment and floors, (9) conduct postsanitation verification, (10) dry the floors, and (11) clean and put away supplies. Some equipment may require disassembling prior to cleaning and sanitizing and may need to be resanitized after reassembling (Tompkin et al., 1999). Drains in the processing areas are often considered a potential source of (persistent) contamination; therefore, they must be regularly cleaned and sanitized.

Sanitary equipment and plant design represent another critical consideration for *Listeria* control. Equipment must be designed to allow for easy cleaning and sanitation and to minimize sites where microbial contamination can occur (Tompkin et al., 1999). Consideration should also be given to avoiding cleaning techniques that can disperse *Listeria* in the processing environment. Any cleaning techniques that would be likely to disperse bacteria throughout the environment via air or water droplets (e.g., high-pressure cleaning equipment) should generally be avoided, particularly in finished product handling areas.

Processing Policies

The processing of smoked fish and other minimally processed seafoods generally requires significant employee handling of finished products. Expense and effort to reduce contamination via sanitation of the processing equipment are lost when employees are poorly trained or are not supplied with the proper tools or equipment to prevent the contamination of product during processing or packaging. Using clean gloves, smocks, and aprons is vital to prevent product contamination. To reduce the risk of cross-contamination, processing activities should be separated so that employees do not move from areas containing raw materials to areas where finished products are handled or stored; this can be an even larger problem when employees rotate duties during the day (Rørvik et al., 1997).

A useful strategy to reduce cross-contamination includes color-coding garments, which allows differentiation of garments and crews designated for processing areas from those designated for postprocessing areas. Disposable gloves and aprons should be used wherever possible, particularly when handling finished product, since reusable garments are frequently worn many times and by multiple individuals before they are cleaned (Anonymous, 1999b; Tompkin et al., 1999).

In addition to cross-contamination, employees are a source of contamination to the operation by introducing bacteria into the processing environment on their hands, clothes, and shoes. Footbaths are frequently used to sanitize shoes of employees as they enter food processing areas. Footbaths are also crucial to minimize introduction of *L. monocytogenes* into the processing environment and the finished product area through other vehicles, such as carts. However, if footbaths are not properly maintained they can become a source of cross-contamination (IFT, 2001; Anonymous, 1999b). Employee hygiene, particularly hand washing before production and after breaks, is an important *L. monocytogenes* control point and should be included in employee training — particularly because 1 to 10% of the population appear to be carriers of *L. monocytogenes* (Notermans et al., 1998).

Raw Material Specifications

Contamination of minimally processed seafoods with *L. monocytogenes* appears to occur predominantly during processing and from environmental sources; however, control of *L. monocytogenes* in raw materials represents an important component of a comprehensive *Listeria* control program for minimally processed seafood. Sanitary harvest and slaughter conditions and continuous refrigeration from time of harvest to processing are likely to be important measures that can help to minimize *L. monocytogenes* contamination. Some processing plants use raw material specifications (products must be free of *L. monocytogenes* or even *Listeria* spp.) for their suppliers and may test multiple samples of raw materials for each lot received. Thawing raw finfish in chlorine solutions and treatment of raw finfish with chlorine rinses or sprays before processing have been suggested as measures to further reduce microbial loads (IFT, 2001).

Packaging

A number of studies have investigated packaging technologies and food additives that would inhibit the growth of *Listeria* in finished products, including minimally processed seafoods. Most of these studies have focused on cold-smoked salmon. Lactate and sorbate are two chemical additives that have been shown to inhibit the growth of *L. monocytogenes* at refrigerated temperatures; however, the practicality of these substances has been questioned because of technical challenges and possible sensory effects (Pelroy et al., 1994b; El-Shenawy and Marth, 1988). Modified atmosphere packaging using carbon dioxide at levels of 70% and higher has also been shown to inhibit the growth of *L. monocytogenes* in smoked fish at refrigerated temperatures (Hendricks and Hotchkiss, 1997; Nilsson et al., 2000; Svado and Cahill, 1998). Implementation of these packaging methods would be challenging for the industry due to capital costs for new packaging equipment and added costs for transportation of larger volume packages (IFT, 2001). Food-grade microorganisms have also been investigated for their use to outcompete *L. monocytogenes* or to produce inhibitory substances to *L. monocytogenes* on smoked fish. *Carnobacteria* in particular have been shown to be promising by several laboratories (Duffes et al., 1999; Nillson et al., 1999; Palundan-Müller et al., 1998).

While *L. monocytogenes* can grow and multiply at refrigeration temperatures, continuous storage at low temperatures (34 to 36°F) will minimize the growth rate of this organism if it is present in the product. Assuring uninterrupted refrigerated storage and distribution, combined with limiting product shelf-life and the use of time-temperature indicators, will help to limit outgrowth of *Listeria* during distribution. Combined with other *Listeria* control strategies, appropriate refrigerated storage and distribution may help assure that *L. monocytogenes* loads will stay below the regulatory limits established in some countries (see below) and the levels likely to cause human disease. Frozen storage and distribution do not kill or inactivate *L. monocytogenes* but provide another option to inhibit growth of this organism completely after processing.

L. monocytogenes Detection and Characterization

Detection of *L. monocytogenes* and *Listeria* spp. from raw materials, environmental samples, and finished products is an important tool to monitor and verify the effectiveness of control strategies designed to prevent finished product contamination. Subtyping methods to further characterize *L. monocytogenes* isolates can be used to specifically track the sources and spread of contamination. Advances in biotechnology have increased options for tracking and detection of *L. monocytogenes* while reducing the time required for testing.

Sampling Methods and Strategies

An overall *Listeria* or *L. monocytogenes* monitoring program is necessary to assess the efficiency of implemented control strategies and to determine the need for additional or improved pathogen control measures (Tompkin et al., 1999). Environmental monitoring is a crucial component of a *L. monocytogenes* control program, but some processors of minimally processed seafoods may also choose to monitor or occasionally test raw materials (see above) or finished products.

The development of an overall environmental sampling and testing strategy can be an important part of an overall *Listeria* control strategy, and the purpose of the testing and monitoring program should be carefully evaluated. For instance, sampling during a production run may provide important information with respect to overall processing hygiene and risk for final product contamination. On the other hand, sampling before production begins may be more suitable to identify persistent environmental contamination or efficacy of SSOPs. Each plant, product, and process must be evaluated to determine appropriate environmental testing points (Tompkin et al., 1999).

In establishing a sampling plan, it is also important to determine whether set sampling locations will be designated or whether a monitoring plan will include testing of random locations that vary from one sample collection to the next (also known as the "seek and destroy" sampling approach). Establishing static sampling points allows monitoring for change over time, provides a frame of reference, and can be used in a quality assurance program that uses statistical process control (SPC). This sampling approach thus allows monitoring of significant changes in contamination frequency and patterns that need to trigger specific and properly documented remedial actions. Establishment of specific upper control limits, which trigger remedial actions, should be established by processing plants. Many plants may still choose to include some variable sampling points in their sampling plans to identify particularly problematic areas for *Listeria* or *L. monocytogenes* contamination.

It has generally been recommended that food-contact and nonfood-contact surfaces be included in a sampling plan (Tompkin et al., 1999). For most plants, contamination frequency has been found to be lower for food-contact surfaces than for nonfood-contact surfaces, particularly floors and drains (Hoffman et al., 2002). Floor drains represent an almost constant problem in many plants (Tompkin et al., 1999; Norton et al., 2001a), and the value of testing drains has been controversial. Drains may often represent good indicators of overall plant sanitation, though. For example, Rørvik et al. (1997) reported that the presence of *L. monocytogenes* in drains was a sensitive predictor for the presence of *L. monocytogenes* in the finished product. Including a number of drain samples in a sampling plan may thus provide a good indicator of the overall efficiency of the *Listeria* control strategies implemented in a processing plant.

Detection Methods

Although *L. monocytogenes* represents the only human pathogen among the species in the genus *Listeria*, many monitoring programs use the detection of *Listeria* spp. as an indicator of potential *L. monocytogenes* contamination or of conditions that may allow survival or presence of *L. monocytogenes*. Historically, detection of *Listeria* spp. has been more rapid and sensitive than detection of *L. monocytogenes*. In most methods, selective enrichment procedures are used to enrich *Listeria* spp. over other microorganisms, followed by plating on selective and differential media that generally distinguish *Listeria* spp. from other bacteria. If specific detection of *L. monocytogenes* is required, *Listeria*-like colonies are subsequently tested by biochemical tests, including hemolysis testing to differentiate *L. monocytogenes* from other *Listeria* spp.

A variety of enrichment media and procedures has been developed for the detection of *Listeria.* Early techniques relied on cold enrichment to select for *Listeria* spp. (Curtis and Lee, 1995). Because cold enrichment protocols require anywhere between a week and several months of incubation, antibiotics have become the selective agents of choice today, with acriflavin, nalidixic acid, and cycloheximide commonly used as selective agents (Beumer et al., 1996a; Curtis and Lee 1995; Pritchard and Donnelly, 1998). A frequently used enrichment media for the detection of *L. monocytogenes* from seafoods is the *Listeria* enrichment broth (LEB). Commonly used selective plating media include Oxford and Listeria Plating Media (LPM). A detailed description of the standard method for the detection of *L. monocytogenes* from seafoods can be found in the FDA Bacteriological Analytical Manual (Hitchins, 1995); Donnelly (1999) has written a thorough review of detection methods for *Listeria.*

Detection methods have evolved from *Listeria*-selective agars to genetic tests and plating media highly specific for *Listeria* spp. and *L. monocytogenes.* Methods currently entering the market to screen for and identify *L. monocytogenes* have cut detection times from 1 to 2 weeks to 1 to 4 days (Carroll et al., 2000; Entis and Lerner, 1999; Norton et al., 2001a; O'Connor et al., 1999; Simon et al., 1996). Even new methods generally still require enrichment for 1 to 2 days to allow detection of the low numbers of *Listeria* often present in foods. New genetic tests are available that use molecular probes that are highly specific for *Listeria* or *L. monocytogenes.*

Commercial methods such as BAX® and Probelia® are based on DNA amplification by polymerase chain reaction (PCR) of Listerial DNA and have a sensitivity capable of detecting as few as 10^2 CFU/ml in 2 to 3 days (Norton et al., 2000; Hoffman and Wiedmann, 2001; Coquard et al., 1999; Stewart and Gendel, 1998). Cultural methods have evolved from screening for *Listeria* species by esculin hydrolysis on plates or broth to highly specific media that release fluorogenic or chromogenic products when substrates are cleaved by enzymes specific to *L. monocytogenes.* The discriminatory ability of these media represents a significant improvement over traditional media when screening for *L. monocytogenes.* Some of these *L. monocytogenes* differential media are also commercially available (Restaino et al., 1999; Karpiskova et al., 2000; Hoffman and Wiedmann, 2001).

The U.S. FDA currently has a zero tolerance policy for the presence of *L. monocytogenes* in RTE foods (Elliot and Kvenberk, 2000), which requires the absence of *L. monocytogenes* in a 25-g sample of an RTE food. Thus, most *L. monocytogenes* detection methods used in the U.S. are qualitative rather than quantitative and are designed to detect the presence of *L. monocytogenes* in a 25-g sample. Quantitative detection of low levels of *Listeria* or *L. monocytogenes* is generally performed using a 3- or a 5-tube most probable number (MPN) method (Rawles et al., 1995; Tortorello et al., 1997).

Subtyping Methods to Track Contamination Sources

As described previously, subtyping methods, which allow differentiation of *L. monocytogenes* beyond the species level, can provide important information on the sources and spread of *L. monocytogenes* contamination throughout the food chain and in seafood processing. Accurate tracking of the sources of food-borne pathogens is crucial to design appropriate intervention and control strategies. The simple identification of the same genus and even the same species by standard methods in either raw materials or environmental samples and in the finished product does not unequivocally establish a causal relationship. Analytical methods that allow the characterization and precise identification and subtyping of microorganisms are thus critical to determining the relative importance of different contamination sources (e.g., raw materials vs. environmental sources in the processing plant).

For example, in some previous studies discriminatory subtyping methods have allowed identification of the processing environment as a significant source of *L. monocytogenes* isolated from samples during processing and from finished products (Rørvik et al., 1995; Autio et al., 1999). Rørvik et al. (1995) observed the persistence of a single subtype in the processing environment of a smoked salmon processing plant and in finished products, strongly suggesting the environment as a primary source of contamination. Molecular subtyping data also indicated that the contamination of smoked rainbow trout most likely occurred during brining and slicing in a processing plant (Autio et al., 1999). The apparent importance of environmental *L. monocytogenes* sources clearly suggests that sanitation standard operation procedures (SSOPs) may provide a more appropriate approach for control of this organism than definition of specific critical control points in the context of an HACCP plan (Norton et al., 2001a).

In general, bacterial subtyping methods can be divided into (1) conventional and phenotypic and (2) genetic or DNA-based methods. Conventional and phenotypic methods have been used for many years to subtype *L. monocytogenes* and other food-borne pathogens; however, genetic subtyping methods have revolutionized this field. A variety of molecular typing ("fingerprinting") methods allowing sensitive strain differentiation of *L. monocytogenes* have been described. These methods are often superior to classical methods (such as serotyping) because they generally provide more sensitive strain discrimination. Serotyping, one classical strain differentiation method for *L. monocytogenes*, allows discrimination of only 13 subtypes. Thus, this method provides a relatively insensitive tool for epidemiological investigations and is inadequate for tracking *L. monocytogenes* contamination sources.

New and more discriminatory approaches for subtyping *L. monocytogenes* are necessary for accurate and effective tracking of contamination sources. Commonly used DNA-based subtyping approaches for bacterial isolates include random amplification of polymorphic DNA (RAPD), pulsed-field gel electrophoresis (PFGE) (Brosch et al., 1994), ribotyping (Bruce et al., 1995), and, increasingly, DNA sequencing based methods. The most commonly used molecular methods that provide accurate and discriminatory typing results for *L. monocytogenes* include ribotyping and PFGE. Both methods provide more sensitive subtype discrimination (>100 *L. monocytogenes* subtypes) as compared to most classical subtyping methods. The choice of an appropriate subtyping method (or methods) depends significantly on the intended application and the goal of subtyping *L. monocytogenes* isolates. For more in-depth reviews of subtyping methods for *L. monocytogenes*, refer to Graves et al. (1999) and Wiedmann (2002).

CURRENT REGULATORY POLICIES

The levels of *L. monocytogenes* allowed in RTE foods vary by country. As mentioned, the U.S. has a zero tolerance policy for *L. monocytogenes* in a 25-g sample of an RTE food; however, this policy is rare in the international community, with Italy the only other country with a zero tolerance for all RTE foods (Ross et al., 2000). The next most stringent nation is Australia, which has a zero tolerance policy for foods that support the growth of *L. monocytogenes*. Germany and France each have a tolerance limit of less than 100 CFU per gram of food at the point of consumption for RTE foods (Ross et al., 2000; Nørrung, 2000).

Canada has a progressive three-tiered policy based on the shelf-life of the food and whether or not the product has been linked to any cases of listeriosis. The most stringent regulations are placed on "Category 1" products. These are products that have been linked to outbreaks of listeriosis and have a zero tolerance for *L. monocytogenes* (Farber, 2000; Ross et al., 2000). Category 2 includes RTE foods that have a shelf-life greater than 10 days and are capable of supporting the growth of *L. monocytogenes.* Category 2 products receive a lower priority for inspections, and if *L. monocytogenes* is found, a class II recall, but not necessarily a public alert, is required (Farber, 2000). Category 3 foods are foods that do not support the growth of *L. monocytogenes* and foods that have a shelf-life less than 10 days. These products are allowed to have levels of *L. monocytogenes* up to 100 CFU/g (Farber, 2000; Ross et al., 2000). Maintenance of good hygienic conditions and compliance with GMPs are required for all categories of foods.

Current regulatory policies for *L. monocytogenes* in the Netherlands divide food products into six categories, four of which are relevant to this discussion. Heat-treated foods that are handled before packaging and have a shelf-life greater than 1 week (e.g., hot-smoked fish) and lightly preserved RTE foods that have a shelf-life greater than 3 weeks (e.g., cold-smoked fish) have a zero-tolerance limit. Conversely, these products are placed in separate categories and have a 100 CFU/g tolerance limit if they are stabilized to prevent the growth of *L. monocytogenes* or have a shelf-life less than 1 or 3 weeks, respectively (Nørrung, 2000). In situations where testing reveals *L. monocytogenes* at a level between 10 and 99 CFU/g, steps must be taken to improve the situation, but no recall is required (Nørrung, 2000).

RISK ASSESSMENT

Risk assessments provide a tool to better quantify the risk of food-borne disease transmission associated with specific foods and production and distribution practices, to subsequently target intervention and control strategies, and to identify data gaps. Due to the severity of human listeriosis, *L. monocytogenes* represents a serious public health concern. To be able to address this pathogen better, a hazard identification and hazard characterization of *L. monocytogenes* in RTE foods, as well as an exposure assessment, have been performed by FAO (http://www.fao.org/es/esn/pagerisk/techdocs.htm). The U.S. FDA and USDA also performed a risk characterization of listeriosis caused by consumption of different RTE foods, which was published in draft form in January 2001 (Anonymous, 2001; http://www.foodsafety.gov/

~dms/lmrisk.html). Risk characterization was broken down by the risk to three groups of the population based on age (elderly, perinatal, and intermediate-aged individuals [30 days to 59 years old]). The predicted risk was based on five factors:

- Frequency and extent of *L. monocytogenes* contamination in the food
- Amounts and frequency of consumption of the food
- Potential for growth of *L. monocytogenes* during refrigerated storage
- Duration of refrigerated storage before consumption
- Temperature at which the food is held during refrigerated storage

Based on these factors, FDA estimated and ranked the predicted relative risk of contracting listeriosis from eating different types of RTE foods and adjusted the relative rank of each to reflect current consumption levels for these food products. This study ranked smoked seafood (which included both hot- and cold-smoked products) 6th, cooked RTE crustaceans 9th, preserved fish 13th, and raw seafood 17th, in terms of their relative risk for causing listeriosis on a per annum basis in intermediate-aged individuals. The five product categories with the highest risk for this age group were deli meats, deli salads, pasteurized milk, frankfurters, and "miscellaneous dairy products (butter, yogurt, cream)." Ranked 7th and 8th directly behind smoked fish were pâté and soft cheese. In the FDA assessment, products with known moderate to high contamination rates are not necessarily the highest risk products if consumed less frequently, in smaller amounts, or by fewer people. Deli meats ranked first in relative risk per annum risk because of a moderate to high contamination rate (postprocessing contamination) coupled with favorable growth conditions and broad exposure to the general population with billions of servings per year.

The FDA/USDA risk assessment also addresses questions on the human health risk associated with consumption of low numbers of *L. monocytogenes*. In their conclusions, the FDA/USDA risk assessment stated the following regarding their assessment:

> The risk assessment reinforces past epidemiological conclusions that food-borne listeriosis is a moderately rare although severe disease. Although the exposure assessment suggests that U.S. consumers are exposed to low levels of L. monocytogenes on a regular basis, the likelihood of acquiring listeriosis is very small.

Other authors also have noted that the average person consumes a dose of 5 x 10^5 *L. monocytogenes*/serving around 4 times per year, but only a small fraction of individuals contract this illness (Nørrung, 2000; Notermans et al., 1998). This suggests that the presence of low levels of *L. monocytogenes* in foods at point of consumption represents a minimal health risk and that tolerance levels for the presence of *L. monocytogenes* in RTE foods (e.g., <100 CFU/g at point of consumption or in foods that do not allow growth of *L. monocytogenes*) are scientifically justifiable.

CONCLUSIONS

Production of some minimally processed seafood (products such as cold-smoked fish) does not include a kill step, which would eliminate *L. monocytogenes* if it were present on raw materials. Other products, such as cooked crustaceans, have a thermal kill step in their process but may be susceptible to postprocessing contamination. Even though it has been demonstrated that raw seafood products are occasionally contaminated with *Listeria*, finished product contamination — when it occurs — appears to originate predominantly from environmental sources. This is similar to what has been found for other RTE foods that include a lethal kill step for *Listeria* in the production process. Based on the available data, a combination of stringent microbial quality criteria for raw materials and adherence to SSOPs and GMPs should assure production of safe minimally processed seafoods. Although an HACCP program is required for seafood processing plants in the U.S., control of *L. monocytogenes* cannot be addressed through critical control points but should be addressed in the prerequisite programs (SSOP, GMP) (Elliot and Kvenberk, 2000).

In general, total elimination of *L. monocytogenes* from the food supply is not feasible. However, the reduction and management of risk associated with *L. monocytogenes* based on sound scientific reasoning is possible (Farber, 2000). Reduction and exclusion of *L. monocytogenes* at the source, during processing, and during handling have been and should continue to be a priority for industry. Increasing evidence from risk assessments indicates that, while most people regularly are exposed to significant numbers of *L. monocytogenes,* only few human listeriosis cases occur. This indicates that the feasibility and costs associated with a general zero tolerance policy for *L. monocytogenes* in RTE foods should be reevaluated and revised. Specific risk groups (including the elderly, immunocompromised people, and pregnant women) account for the vast majority of human listeriosis cases. As part of a comprehensive strategy to minimize food-borne listeriosis, limiting product shelf-life, using alternative distribution methods such as distributing products in the frozen state, using warning labels on packaging materials for specific foods, and using specific antimicrobial additives to minimize growth represent options that could help to minimize exposure for consumers at risk.

ACKNOWLEDGMENTS

Research in the authors' laboratory is funded by the National Oceanic and Atmospheric administration award NA86RG0056 to the Research Foundation of State University of New York for New York Sea Grant (M.W.) and by the Cooperative State Research, Education, and Extension Service, U.S. Department of Agriculture, under Agreement No. 00–51110–9769 (M.W. and K. G.). The U.S. government is authorized to produce and distribute reprints for governmental purposes, notwithstanding any copyright notation that may appear hereon. The views expressed herein are those of the authors and do not necessarily reflect the views of NOAA or any of its subagencies. Any opinions, findings, conclusions, or recommendations expressed in this publication are those of the authors and do not necessarily reflect the view of the U.S. Department of Agriculture.

REFERENCES

Anonymous, 1998, Multistate outbreak of listeriosis — United States, 1998, MMWR 47:1085–1086.

Anonymous, 1999a, Update: multistate outbreak of listeriosis — United States, 1998–1999, MMWR 47:1117–1118.

Anonymous, 1999b, FAO expert consultation on the trade impact of *Listeria* in fish products, Food and Agriculture Organization, Amherst, MA.

Anonymous, 2001, Draft assessment of the relative risk to public health from foodborne *Listeria monocytogenes* among selected categories of ready-to-eat foods, http://www.foodsafety.gov/~dms/lmrisk.html.

AFDO, 1991, Cured, salted, and smoked fish establishments good manufacturing practices, An Association of Food and Drug Officials model code adopted June 1991.

Autio, T. et al., 1999, Sources of *Listeria monocytogenes* contamination in a cold-smoked rainbow trout processing plant detected by pulsed-field gel electrophoresis typing, *Appl. Environ. Microbiol.*, 65:150–155.

Arimi, S.M. et al., 1997, Diversity of *Listeria* ribotypes recovered from dairy cattle, silage, and dairy processing environments, *J. Food. Prot.*, 60:811–816.

Beumer, R.R. et al., 1996a, The effect of acriflavin and nalidixic acid on the growth of *Listeria* spp. in enrichment media, *Food Microbiol.*, 13:137–148.

Beumer, R.R. et al., 1996b, *Listeria* species in domestic environments, *Epidemiol Infect.*, 117:437–442.

Brosch, R., Chen, J., and Luchansky, J.B., 1994, Pulsed-field fingerprinting of listeriae: identification of genomic divisions for *Listeria monocytogenes* and their correlation with serovar, *Appl. Environ. Microbiol.*, 60:2584–2592.

Bruce J.L. et al., 1995, Sets of EcoRI fragments containing ribosomal RNA sequences are conserved among different strains of *Listeria monocytogenes*, *Proc. Natl. Acad. Sci.* (USA), 92:5229–5233.

Carroll, S.A. et al., 2000, Development and evaluation of a 24-hour method for the detection and quantification of *Listeria monocytogenes* in meat products, *J. Food Prot.*, 63:347–353.

Coquard, D., Exinger, A., and Jeltsch, J.M., 1999, Routine detection of Salmonella species in water: comparative evaluation of the Iso and PROBELIA polymerase chain reaction methods, *J. AOAC Int.*, 82:871–976.

Curtis, G.D.W. and Lee, W.H., 1995, Culture media and methods for the isolation of *Listeria monocytogenes*, *Int. J. Food Microbiol.*, 26:1–13.

Dauphin, G., Ragimbeau, C., and Malle, P., 2001, Use of PFGE typing for tracing contamination with *Listeria monocytogenes* in three cold-smoked salmon processing plants, *Int. J. Food Microbiol.*, 64:51–61.

Donnelly, C.W., 1999, Conventional methods to detect and isolate *L. monocytogenes*, in *Listeria, Listeriosis, and Food Safety*, 2nd ed., Ryser, E.T. and Marth, E.H., Eds., New York: Marcel Dekker, Inc., 225–260.

Dore, I., 1993, *The Smoked and Cured Seafood Guide*, Toms River, NJ: Urner Barry Publications, Inc., 114–115.

Duffes, F. et al., 1999, Inhibition of *Listeria monocytogenes* by *Carnobacterium* spp. strains in a simulated cold smoked fish system stored at 4°C, *Int. J. Food Microbiol.*, 47:33–42.

Eklund, M.W. et al., 1995, Incidence and sources of *Listeria monocytogenes* in cold-smoked fishery products and processing plants, *J. Food Prot.*, 58:502–508.

Elliot, E.L. and Kvenberg, J.E., 2000, Risk assessment used to evaluate the U.S. position on *Listeria monocytogenes* in seafood, *Int. J. Food Microbiol.*, 62:253–260.

El-Shenawy, M.A. and Marth, E.H., 1988, Inhibition and inactivation of *Listeria monocytogenes* by sorbic acid, *J. Food Prot.*, 51:842–847.

Embarek, B., 1994, Presence detection and growth of *Listeria monocytogenes* in seafoods: a review, *Int. J. Food Microbiol.*, 23:17–34.

Entis, P. and Lerner, I., 1999, Twenty-four-hour direct presumptive enumeration of *Listeria monocytogenes* in food and environmental samples using the ISO-GRID method with LM-137 agar, *J. Food Prot.*, 63:354–363.

Farber, J.M., 2000, Present situation in Canada regarding *Listeria monocytogenes* and ready-to-eat seafood products, *Int. J. Food. Mirobiol.*, 62:247–251.

Farber, J.M. and Peterkin, P.I., 1991, *Listeria monocytogenes*, a food-borne pathogen, *Microbiol. Rev.*, 55:76–511.

Gall, K., Reddy, K.P., and Regenstein, J.M., 2000, Specialty seafood products, in *Marine and Freshwater Products Handbook,* Martin, R.E. et al., Eds., Inc., Lancaster, PA: Technomic Publishing Co., 403–416.

Gram, L. and Huss, H.H., 2000, Fresh and processed fish and shellfish, in *The Microbiological Safety and Quality of Food*, Lund, B.M., Baird-Parker, T.C., and Gould, G.W., Eds., Gaithersburg, MD: Aspen Publishers, Inc., 472–506.

Gravani, R., 1999, Incidence and control of *Listeria* in food-processing facilities, in *Listeria, Listeriosis, and Food Safety,* 2nd ed., Ryser, E.T. and Marth, E.H., Eds., New York: Marcel Dekker, Inc., 657–709.

Graves, L.M., Swaminathan, B., and Hunter, S.B., 1999, Subtyping *L. monocytogenes, Listeria, Listeriosis, and Food Safety,* 2nd ed., Ryser, E.T. and Marth, E.H., Eds., New York: Marcel Dekker, Inc., 279–297.

Heinitz, M.L. and Johnson, J.M., 1998, The incidence of *Listeria* spp., Salmonella spp., and *Clostridium botulinum* in smoked fish and shellfish, *J. Food Prot.*, 61:318–323.

Hendricks, M.T. and Hotchkiss, J.H., 1997, Effect of carbon dioxide on the growth of *Pseudomonas fluorescens* and *Listeria monocytogenes* in aerobic atmospheres, *J. Food Prot.*, 60:1548–1552.

Hitchins, A.D., 1995, *Listeria monocytogenes*, in *Food and Drug Administration Bacteriological Analytical Manual*, 8th ed., Gaithersburg, MD: AOAC International, 10.01–10.13.

Hoffman, A.D., and Wiedmann, M., 2001, Comparative evaluation of culture- and PCR-based detection methods for *Listeria monocytogenes* in environmental and raw fish samples, *J. Food Prot.,* 64:1521–1526.

Hoffman, A.D. et al., 2002, *Listeria monocytogenes* contamination patterns in the smoked fish processing environment and raw fish, *J. Food Prot.,* (in press).

Hudson, J.A. and Mott, S.J., 1993, Presence of *Listeria monocytogenes*, motile aeromonads and *Yersinia entercolitica* in environmental samples taken from a supermarket delicatessen, *Int. J. Food Microbiol.*, 18:333–337.

Institute of Food Technologists (IFT), 2001, Processing parameters needed to control pathogens in cold smoked fish, *J. Food Sci.*, 66:51055–51133.

Jarvis, N.R., 1987, *Curing of Fishery Products*, Kingston, MA: Teaparty Books.

Jinneman, K.C., Wekell, M.M., and Eklund, M.W., 1999, Incidence and behaviour of *Listeria monocytogenes* in fish and seafoods, in *Listeria, Listeriosis, and Food Safety*, 2nd ed., Ryser, E.T. and Marth, E.H., Eds., New York: Marcel Dekker, Inc., 601–630.

Jemmi, T., 1990, Zum Vorkommen von *Listeria monocytogenes* n importierten geraeucherten and fermentierten Fischen, *Arch. Lebenmittelhyg.*, 41:107–109.

Jemmi, T. and Keush, A., 1992, Behavior of Listeria monocytogenes during processing and storage of experimentally contaminated hot-smoked trout, *Int. J. Food Microbiol.*, 15:339–346.

Jeyasekaran G., Karunasagar, I., and Karunasagar, I., 1996, Incidence of *Listeria* spp. in tropical fish. 1996, *Int. J. Food Microbiol.*, 31:333–340.

Karpiskova, R. et al., 2000, Application of a chromogenic medium and the PCR methods for the rapid confirmation of *L. monocytogenes* in foodstuffs, *J. Microbiol. Methods*, 41:267–271.

Lawrence, L.M. and Gilmour, A., 1995, Characterization of *Listeria monocytogenes* isolated from poultry products and from the poultry-processing environment by random amplification of polymorphic DNA and multilocus enzyme electrophoresis, *Appl. Environ. Microbiol.*, 61:2139–2144.

McCarthy, S.A., 1997, Incidence and survival of *Listeria monocytogenes* in ready-to-eat seafood products, *J. Food Prot.*, 60:372–376.

Mead, P.S. et al., 1999, Food related illness and death in the United States, *Emerging Infect. Dis.*, 5:607–625.

Nesbakken, T., Kapperud, G., and Caugant, D.A., 1996, Pathways of *Listeria monocytogenes* contamination in the meat processing industry, *Int. J. Food Microbiol.*, 31:161–171.

Nilsson, L., Gram, L., and Huss, H.H., 1999, Growth control of *Listeria monocytogenes* on cold-smoked salmon using a competitive lactic acid bacteria flora, *J. Food Prot.*, 62:336–342.

Nilsson L. et al., 2000, Carbon dioxide and nisin act synergistically on *Listeria monocytogenes*, *Appl. Environ. Microbiol.*, 66:769–774.

Nørrung, B., 2000, Microbiological criteria for *Listeria monocytogenes* in foods under special consideration of risk assessment approaches, *Int. J. Food Microbiol.*, 62:217–221.

Norton, D.M. et al., 2000, Application of the BAX for screening/genus *Listeria* polymerase chain reaction system for monitoring *Listeria* species in cold-smoked fish and in the smoked fish processing environment, *J. Food Prot.*, 63:343–346.

Norton, D.M. et al., 2001a, Molecular studies on the ecology of *Listeria monocytogenes* in the smoked fish processing industry, *Appl. Environ. Microbiol.*, 67:198–205.

Norton, D.M. et al., 2001b, Characterization and pathogenic potential of *Listeria monocytogenes* isolates from the smoked fish industry, *Appl. Environ. Microbiol.*, 67:646–653.

Notermans, S. et al., 1998, Studies on the risk assessment of *Listeria monocytogenes*, *J. Food Prot.*, 61:244–348.

Notermans, S. and Hoornstra, E., 2000, Risk assessment of *Listeria monocytogenes* in fish products: some general principles, mechanism of infection and the use of performance standards to control human exposure, *Int. J. Food Microbiol.*, 62:223–229.

O'Connor, L. et al., 1999, Rapid polymerase chain reaction/DNA probe membrane-based assay for the detection of *Listeria* and *Listeria monocytogenes* in food, *J. Food Prot.*, 63: 337–342.

Palundan-Müller, C. et al., 1998, Evaluation of the role of *Carnobacterium pisciola* in spoilage of vacuum- and modified packed cold-smoked salmon stored at 5 degrees C, *Int. J. Food Microbiol.*, 39:155–166.

Pelroy, G.A. et al., 1994a, Inhibition of *Listeria monocytogenes* in cold-process (smoked) salmon by sodium nitrite and packaging method, *J. Food Prot.*, 57:114–119.

Pelroy, G.A. et al., 1994b, Inhibition of *Listeria monocytogenes* in cold-process (smoked) salmon by sodium lactate, *J. Food Prot.*, 57: 108–113.

Pritchard, T.J. and Donnelly, C.W., 1998, Combined secondary enrichment of primary enrichment broths increases *Listeria* detection, *J. Food Prot.*, 62:532–535.

Rawles, D. et al., 1995, *Listeria monocytogenes* occurrence and growth at refrigeration temperatures in fresh blue crab (*Callinectes sapidus*) meat, *J. Food Prot.*, 58:1219–1221.

Restaino, L. et al., 1999, Isolation and detection of *Listeria monocytogenes* using fluorogenic and chromogenic substrates for phosphatidylinositol-specific phospholipase C, *J. Food Prot.*, 62:244–251.

Rocourt, J., Jacquet, C., and Reilly, A., 2000, Epidemiology of human listeriosis and seafoods, *Int. J. Food Microbiol.*, 62:197–209.

Rørvik, L.M., Caugant, D.A., and Yndestad, M., 1995, Contamination pattern of *Listeria monocytogenes* and other *Listeria* spp. in a salmon slaughterhouse and smoked salmon processing plant, *Int. J. Food Microbiol.*, 25:19–27.

Rørvik, L.M. et al., 1997, Risk factors for contamination of smoked salmon with *Listeria monocytogenes* during processing, *Int. J. Food Microbiol.*, 37:215–219.

Rørvik, L.M. et al., 2000, Molecular epidemiological survey of *Listeria monocytogenes* in seafoods and seafood-processing plants, *Appl. Environ. Microbiol.*, 66:4779–4784.

Ross, T., Todd, E., and Smith, M. 2000, Joint FAO/WHO activities on risk assessment of microbiological hazards in foods risk assessment: *Listeria monocytogenes* in ready-to-eat foods, http://www.fao.org/es/esn/pagerisk/techdocs.htm.

Sabanadesan S., Lammerding, A.M., and Griffiths, M.W., 2000, Survival of *Listeria innocua* in salmon following cold-smoke application, *J. Food Prot.*, 63:715–720.

Schuchat, A., Swaminathan, B., and Broome, C.W., 1991, Epidemiology of human listeriosis, *Clin. Microbiol. Rev.*, 4:169–183.

Simon, M.C., Gray, D.I., and Cook, N., 1996, DNA extraction and PCR methods for the detection of *Listeria monocytogenes* in cold-smoked salmon, *Appl. Environ. Microbiol.*, 62:822–824.

Stewart, D. and Gendel, S.M., 1998, Specificity of the BAX polymerase chain reaction system for detection of the foodborne pathogen *Listeria monocytogenes*, *J. AOAC Int.*, 81: 817–824.

Svado, E.A. and Cahill, M.E., 1998, The combined affects of modified atmosphere, temperature, nisin and ALTA™ 2341 on the growth of *Listeria monocytogenes*, *Int. J. Food Microbiol.*, 43:21–31.

Thurette, J. et al., 1998, Behavior of *Listeria* spp. in smoked fish products affected by liquid smoke, NaCl concentration, and temperature, *J. Food Prot.*, 61:1475–1479.

Tompkin, R.B. et al., 1999, Guidelines to prevent post-processing contamination from *Listeria monocytogenes*, *Dairy, Food Environ. Sanit.*, 19:551–562.

Tortorello, M.L., Reineke, K.F., and Stewart, D.S., 1997, Comparison of antibody-direct epifluorescent filter technique with the most probable number procedure for rapid enumeration of Listeria in fresh vegetables, *J. AOAC Int.*, 80:1208–1214.

Uyttendaele, M., De-Troy, P., and Debevere, J., 1999, Incidence of *Listeria monocytogenes* in different types of meat products on the Belgian retail market, *Int. J. Food Microbiol.*, 53:75–80.

Wiedmann, M. et al., 1997, Ribotypes and virulence gene polymorphisms suggest three distinct *Listeria monocytogenes* lineages with differences in pathogenic potential, *Infect. Immun.*, 65:2707–2716.

Wiedmann, M., 2001, Molecular subtyping methods for *Listeria monocytogenes*, *J. AOAC Int.*, 85:524–531.

4 Fate of *Clostridium Perfringens* in Cook–Chill Foods*

John S. Novak

CONTENTS

* Mention of brand or firm name does not constitute an endorsement by the U.S. Department of Agriculture above others of a similar nature not mentioned.

1-58716-041-2/03/$0.00+$1.50

INTRODUCTION

Reduced stringency in food processing preparations has created prime environments for proliferation of food-borne pathogens. With respect to the increased availability of minimally processed foods in supermarket deli sections in recent years, conditions exist that may enable food-borne outbreaks associated with the pathogen *Clostridium perfringens*. This chapter highlights features of *C. perfringens* that make it a food safety concern, especially under the unique specifications of minimally processed foods. Recognition of the increased potential for food-borne illness can be expected to impact care in food handling practices so as to avert future disease-related problems in the marketplace.

BACKGROUND INFORMATION

CURRENT STATUS OF FOOD-BORNE ILLNESS ASSOCIATED WITH *C. PERFRINGENS* IN THE U.S.

Recently reported demographic and lifestyle changes in the U.S. have shown decreases in family size and increases in U.S. population (Knabel, 1995). Increases are noted in dual-income families, single-parent households, a greater proportion of elderly individuals, and increased food preparation by "latch-key" children out of necessity (Knabel, 1995). As a result, time constraints have produced decreased devotion to home meal preparations and the perception that these time-energy investments are not prudent. Coupled with consumer demands for increased varieties of healthful, fresh-tasting, low-preservative foods, this trend has resulted in prepared meals dominating grocery, supermarket, and convenience store deli sections. If food handling and temperature regulations become lax at the same time, conditions become advantageous for the growth of *C. perfringens.*

C. perfringens food poisoning is the third most common cause of food-borne disease in the U.S., following *Campylobacter* and *Salmonella* spp.; it results from the anaerobic spore former, *C. perfringens* (Mead et al., 1999). The Centers for Disease Control and Prevention (CDC) estimate a quarter of a million food-borne illnesses annually due to *C. perfringens*–contaminated foods, resulting in 41 hospitalizations and 7 deaths (Mead et al., 1999). A 1995 economic estimate calculated costs associated with 10,000 cases of *C. perfringens*–incurred food-borne illnesses in the U.S. to reach $500 million (Buzby and Roberts, 1997). That sum includes the total cost of lost productivity for all individuals affected including victims or, in the cases of children and the elderly, their parents or paid caretakers (Buzby and Roberts, 1997). These consequences are realized in lost work hours and profits regardless of degree of illness severity.

PROPERTIES OF COOK–CHILL FOODS AND SAFETY CONCERNS

The market response for home meal replacements has consisted of cook–chill or frozen foods. Cook–chill foods include foods that are mildly precooked, rapidly chilled, portioned, and subsequently distributed for refrigerated storage prior to

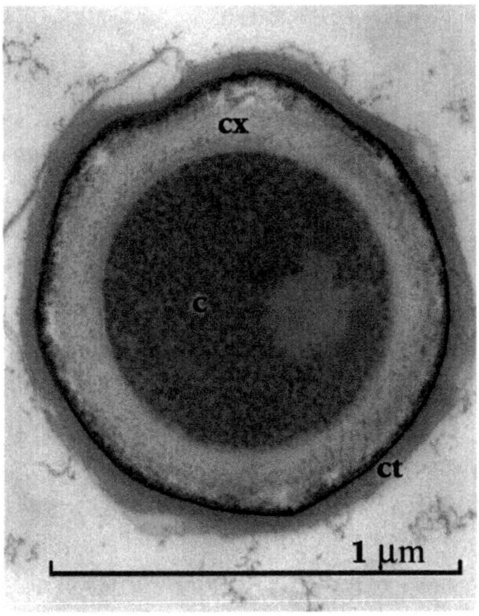

FIGURE 4.1 Transmission electron micrograph depicting the ultrastructure and layers within a spore of *C. perfringens*. The central dehydrated core region (c) contains ribosomes and nuclear material complexed with calcium DPA. This is surrounded by a thick cortex (cx) composed largely of peptidoglycan, minerals such as calcium, DAP, mucopeptides, and some DPA. Numerous proteinaceous coat layers (ct) can be seen at the outer periphery of the spore. The outermost exosporium layer has been lost in this example following spore harvesting washes. The bar represents 1.0 μm in size.

reheating and consumption by the consumer (Sandys and Wilkinson, 1988). Unlike frozen foods that undergo a rapid phase change, cook–chill foods are more susceptible to temperature abuse (i.e., temperatures above 10°C for extended periods of time), microbial contamination, pathogen growth, and food-borne disease. At the same time, cook–chill foods may be more desirable because they are more uniformly heated to safe temperatures and retain many of their fresh-like qualities. Their popularity arises from the flexibility of menu choices, the time efficiency of food production, and the economy of energy and labor (Sandys and Wikenson, 1988). A consequence of additional handling and storage stages is the increased potential for growth from spores of *C. perfringens* because probability of their presence is high.

Spores are specialized survival cells whose formation may be catalyzed by restrictive cellular growth conditions. Structurally, they contain numerous outer protective layers, proteinaceous coats, and membranes (Figure 4.1). The centrally located core region of the spore contains ribosomes and nucleic acids complexed with calcium dipicolinic acid (DPA) in a dehydrated metabolically dormant state that provides protection from exposure to 100°C for up to 60 min. A cortex consisting mainly of cross-linked peptidoglycan, diaminopimelic acid (DAP), mucopolypeptides, and some DPA surrounds and is believed to regulate the hydration of the core. Once conditions become favorable for growth, as occurs in temperature-abused foods, the spores germinate and the microorganism proliferates. The times and temperatures necessary to destroy all spores completely would also damage the food products, making them less natural and wholesome in the process.

A Tolerable Form of Food Poisoning

Non-life-threatening in healthy individuals, the symptoms associated with *C. perfringens* food poisoning are characterized by acute diarrhea and severe abdominal pain 8 to 24 h following ingestion of foods contaminated with large numbers of *C. perfringens* bacteria (Brown, 2000). Full recovery is normal within 24 to 48 h and, as a result of the comparatively common symptoms, cases are frequently not reported (Brown, 2000; Mead et al., 1999). Because of this under-reporting, it has been argued that there was a ten-fold greater occurrence of *C. perfringens*–related food poisoning than the average annual number of outbreak-related cases reported to the CDC from 1983 through 1992 (Mead et al., 1999).

C. perfringens food-borne illnesses can be extremely serious in immunocompromised, very young, or elderly individuals. This is apparent in nursing homes, where the death rate is ten times greater than in the general population (Smith, 1998). For the period from 1988 to 1992 alone, *C. perfringens* was responsible for 1.6% of total food-borne disease outbreaks, 4.9% of total food-borne disease cases, and 1.4% of total food-borne disease deaths in the U.S. (Smith, 1998). In nursing homes during the same time period, *C. perfringens* resulted in 5.2% of such outbreaks, 10% of such cases, and 3.9% of such deaths (Smith, 1998). All of these tragedies could most probably have been prevented through proper food-handling practices and preparation temperatures.

Regulations and Recommendations Regarding *C. perfringens* and Foods

Guidelines and acceptable limits have been established to control the inevitable presence of *C. perfringens* in prepared foods. It is generally accepted that this ubiquitous *Clostridium* species is present in the air, water, and soil. A typical soil sample can contain 10^3 to 10^4 viable cells per gram and it can be assumed that at least 50% of all raw or frozen meat contains *C. perfringens* (McClane, 1997). Human feces from healthy individuals can contain up to 10^3 to 10^6 spores per gram (McClane, 1997). However, illness is often present in individuals from whom greater than 10^6 spores per gram of feces are isolated (Labbe and Juneja, 2001). The heat-labile enterotoxin produced by *C. perfringens* sporulating cells results in disease symptoms; therefore, large numbers of cells undergoing sporulation are necessary to produce the 8 to 10 mg of toxin required to induce symptoms (Labbe and Juneja, 2001). For this reason, restrictions have been imposed to limit the growth of *C. perfringens* present in foods.

The Food Safety Inspection Service (FSIS) of the U.S. Department of Agriculture (USDA) has issued a directive regarding stabilization performance standard requirements in meat products (FSIS, 1999). With respect to the inevitable presence of *C. perfringens* in foods (10^4 *C. perfringens* cells per gram can occur in raw products on occasion), the directive restricts growth of the spore former to no more than a 1 $\log_{10}$ increase or tenfold increases from no more than 10^4 to 10^5 viable cells per gram (FSIS, 1999) (Figure 4.2). This leaves little room for deviations from prescribed food-handling guidelines, considering that an illness outbreak is confirmed upon the isolation limits of greater than 10^6 spores per gram of feces (Labbe and Juneja, 2001).

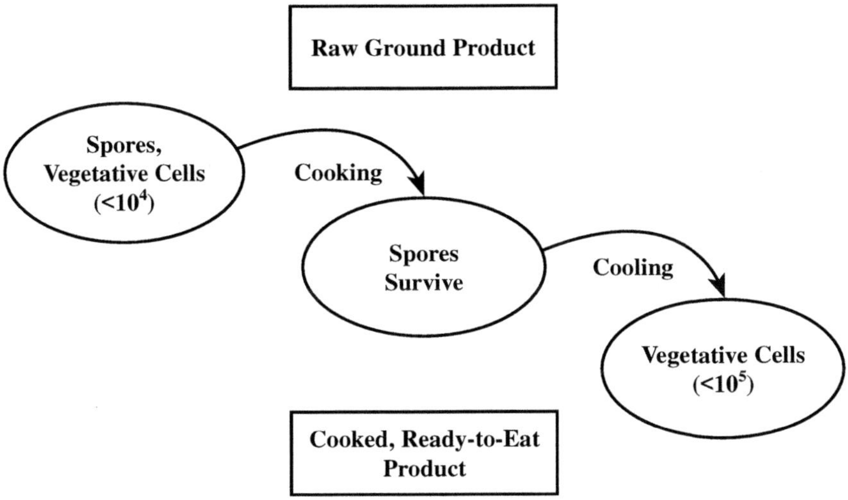

FIGURE 4.2 Germination of spores and growth of resulting vegetative cells of *C. perfringens* for cooked ready-to-eat meat and poultry products. From Marks, H. and Coleman, M., *J. Food Prot.*, 61:1535–1540, 1998.

In the U.K., recommendations for precooked, chilled foods require that *C. perfringens* be present in less than 10^2 CFU per gram and, to ensure safety, that food is chilled from 70 to 3°C within 90 min, followed by storage between 0 and 3°C for no longer than 5 days, including dates of production and consumption (Sandys and Wilkinson, 1988). In the U.S., "sell by" dates on cook–chilled food products are listed, but product expiration is often at the discretion of the consumer. Compliance guidelines on stabilization recommend that cooked meats be cooled from 130 to 80°F in 5 h and from 80 to 45°F in 10 h, for a total of 15 h of cooling time (FSIS, 1999). Because *C. perfringens* is the fastest growing species of *Clostridia*, controlling its growth would be expected to arrest growth of more dangerous spore-formers such as *C. botulinum*, with the exception of non-proteolytic strains that can grow at refrigeration temperatures (Marks and Coleman, 1998).

CLOSTRIDIUM PERFRINGENS

Growth Characteristics

Cook–chill foods provide a particularly suitable environment for the survival and propagation of an opportunistic pathogen such as *C. perfringens.* In order to comprehend the potential for an illness fully, one needs to examine the biological criteria that enable the food-borne pathogen to survive in that particular environment. The fate of *C. perfringens* in minimally processed, cook–chill foods is determined by the microorganism's specific growth characteristics, requirements, and product handling history following refrigeration.

C. perfringens is an anaerobe that can tolerate aerobic conditions. The vegetative cells are Gram-positive, nonmotile, encapsulated rods (Rhodehamel and Harmon,

1998). Microbiological testing shows the microorganism capable of hydrolyzing gelatin, reducing nitrate, and fermenting lactose (Rhodehamel and Harmon, 1998). At the optimal growth temperature (43 to 45°C), generation times are often less than 10 min (Labbe and Juneja, 2001). Growth has been shown at temperatures as high as 50°C, while slowly arresting near 6°C and below (McClane, 1997). In addition, *C. perfringens* is capable of vegetative growth at relatively low water activities (a_w) in the range of 0.93 to 0.97 (Bartsch and Walker, 1982). Growth of *C. perfringens* is limited by a number of environmental and physiological conditions, including pH (<5.0 or >8.0), NaCl (>8%), $NaNO_3$ (>1000 ppm), and $NaNO_2$ (>400 ppm) (McClane, 1997). *C. perfringens* is mostly, but not exclusively, found in meat products due to an auxotrophic requirement for 13 different amino acids that the microorganism cannot synthesize (Andersson et al., 1995). Although extreme conditions may limit vegetative growth, they do not ensure pathogen inactivation due to the enhanced resistance of the metabolically dormant spores.

Enterotoxin Production

The resilience of *C. perfringens*, documented to survive stressful conditions and temperatures as high as 100°C for more than 1 h, is attributed to the formation of heat-resistant spores (Rhodehamel and Harmon, 1998). That heat resistance is attained coincident with the formation of the spore coat layers (Labbe and Duncan, 1977). The *C. perfringens* enterotoxin (CPE) produced during sporulation and formation of the coat layers is believed to bind and affect the villus tip cells of the small intestine, resulting in a disruption of the ability to maintain membrane ionic balance and resultant diarrheal symptoms (McClane, 1997). However, the diarrhea is also a remedy to the problem because it flushes unbound CPE and many *C. perfringens* cells from the small intestine (McClane, 1997).

Large numbers of vegetative cells (greater than 10^6 per gram of ingested food product) are required to elicit illness symptoms because many cells are killed by exposure to stomach acid (McClane, 1997). Ironically, it is believed that the acidic conditions encountered upon passage through the gastrointestinal tract actually trigger the sporulation of vegetative cells (Wrigley et al., 1995). Although there is evidence of "leaky" gene regulation of CPE production in *C. perfringens* cells, a 1500-fold increase in enterotoxin is associated with sporulating cells (McClane, 1997). Preformed CPE in foods is often not implicated in food-borne illness because heating for 5 min at 60°C will inactivate the enterotoxin (McClane, 1997). The toxinogenic typing (A, B, C, D, or E) of *C. perfringens* is not based on the serologic specificity of CPE-related food-borne illness but on many other exotoxins produced by the microorganism and designated α, β, ε, and τ (Petit et al., 1999; Brown, 2000). These toxins do not create a food-borne illness concern in the U.S. and will not be given further attention in this chapter.

An Opportunistic Pathogen under Conditions of Temperature Abuse

The combination of rapid growth, an ability to form a dormant stage (the spore) that enables survival after thermal processing, and the characteristics of minimally

TABLE 4.1
Maximum Temperature (°C) in Products after Transportation for 1 h in the Trunk of a Car, without Protection or within a Cooled Insulated Container

Product	Unprotected	Cool Box
Beef pie	24	7
Chicken sandwich	32	10
Cooked chicken	28	12
Minced beef	18	9
Prepared salad	29	14
Quiche	26	18
Sausage (raw)	28	15
Smoked ham	30	14
Trout	28	5
Brie cheese	28	11
Coleslaw	30	14
Lasagne	21	6
Pate	25	13
Prawns	37	14
Raw chicken	24	4
Sausage roll	28	12
Smoked salmon	38	18

Reprinted from James, S.J. and Evans, J., *Int. J. Refrig.*, 15:299–306, 1992a. With permission.

processed, cook–chill foods create a niche suitable for the opportunistic pathogen *C. perfringens*. Temperature abuse becomes a critical factor in the control of *C. perfringens*. The length of exposure of any food product to temperatures between 10°C (50°F) and 47°C (117°F) must be limited, even during transport from the supermarket by the consumer. Table 4.1 lists the maximum temperatures recorded in various foods transported for 60 min in the trunk of a car when the ambient temperature ranged from 23°C (73°F) to 27°C (81°F) (James and Evans, 1992a). After 60 min, all of the products were in the growth-permissive range for *C. perfringens,* and a few exceeded temperatures of 15°C (59°F) when kept within a cooled, insulated container.

It is often assumed that refrigeration is enough to limit growth of *C. perfringens*. Unfortunately, home refrigeration temperatures are seldom set with a calibrated thermometer and are routinely regulated at the discretion of the consumer. A survey of 150 domestic refrigerators showed a wide distribution of average temperatures, including 7.3% of those above 10°C (50°F) (Figure 4.3) (Flynn et al., 1992). Even refrigerators properly set and capable of maintaining temperatures below 4°C (40°F) can reach temperatures well above those recommended following repeated door openings (Figure 4.4) (James and Evans, 1992b). Therefore, the avoidance of temperature abuse in food products is the responsibility not only of the retailer but also of the consumer who purchases those products.

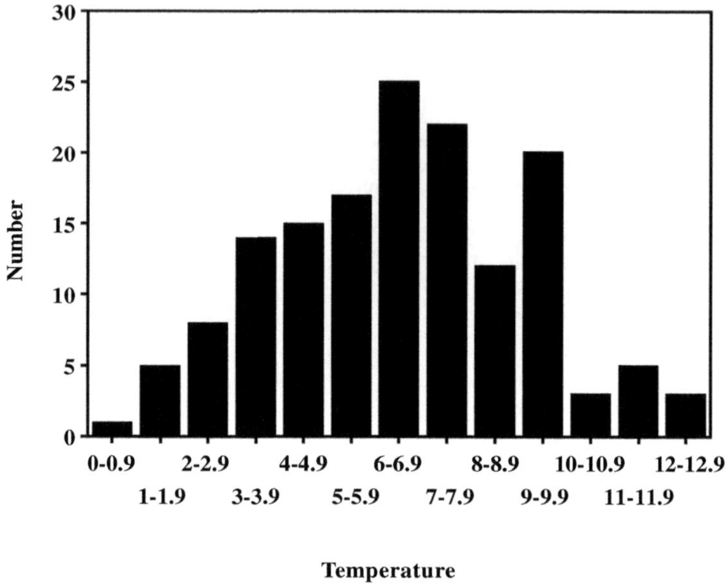

FIGURE 4.3 Mean temperature (°C) frequency distribution for all sampled refrigerators. Reprinted from Flynn, O.M.J., Blair, J.I., and McDowell, D., *Int. J. Refrig.*, 15:307–312, 1992.

RESEARCH IN THE FOOD LABORATORY

Variations in Heating Efficiency Dependent upon Food Chemistry

Food research is necessary to ensure that conditions are not conducive for allowing growth beyond 1 $\log_{10}$ unit increase within a food product. Typical growth parameters for vegetative cells of *C. perfringens* are depicted in Tables 4.2 and 4.3. In thioglycolate broth, a common, complex laboratory medium used in culturing anaerobes, growth of *C. perfringens* was optimal in the 35 to 45°C range, with generation times approaching 19 min (Park and Mikolajcik, 1979). When the medium was switched to ground-beef broth, the generation time increased to 156 min (Park and Mikolajcik, 1979); the chemical composition of the ground-beef broth was less favorable for rapid growth of *C. perfringens*. However slowed, the microorganism was still capable of growth in the less optimal food product. Vegetative cell growth slowed at 15°C in thioglycolate broth, and a 2 $\log_{10}$ decrease in cellular viability during a 6-h time period was recorded (Park and Mikolajcik, 1979). This reemphasized the importance of temperature control during food handling to limit pathogen growth. As general as these control measures appear, pathogen responses in cook–chill foods would be simplified if they were dependent only upon the food composition and growth temperatures.

Strain Variability within a Species

NCTC 8238 and NCTC 8798, two strains of *C. perfringens* that produce CPE, were evaluated in ground beef following growth to stationary phase at constant temper-

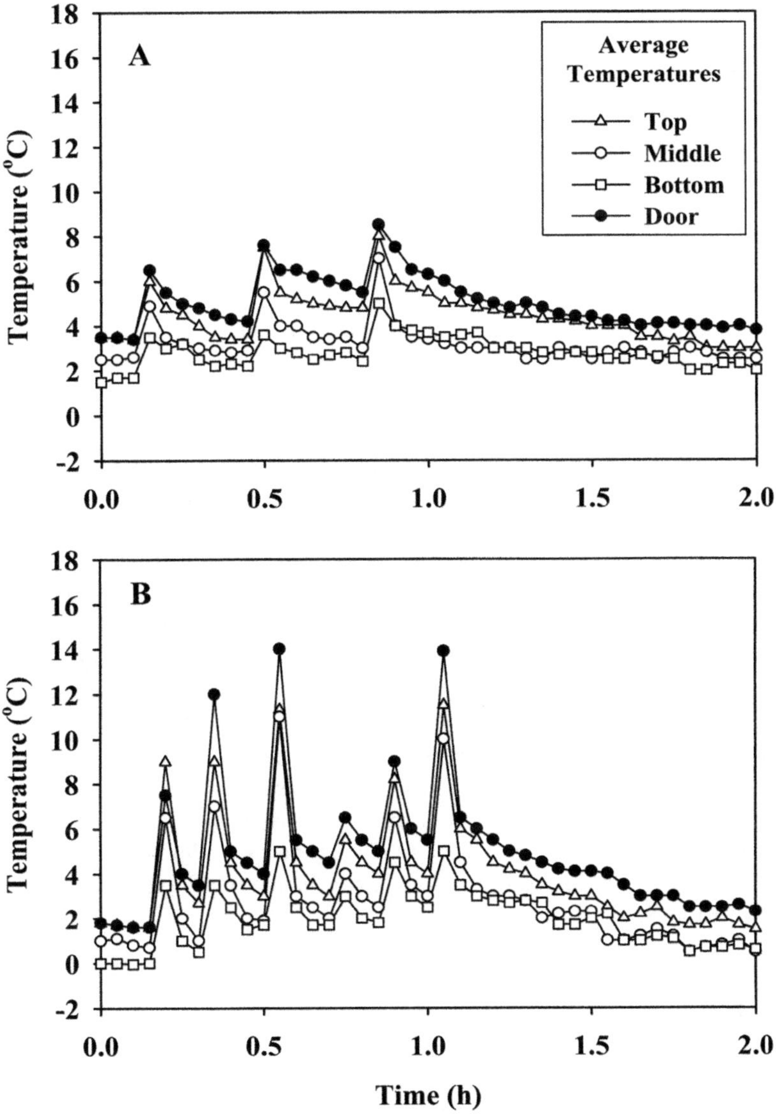

FIGURE 4.4 Temperature response of a refrigerator to repeated door opening: (A) three 1-min openings at 20-min intervals; (B) six 1-min openings at 10-min intervals. Reprinted from James, S.J. and Evans, J., *Int. J. Refrig.*, 15:313–319, 1992b. With permission.

TABLE 4.2
***C. perfringens* Vegetative Cell Growth**

Temperature (°C)	Generation Time (min)	Growth Medium
15	ND[a]	Thioglycolate
25	61.5	Thioglycolate
35	19.3	Thioglycolate
35	156.0	Ground-beef broth
45	19.7	Thioglycolate
50	22.6	Thioglycolate

[a] Not determined because there was a 2 $\log_{10}$ decrease in 6 h at 15°.

Adapted from Park, Y. and Mikolajcik, E.M., *J. Food Prot.*, 42:848–851, 1979.

ature (37, 41, 45, or 49°C) and growth at rising temperature from 25 to 50°C at rates of 4, 6, and 7.5°C per h (Roy et al., 1981). Heat treatment is the most common food preservation process in use today (Juneja and Thayer, 2001). D-values necessary to result in a 90% or 1-$\log_{10}$ decrease in the viable cell population were calculated as the time in minutes at a specific temperature (57 or 59°C in this experiment) and used as the standard measure of pathogen thermal inactivation. The results depicted in Table 4.3 (Roy et al., 1981) validate the hypothesis that vegetative cells surviving mild heat exposure, as might be expected with minimally processed foods, can acquire increased heat resistance, as exemplified by high D-values for growth in ground beef at increasing temperatures. Adaptations to growth at 49°C were greater for strain NCTC 8798 compared to strain NCTC 8238. Growth to stationary phase at 6°C per h increasing temperature produced more heat-resistant *C. perfringens* NCTC 8238 vegetative cells, whereas growth at 4°C per h increasing temperatures from 25 to 50°C best adapted *C. perfringens* NCTC 8798 to survival from additional heat treatment. This enhanced thermal tolerance of pathogen cells surviving an initial mild heating becomes worrisome when evaluating the safety of cook–chill foods.

TABLE 4.3
D_{59}-Values for *C. perfringens* Vegetative Cells Following Growth at Constant Temperatures (37, 41, 45, and 49°C) or Following Growth at Rising Temperatures (25–50°C) at a Rate of 4, 6, and 7.5°C per h

C. perfringens Strains	37°C	41°C	45°C	49°C	4°C/h	6°C/h	7.5°C/h
NCTC 8238	2.3	2.8	4.1	6.9	7.7	12.5	6.9
NCTC 8798	3.1	4.4	7.2	10.6	11.0	8.5	7.6

Adapted from Roy, J.R. et al., *J. Food Sci.*, 46:1586–1591, 1981.

The ability of *C. perfringens* vegetative cells to survive refrigerated temperatures used to chill foods can also be strain dependent. *C. perfringens* vegetative cells were reported to reproduce more rapidly in frankfurters at 23 to 37°C compared to 12 to 15°C but not at all during a 2- to 4-week storage at 0 to 10°C (Solberg and Elkin, 1970). More meticulous examination demonstrated that *C. perfringens* strain 1362 vegetative cells decreased 0.5 $\log_{10}$ in 169 days at 5°C, whereas strain S80 vegetative cells decreased 3 logs in 83 days under the same conditions (Solberg and Elkin, 1970).

When turkey or beef casserole was the contaminated food source, 5 to 10°C for 48 h produced a similar stabilization or decrease of vegetative cell numbers, whereas 24°C allowed multiplication of cells that remained viable, possibly through sporulation at temperatures as high as 68°C for 6 h (Strong and Ripp, 1967). It is generally accepted that temperature-abused, precooked foods, if not reheated adequately before consumption, would be a potential source for food poisoning but that reheating to 65°C would be sufficient to kill *C. perfringens* (Juneja et al., 1994). The increased thermal stability and survivability of spores that this microorganism produces often complicate the problem.

Increased Heat Resistance and Spores

Studies have been done with *C. perfringens* vegetative cells that record the optimal temperature for sporulation at 37°C in comparison with 5, 22, and 46°C (Kim et al., 1967). Sporulation was evident by 6 h and was complete after 20 h at 37°C (Kim et al., 1967). Although *C. perfringens* is an anaerobe, the type of atmosphere was found to have minimal, if any, influence on sporulation temperatures above 28°C (Juneja et al., 1994). There is concern that a heat-shocking condition may be created in cook–chill processing, potentially facilitating an increase in the heat resistance of pathogens and their spores (Juneja and Thayer, 2001).

A direct relationship has been shown to exist between spore heat resistance and the temperature at which spores are produced (Garcia-Alvarado et al., 1992). The implication is that spores, like vegetative cells, can become adapted to survival following sublethal high temperature exposures. A sublethal heat shock at 55°C for 30 min, applied as *C. perfringens* spores were in the process of forming, resulted in spores with increased heat resistance (Heredia et al., 1997, 1998). The acquired thermotolerance or adaptation was maintained transiently for 2 h in metabolically active vegetative cells, which may indicate metabolic turnover events such as the production and degradation of protein products (Heredia et al., 1997). The duration to which dormant spores maintained the heat adaptation was not evaluated in these studies, and the possibility exists that spores remain heat adapted until germination.

C. perfringens spores preheated to 100°C for 60 min were found to be heat resistant, but upon germination the resultant vegetative cells sporulated poorly, whereas spores heated to 70°C for 10 min were far less heat resistant but had subsequently increased sporulation potential of the vegetative cells (Nishida et al., 1969). Commercial cooking frequently uses long-time, low-temperature cooking, with temperatures increased gradually from 40 to 60°C in 4 h (Smith et al., 1980). A population of *C. perfringens* inoculated into ground beef was found to increase

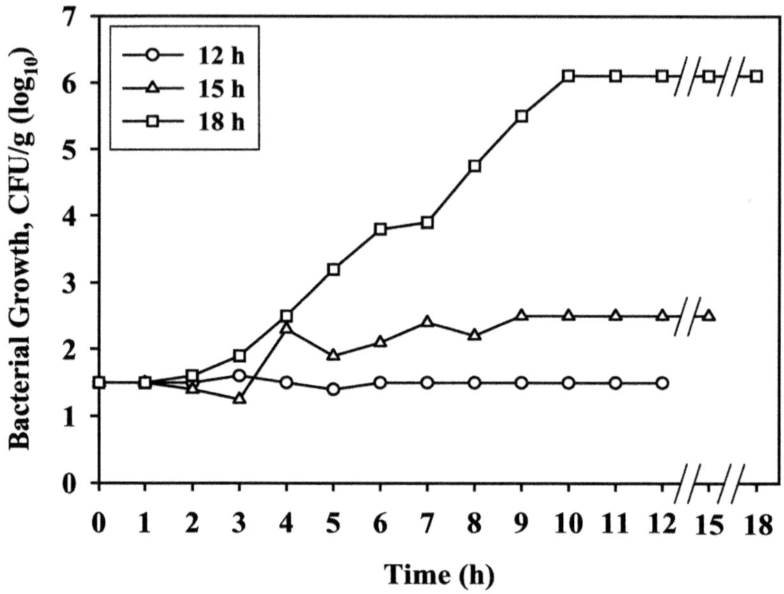

FIGURE 4.5 Fate of three-strain composite of *C. perfringens* spores in cooked ground beef cooled through the temperature range of 54.4 to 7.2°C in 12, 15, and 18 h. From Juneja, V.K. et al., *J. Food Prot.*, 57:1063–1067, 1994.

at 42°C, reaching maximum numbers at 55°C following 9.5 h of increasing temperature (Smith et al., 1980). Further heating for an hour at 56°C resulted in a dramatic decrease in viable cells (Smith et al., 1980). This does not imply that cooking foods to 60°C will ensure safety from *C. perfringens* outbreaks; the cooking temperatures may kill vegetative cells, but the same temperatures can also serve to activate dormant spores that will germinate and multiply if cooking temperatures fall (Adams, 1973).

Cooling Studies Following Heat Treatment of Foods

C. perfringens inoculated at 1.5 $\log_{10}$ spores per gram in cooked beef was found to be capable of germinating and growing to potentially hazardous infectious dose levels greater than 6 $\log_{10}$ cells per gram if cooled from 54.4 to 7.2°C in longer than 18 h (Figure 4.5) (Juneja et al., 1994). Spores germinating in cooked chili resulted in observed growth if the cooling period between 48.9 and 37.8°C lasted more than 2 h (Blankenship et al., 1988). Therefore, even though a food product may be thoroughly cooked, it is important that the cooling stage be as short as possible — even more so for minimally processed foods.

Evolution and Adaptability of *C. perfringens* Food Poisoning

Evidently the environment and the specific composition of a food product play a role in heat resistance. *C. perfringens* spores produced in Ellner's medium were found to be less heat resistant than the same spores in physiological saline (Weiss and Strong, 1967). Likewise, spores from different strains of *C. perfringens* can exhibit different tolerances to refrigeration temperatures. It has already been shown that chilled temperatures arrest vegetative cell growth in *C. perfringens*; although not a psychrotroph, *C. perfringens* is capable of limited survival in chilled foods (Goepfert and Kim, 1975). Following germination, *C. perfringens* strain 1362 spores increased in number or showed no change in viability when stored in water for 169 days at 5°C (Solberg and Elkin, 1970). S-80 spores, however, decreased 1.5-fold in 83 days under the same conditions (Solberg and Elkin, 1970). Therefore, all possible variables must be examined in all likely scenarios for a given food product containing a potential pathogen.

With respect to heat adaptations and pathogen toxicity, it is an interesting finding that only *C. perfringens* strains exposed to heat produce enterotoxin in sufficient amounts to cause food poisoning (Granum, 1990). CPE has been believed to be a structural component of the spore coat but, amazingly, strains of *C. perfringens* associated with food poisoning can produce 2000 times more enterotoxin than is found during normal spore coat production (Granum, 1990). Therefore, it has been suggested that all *C. perfringens* strains can be transformed to enterotoxin-positive strains through repeated heat treatments similar to those for cook–chill foods (Granum, 1990). Recently it has been reported that the location of genetic determinants for CPE on the chromosome or on a plasmid may play a significant role in determining whether *C. perfringens* heat resistance is high and whether the isolate is capable of causing food poisoning (Collie and McClane, 1998; Sarker et al., 2000).

APPLICATION AND TECHNOLOGY TRANSFER

Limitations to Growth

Growth-limiting parameters need thorough evaluation in cook–chill foods in order to ensure safety in limiting increases in *C. perfringens* cell numbers under all potential conditions to which the foods will be exposed. A few of these have been extensively studied, such as water activity, redox potential, pH, salt, and the use of other preservatives. While mechanisms of heat resistance in vegetative cells have concentrated on repair of DNA and synthesis of protective proteins, spore heat resistance properties have been theorized to result from mineralization or combinations of systems to maintain a dehydrated core (Miyata et al., 1997). It is generally accepted that the heat resistance in the spore is a consequence of low water activity (a_w) in the spore core, but it is acknowledged that other mechanisms must be in effect to protect germination enzymes present in peripheral locations outside the core (Miyata et al., 1997). Levels of a_w in foods or media could be controlled with NaCl, KCl, or glucose. A trend was established for *C. perfringens* that, as the a_w level was lowered, the rate and amount of growth were lessened (Strong et al., 1970). The use of glucose to lower a_w resulted in growth at lower a_w levels (0.96) compared to NaCl (0.975), whereas KCl was most effective comparatively in producing long lag times and the least amount of measurable *C. perfringens* growth (Strong et al., 1970). The method of handling rehydrated dried foods is considered important with respect to *C. perfringens* because the pathogen has been isolated from a variety of food products of limited water content, such as dried soup and sauce mixes, due to the resilient spores (Strong et al., 1970).

Inhibition by Air

Spores enable an anaerobe such as *C. perfringens* to survive until conditions become favorable for growth. Quite often, foods provide constantly changing environments of which versatile pathogens can take advantage. During the germination and sprouting of mung beans, *C. perfringens* was capable of growth at an unusually low pH of 3.7 as the oxygen concentration decreased (deJong, 1989). This anaerobic environment was attributed not only to the respiration of the sprouts but also to actively growing *Enterobacteriaceae* and lactobacilli (deJong, 1989). When aerobicity was increased, the growth of *C. perfringens* was inhibited (deJong, 1989). In cooked turkey, the growth of *C. perfringens* can be slowed using a modified atmosphere of 25 to 50% CO_2 and 20% O_2 with the balance N_2 (Juneja et al., 1996a). However, growth of *C. perfringens* in the turkey meat could be completely inhibited at 4°C, regardless of anaerobicity (Juneja et al., 1996a). Thus, the importance of refrigerated storage of foods cannot be overstated.

The oxidation-reduction or redox potential (Eh) parameter of foods may not be an effective means for controlling anaerobic bacteria. A positive oxidation-reduction potential may not always be indicative of the presence of oxygen. The addition of 10% potassium ferricyanide to anaerobic medium, inoculated with *C. perfringens,* over a 10-h period maintained an average Eh of +325 mV and growth identical to controls (Walden and Hentges, 1975). An aerated culture with a negative Eh (–50 mV) inhibited growth of *C. perfringens*, as did a culture aerated to an Eh of +500 mV (Walden and Hentges, 1975). It was concluded that the presence of oxygen was inhibitory to growth of anaerobes such as *C. perfringens* regardless of the Eh value (Walden and Hentges, 1975). It has been of concern that the Eh of many raw meats and gravies is low enough to permit growth of *C. perfringens* (McClane, 1997). The usefulness of Eh as a predictive measure for assessing the fate of *C. perfringens* in foods is questionable because the microorganism is known to modify the Eh of its surrounding environment by producing reducing molecules such as ferredoxin (McClane, 1997).

Synergistic Effects of Stress Hurdles

The "hurdle" concept of limiting pathogen growth involves the combinations of minimally inhibitory factors, or additive hurdles, that alone do not preclude growth but together provide enough accumulative stress on the microorganism that survival is not possible. This approach is preferable for minimally processed foods because inhibitory components are not added to excess. The result of this is the maintenance of the organoleptic properties of foods with a safer, fresh-like quality.

Temperature abuse of ground beef inoculated with 2 $\log_{10}$ *C. perfringens* spores per gram at 15°C was examined by varying pH and salt concentration (Figure 4.6) (Juneja and Majka, 1995). Sodium pyrophosphate was added to 0.3% (w/v) in order to serve as a typical preservative used in packaged meats. Neither a lowered pH of

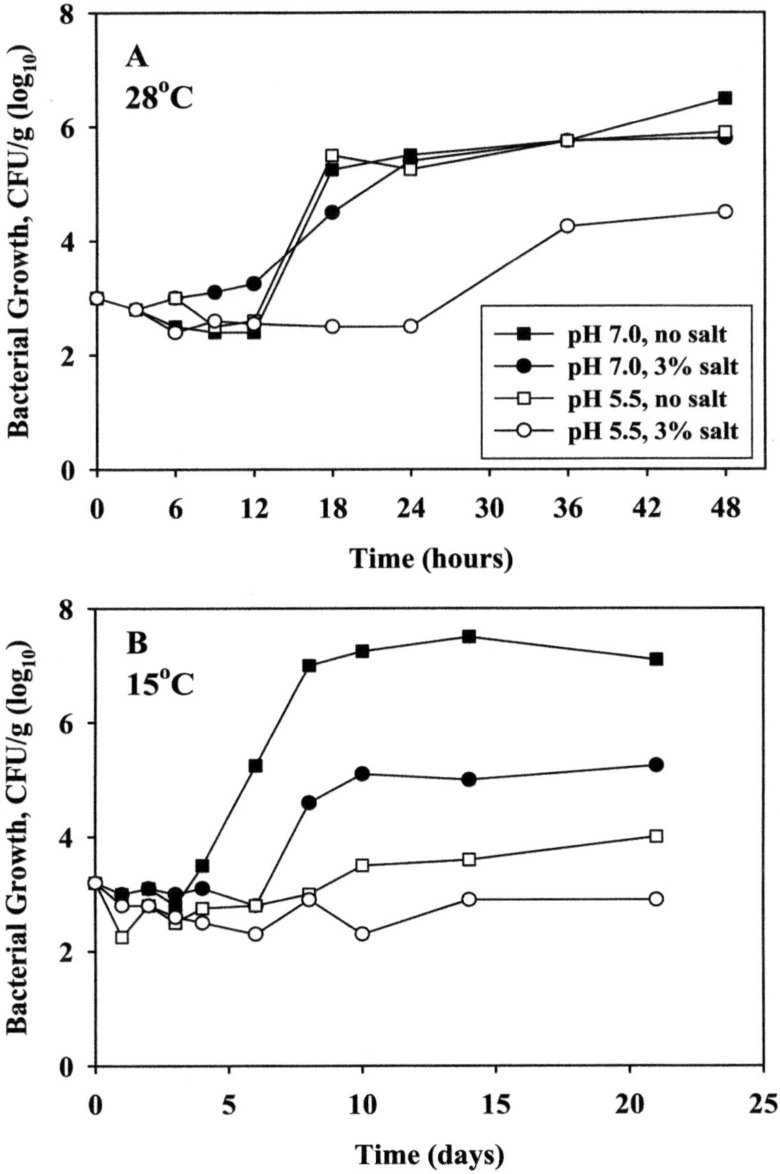

FIGURE 4.6 The effect of temperature abuse (storage at 15 or 28°C) on growth of *Clostridium perfringens* from a spore inoculum in vacuum-packaged, cook-in-bag ground beef that included 0.3% sodium pyrophosphate at pH 5.5 or 7.0 and salt levels 0 or 3%. From Juneja, V.K. and Majka, W.M., *J. Food Safety*, 15:21–34, 1995.

5.5 nor 3% (w/v) salt was capable of completely inhibiting *C. perfringens* growth after 1 to 3 weeks at 15°C when used alone. When the pH was lowered to 5.5 in combination with 3% salt, cell numbers for *C. perfringens* did not increase after 20 days. If the low temperature hurdle was removed to 28°C, then growth was evident after only 36 h at pH 5.5 and 3% salt. Lower concentrations of salt may provide a protective effect to *C. perfringens,* as evidenced by decreasing germination times or faster growth rates between 0 and 3% NaCl (Juneja and Marmer, 1996).

Predictive models based on experimental data can be used to study *C. perfringens* growth and the interactions of additional variables or hurdles (Juneja et al., 1996b, 1999). Experimental data from the growth characteristics of *C. perfringens* have recently been incorporated into the USDA's pathogen modeling program and can be accessed at www.arserrc.gov/mfs/pmparameters.htm for more information.

CONCLUSIONS

It has been demonstrated that cook–chill foods subjected to temperature-abuse conditions provide a suitable environment for propagation of *C. perfringens,* thereby increasing risk of food-borne illness. The foods are not limited to the marketplace because hospital, work, and school cafeterias also contain preprepared chilled foods. Other representative examples include meals served as part of a transportation or catering service. A study was recently conducted on the microbial quality of meals served on aircraft used for public transportation (Hatakka, 1998). It was concluded that during the period from 1991 to 1994 the frequency of pathogenic *C. perfringens* in airline meals was 1.0% of the total meals served and that 0.7% of the meals exceeded the Association of European Airlines' (AEA) acceptable standard for *C. perfringens* (1×10^3 CFU per gram) (Hatakka, 1998). Ultimately, consumer education plays a crucial role in limiting *C. perfringens*–associated food-borne illness. Proper chilling and sufficient heat treatment for the food and portion size will increase food safety considerably with respect to this common but avoidable pathogen.

ACKNOWLEDGMENTS

The author is grateful to Dr. James L. Smith, USDA Agricultural Research Service; Dr. John H. Hanlin, McCormick & Co., Inc.; and Dr. Thomas M. Wahlund, California State University at San Marcos for their critical assessments and helpful peer reviews of this chapter.

REFERENCES

Adams, D.M., 1973, Inactivation of *Clostridium perfringens* type A spores at ultrahigh temperatures, *Appl. Microbiol.*, 26:282–287.

Andersson, A., Ronner, U., and Granum, P.E., 1995, What problems does the food industry have with the spore-forming pathogens *Bacillus cereus* and *Clostridium perfringens*?, *Int. J. Food Microbiol.*, 28:145–155.

Bartsch, A.G. and Walker, H.W., 1982, Effect of temperature, solute, and Ph on the tolerance of *Clostridium perfringens* to reduced water activities, *J. Food Sci.*, 47:1754–1755.

Blankenship, L.C. et al., 1988, Growth of *Clostridium perfringens* in cooked chili during cooling, *Appl. Environ. Microbiol.*, 54:1104–1108.

Brown, K.L., 2000, Control of bacterial spores, *Br. Med. Bull.*, 56:158–171.

Buzby, J.C. and Roberts, T., 1997, Economic costs and trade impacts of microbial foodborne illness, *World Health Stat. Q.*, 50:57–66.

Collie, R.E. and McClane, B.A., 1998, Evidence that the enterotoxin gene can be episomal in *Clostridium perfringens* isolates associated with non-food-borne human gastrointestinal diseases, *J. Clin. Microbiol.*, 36:30–36.

deJong, J., 1989, Spoilage of an acid food product by *Clostridium perfringens*, *C. barati*, and *C. butyricum*, *Int. J. Food Microbiol.*, 8:121–132.

Flynn, O.M., Blair, J.I., and McDowell, D., 1992, The efficiency and consumer operation of domestic refrigerators, *Rev. Int. Froid.*, 15:307–312.

FSIS, 1999, Directive 7111.1, Performance standards for the production of certain meat and poultry products, United States Department of Agriculture, Washington, D.C.: Food Safety and Inspection Service, Federal Register.

Garcia-Alvarado, J.S., Labbe, R.G., and Rodriguez, M.A., 1992, Sporulation and enterotoxin production by *Clostridium perfringens* type A at 37 and 43°C, *Appl. Environ. Microbiol.*, 58:1411–1414.

Goepfert, J.M. and Kim, H.U., 1975, Behavior of selected food-borne pathogens in raw ground beef, *J. Milk Food Technol.*, 38:449–452.

Granum, P.E., 1990, *Clostridium perfringens* toxins involved in food poisoning, *Int. J. Food Microbiol.*, 10:101–112.

Hatakka, M., 1998, Microbiological quality of hot meals served by airlines, *J. Food Prot.*, 61:1052–1056.

Heredia, N.L. et al., 1997, Elevation of the heat resistance of vegetative cells and spores of *Clostridium perfringens* type A by sublethal heat shock, *J. Food Prot.*, 60:998–1000.

Heredia, N.L., Labbe, R.G., and Garcia-Alvarado, J.S., 1998, Alteration in sporulation, enterotoxin production, and protein synthesis by *Clostridium perfringens* type A following heat shock, *J. Food Prot.*, 61:1143–1147.

James, S.J. and Evans, J., 1992a, Consumer handling of chilled foods: temperature performance, *Rev. Int. Froid.*, 15:299–306.

James, S.J. and Evans, J., 1992b, The temperature performance of domestic refrigerators, *Rev. Int. Froid.*, 15:313–319.

Juneja, V.K. and Majka, W.M., 1995, Outgrowth of *Clostridium perfringens* spores in cook-in-bag beef products, *J. Food Safety*, 15:21–34.

Juneja, V.K. and Marmer, B.S., 1996, Growth of *Clostridium perfringens* from spore inocula in *sous-vide* turkey products, *Int. J. Food Microbiol.*, 32:115–123.

Juneja, V.K., Marmer, B.S., and Call, J.E., 1996a, Influence of modified atmosphere packaging on growth of *Clostridium perfringens* in cooked turkey, *J. Food Safety*, 16:141–150.

Juneja, V.K. et al., 1996b, Interactive effects of temperature, initial Ph, sodium chloride, and sodium pyrophosphate on the growth kinetics of *Clostridium perfringens*, *J. Food Prot.*, 59:963–968.

Juneja, V.K., Snyder, O.P., Jr., and Cygnarowicz-Povost, M., 1994, Influence of cooling rate on outgrowth of *Clostridium perfringens* spores in cooked ground beef, *J. Food Prot.*, 57:1063–1067.

Juneja, V.K. and Thayer, D.W., 2001, Irradiation and other physically based control strategies for foodborne pathogens, in *Microbial Food Contamination,* Wilson, C.L. and Droby, S., Eds., Washington, D.C.: CRC Press, 171–186.

Juneja, V.K. et al., 1999, Predictive model for growth of *Clostridium perfringens* at temperatures applicable to cooling of cooked meat, *Food Microbiol.*, 16:335–349.

Kim, C.H., Cheney, R., and Woodburn, M., 1967, Sporulation of *Clostridium perfringens* in a modified medium and selected foods, *Appl. Microbiol.*, 15:871–876.

Knabel, S.J., 1995, Foodborne illness: role of home food handling practices, *Food Technol.*, 49:119–131.

Labbe, R.G. and Duncan, C.L., 1977, Spore coat protein and enterotoxin synthesis in *Clostridium perfringens*, *J. Bacteriol.*, 131:713–715.

Labbe, R.G. and Juneja, V.K., 2001, *Clostridium perfringens*, in *Foodborne Infections and Intoxications*, Riemann, H. and Bryan, F.L., Eds., New York: Academic Press, in press.

Marks, H. and Coleman, M., 1998, Estimating distributions of organisms in food products, *J. Food Prot.*, 61:1535–1540.

McClane, B.A., 1997, *Clostridium perfringens*, in *Food Microbiology: Fundamentals and Frontiers*, Doyle, M.P., Beuchat, L.R., and Montville, T.J., Eds., Washington, D.C.: ASM Press, 305–326.

Mead, P.S. et al., 1999, Food-related illness and death in the United States, *Emerging Infect. Dis.*, 5:607–625.

Miyata, S. et al., 1997, Localization of germination-specific spore-lytic enzymes in *Clostridium perfringens* S40 spores detected by immunoelectron microscopy, *FEMS Microbiol. Lett.*, 152:243–247.

Nishida, S.N. Seo and Nakagawa, M., 1969, Sporulation, heat resistance, and biological properties of *Clostridium perfringens*, *Appl. Microbiol.*, 17:303–309.

Park, Y. and Mikolajcik, E.M., 1979, Effect of temperature on growth and alpha toxin production by *Clostridium perfringens*, *J. Food Prot.*, 42:848–851.

Petit, L., Gibert, M., and Popoff, M.R., 1999, *Clostridium perfringens*: toxinotype and genotype, *Trends Microbiol.*, 7:104–110.

Rhodehamel, J. and Harmon, S.M., 1998, *Clostridium perfringens*, in *FDA Bacteriological Manual*, 8th ed., Bennett, R.W., Ed., Gaithersburg, MD: AOAC International, 16.01–16.06.

Roy, J.R., Busta, F.F., and Thompson, D.R., 1981, Thermal inactivation of *Clostridium perfringens* after growth at several constant and linearly rising temperatures, *J. Food Sci.*, 46:1586–1591.

Sandys, G.H., and Wilkinson, P.J., 1988, Microbiological evaluation of a hospital delivered meals service using precooked chilled foods, *J. Hosp. Infect.*, 11:209–219.

Sarker, M.R. et al., 2000, Comparative experiments to examine the effects of heating on vegetative cells and spores of *Clostridium perfringens* isolates carrying plasmid genes vs. chromosomal enterotoxin genes, *Appl. Environ. Microbiol.*, 66:3234–3240.

Smith, J.L., 1998, Foodborne illness in the elderly, *J. Food Prot.*, 61:1229–1239.

Smith, L.B., Busta, F.F., and Allen, C.E., 1980, Effect of rising temperatures on growth and survival of *Clostridium perfringens* indigenous to raw beef, *J. Food Prot.*, 43:520–524.

Solberg, M. and Elkin, B., 1970, Effect of processing and storage conditions on the microflora of *Clostridium perfringens*-inoculated frankfurters, *J. Food Sci.*, 35:126–129.

Strong, D.H., Foster, E.F., and Duncan, C.L., 1970, Influence of water activity on the growth of *Clostridium perfringens, Appl. Microbiol.*, 19:980–987.

Strong, D.H. and Ripp, N.M., 1967, Effect of cookery and holding on hams and turkey rolls contaminated with *Clostridium perfringens*, *Appl. Microbiol.*, 15:1172–1177.

Walden, W.C. and Hentges, D.J., 1975, Differential effects of oxygen and oxidation-reduction potential on the multiplication of three speciecs of anaerobic intestinal bacteria, *Appl. Microbiol.*, 30:781–785.

Weiss, K.F. and Strong, D.H., 1967, Effect of suspending medium on heat resistance of spores of *Clostridium perfringens*, *Nature*, 215:530–531.

Wrigley, D.M., Hanwella, H.D.S.H., and Thon, B.L., 1995, Acid exposure enhances sporulation of certain strains of *Clostridium perfringens*, *Anaerobe*, 1:263–267.

5 *Sous-Vide* Processed Foods: Safety Hazards and Control of Microbial Risks

Vijay K. Juneja

CONTENTS

INTRODUCTION

Consumers have been demanding fresh-tasting, high-quality, low-salt, and preservative-free meals that can be microwaved, have a high degree of convenience, and require minimal preparation time. This demand has resulted in an increased production of minimally processed, ready-to-eat, extended shelf-life refrigerated foods in the North American and European markets. According to the National Food Processors Association (NFPA, 1988), such food products are known as "new generation refrigerated food" and include *sous-vide* (under vacuum) processed food products. *Sous-vide* food processing is a method of cooking whereby fresh food is vacuum sealed in heat-stable, high-barrier, or air-impermeable bags or plastic pouches to remove all of the air and then cooked (pasteurized) to a time and temperature for a specific food.

1-58716-041-2/03/$0.00+$1.50

During the cooking or pasteurization step, any heat-sensitive microorganisms, such as vegetative food-borne pathogens, spoilage microflora, and some spore-formers, are killed. The pasteurized product is chilled rapidly to avoid germination and outgrowth of surviving bacterial spores and then stored and distributed under refrigerated conditions (<4°C). Before consumption, the foods are reheated by a number of different methods such as microwaves, conventional forced air ovens, and hot-water baths, depending upon the type of food. *Sous-vide* products offer several benefits over frozen or canned foods, and this type of processing is used extensively in Europe. However, commercial application of *sous-vide* processed foods in North America has been limited. A large proportion of *sous-vide* products in North America is actually stored frozen.

Sous-vide processed products have a shelf-life of 1 to 6 weeks, depending upon the severity of the cooking or pasteurization step and storage temperatures. For a short shelf-life product (<10 to 14 days), the significant microbiological risk is the presence of vegetative pathogens, and the heat treatment should achieve at least a 6-$\log_{10}$ reduction (70°C for 2 min in the slowest heating point) in the numbers of pathogens. For longer shelf-life products, the thermal process must eliminate any spores capable of germination and outgrowth during prolonged storage and must be at least equal to 6-D for psychrotrophic *Clostridium botulinum* (90°C for 10 min) or greater if spores of psychrotrophic *Bacillus* species also must be eliminated. It is worth mentioning that milder heat treatments to retain the organoleptic attributes or inadequate processing may not ensure proper destruction of potentially pathogenic and highly heat-resistant bacteria that are nonspore forming. Also, storage abuse of *sous-vide* foods could lead to high levels of surviving vegetative or spore-forming food-borne pathogens.

This chapter will deal with the potential threat of food-borne illness through the consumption of *sous-vide* processed foods contaminated with spore-forming pathogens such as *C. botulinum*, *Clostridium perfringens*, and *Bacillus cereus* because of their ability to survive the heat treatment given to these products and their subsequent germination, outgrowth, and multiplication during cooling, storage. and distribution. Nonspore-forming, facultative, psychrotrophic pathogens considered prime hazards in *sous-vide* processed products include *Listeria monocytogenes*, *Yersinia enterocolitica*, and *Aeromonas hydrophila*. These pathogens are capable of growth at refrigeration temperatures under anaerobic conditions (Gill and Reichel, 1989; Hudson et al., 1994) and hence pose a potential threat to consumer safety in *sous-vide* products. Nonspore-forming, mesophilic, facultative anaerobes such as *Salmonella* spp., *Staphylococcus aureus*, or enteropathogenic strains of *Escherichia coli* may pose a risk if foods are stored at abusive temperatures.

All of these vegetative pathogens should be eliminated by the *sous-vide* pasteurization step. However, they pose a health risk if the wide variety of raw ingredients used in *sous-vide* foods are of poor microbiological quality or if pasteurization is inadequate to destroy the high microbial load of nonspore-forming pathogens. Also, these pathogens are considered hazards in cases of postprocess contamination due to imperfectly sealed *sous-vide* packs. Furthermore, these pathogens may be capable of surviving thermal processes designed for the production of these foods if the pathogens are able to synthesize heat-shock proteins and thus exhibit an induced thermotolerance response. Challenge studies conducted with formulated *sous-vide* products to assure safety from spore-forming pathogens and the psychrotrophic pathogen, *L. monocytogenes,* will be discussed.

JUSTIFICATION FOR CONCERN

Preservation of *sous-vide* processed foods is achieved by a combination of vacuum packaging in high-barrier or air-impermeable flexible bags or pouches/packages, mild cooking, rapid chilling, and chilled extended storage. These processing steps may select for survival and growth of psychrotrophic strains of *C. botulinum*. Accordingly, the microbiological safety and preservation of *sous-vide* processed foods are being questioned and warrant a critical evaluation of the process. Concerns have been expressed about the serious public health risks associated with these foods. These concerns are justified for a variety of reasons, including the following.

Though the absence of oxygen inhibits the growth of aerobic microorganisms and prevents oxidative rancidity from occurring, a risk factor associated with these products occurs because vacuum packaging provides a favorable environment for anaerobic pathogens such as *C. botulinum* to grow and produce toxin in the processed product while the food remains edible. Mild heat treatment in combination with vacuum packaging may, in fact, select for *C. botulinum* and increase the potential for botulism.

An additional concern regarding *sous-vide* foods is based on the inactivation of normal and aerobic spoilage microbiota of meat, which are indicators of spoilage, by mild heat treatment. In fact, the aim of food processors in such foods is to destroy the vegetative cells of spoilage and pathogenic bacteria or to reduce the microbial load. The thermal process is not designed to destroy bacterial spores or to result in commercial sterility; thus, spores may survive and persist in the final product. This increasing trend toward the use of low-heat treatments in an attempt to retain product quality provides an ideal substrate for germination and growth of spore formers. More significantly, though, because the spoilage microflora that play a substantial role in causing deterioration and spoilage are inactivated, the foods may become toxic while remaining organoleptically acceptable.

The majority of *sous-vide* products have a low-acid and high-moisture (high a_w) content. In addition, these products are generally formulated with little or no preservatives because of consumer demands not only for fresh or homemade appeal but also for healthier foods with lower salt levels. Attempts to lower salt levels may increase public health risks and hazards associated with spore formers. To assure microbiological safety by guarding against the surviving pathogens and restricting their outgrowth, and to prevent spoilage, temperature control (as low as 3.3°C or below) from production to consumption is the only barrier. Interestingly, sufficient evidence exists to document that temperature abuse is a common occurrence at retail and consumer levels. Considerable fluctuations in temperature are frequently observed in commercial refrigerators (ICMSF, 1998) as well as in home refrigerators (Torrey and Marth, 1977). Tolstoy (1991) concluded that at least some stages of the chill chain are incapable of achieving and maintaining the necessary, very strict temperature control regimes.

According to NFPA (1988), manufacturers should assume that temperature abuse will occur at some point during the distribution of a refrigerated food product. Although it is relatively easy to control temperature in institutional food service settings (e.g., restaurants), the potential for temperature abuse exists when a product is shipped to remote locations, during storage in the retail environment, or during handling by consumers. Surveys of retail food stores and consumer refrigeration units have revealed that holding temperatures of >10°C are common (Daniels, 1991; Hutton et al., 1991). Thus, dependence on temperature control as the single hurdle to microbial growth is unrealistic to assure safety and extend storage life of *sous-vide* processed foods. Furthermore, since consumers view *sous-vide* packaged foods as shelf stable, there is a high risk of consumer temperature abuse, mishandling, and overextending of the product's shelf-life.

Finally, the extended shelf-life of these foods may allow pathogens, including thermally injured cells, to grow to sufficiently high levels. Also, an increased demand for these foods may mean that production of such products moves beyond the relatively well-controlled food service operations and larger retail operations to smaller, less experienced manufacturers.

BOTULISM OUTBREAKS

It is difficult to find details relating to outbreaks resulting from nonproteolytic *C. botulinum* in foods stored at low temperatures because strains and associated foods may be identified but the storage conditions leading to the outbreak are not. No reported cases of botulism have been associated with *sous-vide* foods. Nevertheless, changes in food preparation and in food use make this organism a concern and increase the probability, however low, that an outbreak could occur. The use of imported spices, meats, and other ingredients also increases the probability that future outbreaks may occur. Thus, it is worth emphasizing that the absolute safety of *sous-vide* products can never be guaranteed, but by recognizing and taking steps to reduce them, *sous-vide* products can be produced, distributed, and marketed with acceptable and manageable risk.

CLOSTRIDIUM BOTULINUM

Clostridium botulinum is a Gram-positive, spore-forming, anaerobic, rod-shaped, catalase-negative soil organism that produces a potent neurotoxic protein known as botulinum neurotoxin. This toxin is relatively heat sensitive and can cause various symptoms of paralysis, the most severe of which is respiratory impairment that can result in death. *C. botulinum* is the most hazardous spore-forming food-borne pathogen; it is widely distributed all over the world and is ubiquitous in the soil. Spores of this organism may find their way into processed food through raw materials or by postprocessing contamination of food. It is not possible to be certain that any food will not contain spores of *C. botulinum* other than foods that have been aseptically packed or have received a sporicidal heat treatment. Presence of spores in foods is not a public health hazard unless they can germinate, outgrow, and multiply into toxin-producing vegetative cells.

Proteolytic type A and B strains of *C. botulinum* are more tolerant to environmental stresses, produce highly heat-resistant spores, and have a minimal growth temperature of 10°C. Thus, the spores of proteolytic *C. botulinum* strains will survive the mild heat treatment given to *sous-vide* food products. However, these are of limited significance in properly refrigerated foods. The nonproteolytic *C. botulinum* strains are less tolerant to stresses, form spores that have reduced heat resistance, and can grow and produce toxin at temperatures as low as 3.3°C. Spores of these strains that survive the thermal process would pose a botulism hazard even under proper refrigeration temperatures, if a secondary barrier is not present. Conner et al. (1989) suggested that additional hurdles or barriers must be built into refrigerated, ready-to-eat products. Thus, without additional barriers, heat processing must be sufficient to destroy nonproteolytic *C. botulinum* spores if the food is to be safe. Scott (1989) recommended that challenge studies be conducted to verify the effectiveness of the combination of hurdles.

The following studies confirm the potential hazard due to growth and toxin production by nonproteolytic *C. botulinum* in extended shelf-life foods stored at temperatures above 3°C:

- Meng and Genigeorgis (1994) found that sodium lactate (NaL) significantly delayed toxigenesis of *C. botulinum* nonproteolytic types B and E in commercially prepared *sous-vide* products (beef and salmon homogenates) containing 0, 2.4, and 4.8% (w/w) NaL and chicken containing 0, 1.8, and 3.6% (w/w) NaL. The processed *sous-vide* products were inoculated with about 4 $\log_{10}$ nonproteolytic *C. botulinum* spores/3-g sample, vacuum packaged, and stored at 4, 8, 12, and 30°C for up to 90 days. Sodium lactate at Æ2.4% in beef and Æ1.8% in chicken delayed toxigenesis for Æ40 d at œ12°C (Table 5.1). It was interesting that salmon was the most conducive to toxigenesis and required 4.8% NaL to delay toxigenesis for a similar time period. Increased NaL levels and decreased temperatures resulted in a parallel delay in toxigenesis in the *sous-vide* products. Taking into account the anticipated low levels of *C. botulinum* spores in meat products and the effect of thermal processing, the authors suggested that Æ2.4% NaL and storage at œ12°C can assure inhibition of toxigenesis for time periods well beyond the expected shelf-life of 3 to 6 weeks. *Sous-vide* processing of raw fish, which is known for frequent contamination with *C. botulinum* at high levels, may require specific formulations as well as specific levels of thermal processing and storage temperature standards to obtain extended shelf-life.
- Maas et al. (1989) conducted a challenge study to assess the efficacy of sodium lactate in delaying toxin production by *C. botulinum* in a cook-in-bag turkey product. In this study, comminuted raw turkey supplemented with 1.4% sodium chloride, 0.3% sodium phosphate, and 0 (control), 2.0, 2.5, 3.0, or 3.5% sodium lactate was inoculated with a 10-strain mixture of proteolytic types A and B *C. botulinum* spores. The inoculated turkey was vacuum packaged, cooked in a water bath to an internal temperature of 71.1°C, and then incubated at 27°C for up to 10 days. Processed turkey

TABLE 5.1
Efficacy of Sodium Lactate on Time to Toxin Detection of *Clostridium botulinum* in *Sous-Vide* Products Stored at 4 to 30°C

Product	Sodium Lactate (%, w/w)	Time to Produce Toxin (Days) 4°C	8°C	12°C	30°C
Beef	0.0	90	8	4	1
	2.4	>90	90	>40	3
	4.8	>90	>90	>40	6
Chicken	0.0	90	16	12	2
	1.8	>90	16	12	2
	3.6	>90	>90	>40	6
Salmon	0.0	60	8	4	1
	2.4	90	12	6	2
	4.8	>90	>90	>40	4

From Meng, J. and Genigeorgis, C.A., *Lett. Appl. Microbiol.*, 19:20–23, 1994.

containing 0, 2.0, 2.5, 3.0, or 3.5% sodium lactate was toxic after 3, 4 to 5, 4 to 6, 7, or 7 to 8 days, respectively. Thus, sodium lactate exhibited an antibotulinal effect that was concentration dependent.

- Simpson et al. (1995) carried out challenge studies to evaluate the safety of reformulated *sous-vide* processed spaghetti and meat-sauce product (salt 1 to 3%, pH 4.5 to 6.0) inoculated with *C. botulinum* types A and B spores. Samples were processed at 75°C for 36 min (equivalent to 1.13 $\log_{10}$ CFU reduction for *Streptococcus faecium*), then stored at 15°C for up to 42 days. Toxin was detected in samples of >pH 5.5 after 14 to 21 days and in products of pH 5.25 after 35 days. Toxin was not detected in any sample of <pH 5.25 within 42 days of storage at 15°C. All products of pH 5.75 and 6 were visibly spoiled, as evidenced by swollen pack, prior to toxigenesis. However, for products of pH 5.5 and 5.25, toxigenesis preceded spoilage. Subsequent studies investigated the effect of the addition of salt on the safety of the sample with a pH of 5.5. Toxin production was inhibited throughout the 42-day storage period at 15°C in samples containing >1.5% (w/w) salt. None of the above studies discussed the sensory implications, if any, of the hurdles used. Research is required because sensory acceptability may be a limiting factor in practical use.
- Brown and Gaze (1990) and Brown et al. (1991) investigated the growth of nonproteolytic *C. botulinum* spores inoculated into carrot, cod, and chicken homogenates that were vacuum packaged and then cooked at 70°C for 2 min and stored at 5, 8, and 15°C. Type B toxin was detected in chicken

within 6 to 8 weeks and in cod within 3 to 6 weeks at 8°C. Type E toxin was formed between 3 to 4 weeks in chicken and between 5 days to 3 weeks in cod. At 15°C, toxins were detected within 7 days in chicken and cod.

- Notermans et al. (1990) studied the potential risk of botulism associated with *sous-vide* processed products. When *sous-vide* products were inoculated with nonproteolytic *C. botulinum* spores and stored at 8°C, toxin was detected after 4 weeks in 2 of 11 commercially available *sous-vide* products. It is worth reiterating that the storage temperature of chilled foods in retail and domestic refrigerators is often around 8°C and frequently above this (NFPA, 1988).
- Hyytia-Trees et al. (2000) evaluated the safety of 16 different types of *sous-vide* processed products containing beef, pork, and mixtures of beef, pork, vegetables, rice, and seafood with respect to nonproteolytic *C. botulinum* by using inoculated pack studies. The unprocessed products were inoculated with a mixture of five nonproteolytic *C. botulinum* strains (three of type E, one type B, and one type F) using low (100 spores/Kg) and high (200,000 spores/Kg) inoculation levels, after which products were vacuum packaged and heat processed. In this study, the pasteurization values of the products were calculated using the formula given by Brown (1993), and the T_{ref} and Z-values were 82.2 and 16.5°C, respectively (Brown, 1990). Surprisingly, *C. botulinum* spores could be detected in 11 of the 16 products after processing that included even low inoculum samples. Only 2 of 16 products were negative for botulinal spores and neurotoxin at the sell-by date and 1 week after the sell-by date. Two products at the high inoculum level showed toxigenesis during storage at 8°C, one of them at the sell-by date. Interestingly, the FoodMicro Model (FMM) predictions for the lethal effect of the thermal process and the FMM and USDA-Pathogen Modeling Program (version 5.0) predictions for the heat inactivation and safe storage time or growth after processing were not in agreement with the observed results in a majority of the challenges. This implies that the safety of *sous-vide* products must be carefully evaluated product by product. Time–temperature combinations used in heat treatments should be reevaluated to provide an adequate degree of protection against survival of spore formers. The authors suggested assessing the efficacy of additional antibotulinal hurdles such as biopreservatives and organic acids.
- Crandall et al. (1994) investigated the ability of *Pediococcus pentosaceus* 43200 to inhibit *C. botulinum* growth and toxin production in *sous-vide* beef with gravy at 4 and 10°C and reported that *Pediococcus* spp., used as a protective culture, was not capable of significantly inhibiting toxin production.

Effect of Lysozyme

Lysozyme is present in varying concentrations in a variety of foods of plant and animal origin such as the eggs of birds and reptiles, mammalian tissue and milk, fish, mollusks and crustaceans, and plants (cauliflower, broccoli, cabbage, etc.) (Lund and Peck, 1994). The levels in foods of plant and animal origin are 1.8 to 27.6 μg/g and 20 to 160 μg/g, respectively. Because lysozyme is heat stable and present in a variety of foods, it may remain active in *sous-vide* processed products and may, in turn, affect the safety margin of such foods. Other lytic enzymes that cause germination of heat-damaged spores may also be present in foods.

Researchers have extensively demonstrated that recovery of heated spores is enhanced when lysozyme is supplemented in the recovery medium (Juneja et al., 1995a; Juneja and Eblen, 1995a; Peck et al., 1992; Lund and Peck, 1994). Consequently, an apparent increase in heat resistance is observed. Peck and Fernandez (1995) concluded from their studies that if lysozyme is present at concentrations up to 50 μg/ml in a refrigerated, processed food with an intended shelf-life of 4 weeks, and the food is likely to be exposed to mild temperature abuse of up to 12°C, a heat treatment at 90°C for 19.8 min would be required to reduce the risk of growth of nonproteolytic *C. botulinum* by a factor of 10^6. However, if a longer shelf-life is expected, then higher heat treatment in conjunction with better control of temperature or additional barriers would be required to ensure safety against neurotoxigenesis by nonproteolytic *C. botulinum*. It is likely that heat treatment inactivates the spore germination system, making a fraction of viable spores unable to germinate. Lysozyme may initiate the germination of sublethally injured spores by permeating the spore coat and degrading the cortex, leading to core hydration and, consequently, to spore germination.

Peck and Fernandez (1995) expressed concerns and cautioned that the addition of lysozyme, including a genetically modified lysozyme with increased heat resistance and other lytic enzymes as preservative factors (Cunningham et al., 1991; Proctor and Cunningham, 1988; Nielsen, 1991; Gould, 1992), is likely to increase the risk of germination and growth of nonproteolytic *C. botulinum*. These findings have implications for assessing heat treatments necessary to reduce the risk of nonproteolytic *C. botulinum* survival and growth during extended storage of *sous-vide* foods. Further investigations are warranted to determine the effect of lysozyme on the efficacy of recommended heat processes and especially on its significance in real food systems.

Clostridium perfringens

Clostridium perfringens is considered to be ubiquitous and is found in soil, dust, vegetation, and a variety of animals including cattle, poultry, and humans. The temperature range for growth of *C. perfringens is* 6 to 50°C, with a doubling time as short as 7.1 to 10 min (Johnson, 1990a). Optimum pH for growth is between pH 6.0 to 7.0, and the growth limiting pH ranges from pH 5.5 to 5.8 to pH 8.0 to 9.0. While most strains are inhibited by 5 to 6.5% salt, the organism has been observed to grow at up to 8% NaCl concentration in foods (Johnson, 1990a).

Researchers in recent years have characterized the behavior of *C. perfringens* in *sous-vide* cooked foods. Juneja and Marmer (1996) investigated the growth potential of *C. perfringens* from a spore inoculum in vacuum-packaged, ground turkey (pH 6) that included 0.3% (w/w) sodium pyrophosphate and sodium chloride at 0,

TABLE 5.2
Mean[a] Generation Times, Lag Times, and D-Values Å Standard Deviation at 99°C of Spore Cocktail of *Clostridium perfringens* Strains NCTC 8238, NCTC 8239, and NCTC 10288 in Ground Turkey Containing 0.3% Sodium Pyrophosphate at pH 6 and Salt Levels of 0, 1, 2, and 3% Salt

Product	Generation Times (min)[b]	Lag Times (h)	D-value at 99°C (min)			
			28°C	15°C	28°C	15°C
Turkey (salt 0%)	39.4	300.0	7.3	61.6	23.2	0.2
Turkey (salt 1%)	31.3	398.8	10.6	59.6	21.3	0.8
Turkey (salt 2%)	24.2	238.2	11.6	106.4	19.5	0.8
Turkey (salt 3%)	88.5	nd[c]	8.0	nd[c]	17.7	0.3
Beef (salt 0%; pH 7)	80.1	415.9	11.55	96.06	23.3[b]	1.4
Beef (salt 3%; pH 7)	88.8	439.0	16.58	159.06	19.8[b,c]	2.1
Beef (salt 0%; pH 5.5)	122.1	4640.7	12.83	200.52	17.3[b,c]	0.1
Beef (salt 3%; pH 5.5)	129.2	NG	27.53	NA	14.0[c]	1.7

[a] Mean of two replications.
[b] Generation times calculated from regression lines for exponential growth using the Gompertz equation.
[c] Not determined.

Sources: Juneja, V.K. and Majka, W.M., *J. Food Safety*, 15:21–34, 1995a, and Juneja, V.K. and Marmer, B.S., *Int. J. Food Microbiol.*, 32:115–123, 1996.

1, 2, or 3% (w/w). The packages were processed to an internal temperature of 71.1°C, ice chilled, and stored at temperatures from 4 to 28°C. *C. perfringens* spores germinated and grew from 2.25 to >5 $\log_{10}$ CFU/g after 16 h in all turkey samples, regardless of the presence or absence of salt at 28°C. The generation times ranged from 39.4 min in salt-free turkey samples to 88.5 min in samples with 3% salt (Table 5.2). The lag times were 7.3 and 8.0 h, respectively. By day 4 at 15°C, *C. perfringens* spores germinated and grew to >5 $\log_{10}$ CFU/g in turkey with no salt.

Although the presence of 3% salt in samples at 15°C completely inhibited germination and subsequent multiplication of vegetative cells, even after 7 days of storage, growth occurred at a relatively slow rate in the presence of 1 to 2% salt. However, the total *C. perfringens* population was consistently lower compared to the levels in turkey with no salt. In contrast to 28°C, *C. perfringens* exhibited 7.5 times longer generation time (300.0 min) and 8 times longer lag time (61.6 h) at 15°C in samples with no salt (Table 5.2). Storage at 4°C inhibited growth, regardless of the presence or absence of salt.

Cyclic and static temperature abuse of *sous-vide* processed refrigerated turkey products may occur during transportation, distribution, storage, display, or consumer handling. When turkey samples stored at 4°C were moved to a 28°C environment for 8 h, *C. perfringens* population in samples did not increase regardless of the presence or absence of salt (Juneja and Marmer, 1996). When samples were transferred to 28°C and held for 12 h, *C. perfringens* spores germinated and grew to >6

$\log_{10}$ CFU/g only in samples with no salt. The numbers of organisms in samples that contained 3% salt did not increase. However, after 20 h at 28°C, turkey samples with 3% salt supported an increase in cell numbers of approximately 2 $\log_{10}$ CFU/g; the levels in all samples were <5$\log_{10}$ CFU/g. When samples stored at 4°C were shifted to 15°C for 24 h, no detectable growth of *C. perfringens* was observed in all samples, regardless of the salt level.

In another study (Juneja and Majka, 1995a), *C. perfringens* growth from a spore inoculum was investigated in vacuum-packaged, cook-in-bag ground beef that included 0.3% (w/w) sodium pyrophosphate, pH 5.5 or 7.0, and sodium chloride at 0 or 3% (w/w). The packages were processed to an internal temperature of 71.1°C, ice chilled, and stored at various temperatures from 4 to 28°C. At 28°C, the generation times ranged from 80.1 min in salt-free beef samples at pH 7.0 to 129.2 min in samples at pH 5.5 with 3% salt (Table 5.2). The lag times were 11.55 and 27.53 h, respectively. In contrast to 28°C, *C. perfringens* exhibited five times longer generation time (415.9 min) and eight times longer lag time (96.06 h) at 15°C in samples at pH 7.0 with no salt. At 15°C, growth occurred within 6 days in samples with pH 7.0 but was delayed until day 8 in the presence of 3% salt at pH 5.5. During storage at 4°C, *C. perfringens* growth from a spore inoculum was not observed in beef samples regardless of pH or salt levels. Cyclic and static temperature abuse of refrigerated *sous-vide* beef products for 12 to 15 h did not permit *C. perfringens* growth. However, temperature abuse of these beef products for periods longer than 15 h in the absence of salt led to growth of *C. perfringens* from a spore inoculum.

The effect of calcium and sodium lactates on growth from spores of *C. perfringens* in a *sous-vide* beef goulash stored at high temperature was assessed by Aran (2001). In this study, calcium lactate was more effective at inhibiting *C. perfringens* growth. Although calcium lactate as low as 1.5% arrested growth at 15, 20, or 25°C, the increased levels of 3% sodium lactate inhibited growth only at 15°C.

Chavez-Lopez et al. (1997) investigated the microbial safety and shelf-life of different vacuum-cooked pasteurized vegetables, pilaf rice, and turkey by challenge tests using a cocktail of *C. perfringens*, *Bacillus cereus*, or *Bacillus licheniformis* spores. None of the treatments achieved a reduction of *C. perfringens* by more than 4 $\log_{10}$ CFU/g. The spore counts of *C. perfringens* continued to decrease during storage at 4 and 15°C for up to 20 days. The reason for the observed decrease in viability of *C. perfringens* is unknown and may be attributed to the presence of competing *Bacillus* species, which may have interfered with viability. Nevertheless, the findings of the challenge study were confirmed by observations that the population densities of *Clostridium* spp. detected in the flora of naturally contaminated *sous-vide* foods increased in lasagna and grilled fillet during storage at 15°C but not in vegetables and roasted beef (Chavez-Lopez et al., 1997).

Table 5.2 shows the thermal resistance of *C. perfringens* spores (expressed as D-values in min) in turkey slurries that included 0.3% sodium pyrophosphate at pH 6.0 and salt levels of 0, 1, 2, or 3%. The D-values at 99°C decreased from 23.2 min (no salt) to 17.7 min (3% salt). In a beef slurry, the D-values significantly decreased ($p < 0.05$) from 23.3 min (pH 7.0, 3% salt) to 14.0 min (pH 5.5, 3% salt) at 99°C. Although addition of increasing levels (1 to 3%) of salt in turkey (Juneja and Marmer, 1996) or a combination of 3% salt and pH 5.5 in beef (Juneja and Majka, 1995a) can result in a parallel increase in sensitivity of *C. perfringens* spores at 99°C, mild heat treatments will not eliminate *C. perfringens* spores in foods that are packaged and then cooked.

In other words, spores are likely to survive normal pasteurization or cooking temperatures applied to *sous-vide* foods. In fact, it is not feasible practically to inactivate the spores by heat: if cooking temperatures are designed to inactivate *C. perfringens* spores, they may negatively impact the product quality; desirable organoleptic attributes of foods are unlikely to be retained. Mild heat treatment given to *sous-vide* foods could serve as an activation step for spores. Thereafter, germination and outgrowth of spores and *C. perfringens* vegetative growth are likely to occur in *sous-vide* foods if the rate and extent of cooling are not sufficient.

Inadequate cooling of foods in retail operations is a major safety problem. Accordingly, to determine a safe cooling rate for cooked beef, Juneja et al. (1994) investigated the potential for *C. perfringens* spores to germinate and grow in cook-in-bag ground beef cooled from 54.4 to 7.2°C using cooling times varying from 6 to 18 h. Vacuum-packaged beef samples inoculated with heat-shocked *C. perfringens* spores were cooked in a stirred water bath to an internal temperature of 60°C in 1 h and then cooled. No outgrowth of *C. perfringens* spores was observed after a cooling period of 6, 9, 12, or 15 h. *C. perfringens* spores germinated and multiplied when an18-h cooling time was followed. The total *C. perfringens* population was 3.16 $\log_{10}$ CFU/g after 5 h and the levels reached 6.0 $\log_{10}$ CFU/g by 10 h (15.6°C) with the 18-h cooling procedure. Rapid growth occurred between 22.1°C (7 h) and 15.6°C (10 h). Juneja et al. (1994) also demonstrated the effectiveness and validity of the "square-root model" under nonisothermal conditions; pasteurized cooked beef must be cooled to 7.2°C in 15 h or less to prevent *C. perfringens* food-borne disease outbreaks.

Juneja et al. (2001) developed a model to predict the growth of *C. perfringens* from spores at temperatures applicable to the cooling of cooked cured meat products. Simulating the conditions that occur in the food industry, the vacuum-packaged bags containing the meat samples inoculated with *C. perfringens* spores were cooked in a water bath programmed to increase linearly to 60°C within 1 h. In this study, *C. perfringens* growth from spores was not observed at 12°C for up to 3 weeks. Germination, outgrowth, and lag (GOL) time and exponential growth rate (EGR) were determined using a function derived from mechanistic and stochastic considerations and the observed relative growths at specified times.

A general model to predict the amount of relative growth for an arbitrary temperature was determined by fitting the exponential growth rates to a square root Ratkowsky function and assuming a constant ratio of GOL and generation times. A closed-form equation was developed to be used to estimate the relative growth for a general cooling scenario and to determine a standard error of the estimate. Applying multivariate statistical procedures, a confidence interval was computed on the prediction of the amount of growth for a given temperature. The model predicts, for example, a relative growth of 3.17 with an upper 95% confidence limit of 8.50 when cooling the product from 51 to 11°C in 8h, assuming log linear decline in temperature with time. The predictive models should aid in evaluating the safety of cooked products after cooling and, thus, with the disposition of products subject to cooling deviations.

Bacillus cereus

Bacillus cereus is widely distributed in nature and causes two different types of food poisoning: the diarrheal type and the emetic type. The organism has been isolated from a variety of foods including many ready-to-serve meals (Harmon and Kautter, 1991). *B. cereus* control is a challenge in *sous-vide* processing because the organism is a facultative anaerobe and a spore former. Although the temperature for growth of *B. cereus* ranges from 15 to 50°C, the organism is recognized as a psychrotrophic pathogen.

Foegeding and Berry (1997) screened a collection of 27 *B. cereus* isolates for the ability to grow at cold temperatures. Of 27 isolates, 19 could grow in brain–heart infusion broth at 7°C if previously adapted to 7°C over a 5-week period. The authors suggested that the cold adaptation response exhibited by *B. cereus* isolates should be considered while assessing shelf-life or safety of foods relative to this organism. However, not all food-borne isolates are psychrotrophic. The doubling time in a nutritive medium at optimum temperature is 18 to 27 min. The growth limiting pH under ideal conditions ranges from the minimum of pH 4.9 to the maximum pH 9.3. The minimum water activity for growth is 0.95 (Johnson, 1990b). Growth depends upon the interactive effects of all the environmental parameters.

The heat resistance of *B. cereus* spores is a concern to the food industry and has been studied extensively. In general, the heat resistance is similar to that of other mesophilic spore formers; however, some strains, referred to as heat-resistant strains, are about 15- to 20-fold more heat resistant than the heat-sensitive strains (Johnson, 1990b). *B. cereus* strains involved in food poisoning have D-values at 90°C ranging from 1.5 to 36 min. It is most likely that the organism will not be completely destroyed by the heat treatment given to most *sous-vide* foods; therefore, the organism must be controlled in these foods by preventing its growth or restricting the shelf-life of the product. As with *C. perfringens*, the risk of food poisoning due to *B. cereus* is relatively low because of the relatively high infective dose, which ranges from 10^5 to 10^7 organisms (total) for the diarrheal type and from 10^5 to 10^8 organisms per gram of food for the emetic syndrome.

A psychrotrophic *B. cereus* strain survived pasteurization and grew at 7°C in *sous-vide* cooked green beans (Knochel et al., 1997). In a challenge study (Chavez-Lopez et at., 1997) using *B. cereus* in vacuum-cooked foods, none of the heat treatments applied was able to inactivate *B. cereus* by more than 2 $\log_{10}$ CFU/g in the tested foods. Counts of this pathogen declined progressively in all products during storage at 4 and 15°C, except for a transient increase to >6 $\log_{10}$ CFU/g observed in rice pilaf at 15°C. Presumably, these observations might be due to the effect of a particular food component.

In another study (Aran, 2001), no *B. cereus* growth was observed at 10°C, but after 7 days at 15°C, population densities increased by 1 $\log_{10}$ CFU/g in *sous-vide* beef goulash samples. Calcium lactate at concentrations of 1.5% in beef goulash completely inhibited *B. cereus* growth at 20°C, but the level of sodium lactate required to inhibit growth was 3%.

Turner et al. (1996) assessed the safety of *sous-vide* chicken breast with respect to *B. cereus* and evaluated the effect of processing parameters on natural microflora.

The product was inoculated and processed to 77 or 94°C. *B. cereus* populations were reduced by 0.5 to 1.0 $\log_{10}$ CFU/g and by 3 $\log_{10}$ CFU/g in products heated to 77 and 94°C, respectively. Spores germinated within 1 day at 10°C, yet detectable changes in populations were not evident through 28 days of storage. Sodium lactate (2%) did not influence *B. cereus* populations or spore germination. The natural microflora was reduced by processing and levels did not increase throughout the 28-day storage at 4 and 10°C. Turner et al. (1996) concluded that the final temperature is important in controlling this organism, even though *B. cereus* populations were reduced by the mildest heat.

LISTERIA MONOCYTOGENES

Listeria monocytogenes continues to be a food-borne pathogen of great concern to the food industry because it is ubiquitous in the environment and in a wide variety of foods. The safety concerns in *sous-vide* processed foods relate to this microorganism's ability to grow rapidly at refrigeration temperatures and the fact that it is more heat resistant than other vegetative pathogens. Moreover, the slow heating rate and long come-up times employed in the production of *sous-vide* foods expose the microbial cells to conditions similar to heat shock, with the possibility of rendering these cells more thermal resistant and thus facilitating a longer survival during low final cooking temperatures.

Stephens et al. (1994) and Kim et al. (1994) reported that heating by slowly raising the temperature of pork exposes *L. monocytogenes* cells to conditions similar to heat stress, thereby enhancing the pathogen's heat resistance. Because recovery of heat-stressed pathogenic bacteria is increased under anaerobic conditions (Knabel et al., 1990; George et al., 1998), possible growth of heat-injured pathogens in *sous-vide* products is certainly a concern. Hansen and Knochel (1996) compared the effect of slow and rapid heating regimes on the heat resistance of *L. monocytogenes* in *sous-vide* cooked beef. The authors found no significant difference between slow (0.3 to 0.6°C/min) and rapid (>10°C/min) heating and the heat resistance of *L. monocytogenes* in low pH (<5.8) *sous-vide* cooked beef prepared at a mild processing temperature. However, they did observe an increase in the heat resistance of *L. monocytogenes* in higher pH (6.2) *sous-vide* beef.

Although processing at slowly rising temperatures may slightly increase the survival of *L. monocytogenes* in cooked beef, no evidence indicated an increase in subsequent growth potential of the surviving cells. Accordingly, Hansen and Knochel (2001) assessed the influence of growth phase, prior heat shock or slowly rising temperature, pH of beef, degree of heat injury, and recovery medium on the potential for resuscitation and growth of heat injured *L. monocytogenes* in *sous-vide* cooked beef. Heat injured, late stationary phase cultures with 95 to 99.9% injured cells in the surviving population did not grow or repair sublethal injuries in *sous-vide* cooked beef at 3°C; repair and growth took place at 10 and 20°C. In logarithmic phase cultures, heat injury occurred very rapidly and Æ99.9% injury was observed in all trials in spite of much lower pasteurization values and fewer $\log_{10}$ reductions compared with late stationary phase cultures. Regardless of growth phase, all cultures in which a high degree of heat injury (Æ 99%) was observed did not subsequently grow in a beef product at 3 or 10°C within 30 days. A longer lag period was observed in beef processed at slowly rising temperatures and in normal pH beef at 10°C.

Quintavala and Campanini (1991) determined the heat resistance of *L. monocytogenes* strain 5S heated at 60, 63, and 66°C in a meat emulsion at a rate of 5°C/min compared to instantaneous heating. The D-values of cells heated slowly were twofold higher than those for cells heated instantaneously at all heating temperatures. Failure to inactivate *L. monocytogenes* during cooking may lead to an unsafe product even if subsequent transportation, distribution, storage, or handling is carried out properly. Although it is universally agreed that proper pasteurization and cooking temperatures will destroy the organism, it is important to ensure that the mild heat treatment designed for *sous-vide* foods provides an adequate degree of protection against *L. monocytogenes*.

Radiation and heat often inactivate microorganisms synergistically. This combined treatment may allow *sous-vide* processing at lower temperatures than individual treatments. Shamsuzzaman et al. (1992) investigated a combined treatment effect of electron-beam irradiation and *sous-vide* cooking of chicken breast meat on survival and growth of *L. monocytogenes*. Chicken breasts were inoculated with the pathogen, vacuum packaged, irradiated with an electron beam to a dose of 2.9 kGy, and cooked to an internal temperature of 65.6°C. *Sous-vide* cooking alone had a marginal effect (only a 0.35 $\log_{10}$ CFU/g reduction). However, after the combined treatment, the pathogen remained undetectable in the product during storage for 8 weeks at 2°C. Parallel studies on uninoculated breast meat revealed that *sous-vide* cooked samples had a shelf-life of less than 6 weeks without irradiation, whereas those samples irradiated before *sous-vide* cooking had a shelf-life of at least 8 weeks.

In another study by Shamsuzzaman et al. (1995), chicken breast meat inoculated with 10^6 CFU/g of *L. monocytogenes* was irradiated with an electron beam at doses up to 3.1 kGy under vacuum in barrier bags, cooked in a boiling water bath for 3 min 45 sec to achieve an internal temperature of 71.1°C, and stored for up to 5 weeks at 8°C to simulate actual display-case temperatures. In this study, *Listeria* was undetectable in samples treated with both *sous-vide* and irradiation at 3.1 kGy, but the organism survived the *sous-vide* treatment without irradiation and multiplied during storage. In the same study, the authors reported that uninoculated chicken breast meat that received both irradiation (3 kGy) and a *sous-vide* treatment had a shelf-life of at least 8 weeks at 8°C, whereas the unirradiated *sous-vide* treated samples spoiled in 16 days.

Listeria were undetectable in combination-treated samples, but some of the nonirradiated *sous-vide* samples tested positive for *Listeria*. Thus, irradiation to 3 kGy can be combined with *sous-vide* treatment to extend the shelf-life of chicken breast and to enhance its microbiological safety without affecting the organoleptic quality. However, the authors reported some loss of thiamine occurred with the combined treatments.

Stillmunkes et al. (1993) evaluated the comparative effects of various additives and storage temperatures on survival of *L. monocytogenes* in *sous-vide* foods. The pathogen was injected into vacuum-packaged, nitrite-free beef roasts with varying concentrations of sodium lactate, glycerol monolaurin, or sodium gluconate and cooked to an internal temperature of 62.8°C. The samples were stored for up to 5 weeks at temperatures simulating wholesale (2°C), retail (7°C), consumer-storage (10°C), and temperature-abuse (25°C) conditions. In this study, sodium lactate (up to 3.5%) resulted in effective inhibition of the pathogen. Glycerol monolaurin demonstrated a concentration-related effect (0 to 3.5%), but the effectiveness was not sustained beyond about 3 weeks of storage. Growth of the pathogen was suppressed by lactate but not by monolaurin in temperature-abused samples. Sodium gluconate did not provide any significant control of the pathogen in cooked uncured beef roasts. In conclusion, the use of sodium lactate and irradiation can increase the microbiological safety of *sous-vide* foods.

Ben Embarek et al. (1994) studied an inhibitory potential of a protective culture of *Enterococcus faecium* against *L. monocytogenes*. The strains of *E. faecium* were isolated from *sous-vide* cooked cod fillets and were characterized as nonhemolytic and sensitive to a range of antibiotics and, therefore, presumably nonpathogenic. In the presence of 7 $\log_{10}$ CFU/ml of *E. faecium*, *L. monocytogenes* (2 $\log_{10}$ CFU/ml) in culture was reduced to <10 CFU/ml after 14 and 4 days at 3 and 15°C, respectively. When cultivated with 4 $\log_{10}$ CFU/ml of *E. faecium*, *L. monocytogenes* was inhibited at 15°C but not at 3 or 5°C.

The limited inhibitory effects observed when *E. faecium* was inoculated at a low level could be attributed to the relatively higher growth rate of *L. monocytogenes* at 3 and 5°C. Spontaneous resistance to an *E. faecium* bacteriocin was observed at 15°C after 11 days. This study suggests that *L. monocytogenes* cells became resistant following a period of effective inhibition. Further investigations are needed to determine the nature and likelihood of resistance phenomena because this could have serious implications for the future development of the use of bacteriocins in *sous-vide* processed foods.

Sous-vide processed fish is subjected to low time and temperature for cooking to retain intrinsic organoleptic attributes. Ben Emarek and Huss (1993) investigated the heat resistance of two strains of *L. monocytogenes* in *sous-vide* cooked fillets of cod and salmon. Pasteurized salmon fillets (10.56 to 17.2%, w/w, fat) had one to four times higher D-values for both strains of *L. monocytogenes* than the lower fat (0.6 to 0.8%, w/w, fat) cod fillets. These findings document the protective effect of fatty materials in the heating medium and the importance of food type on the heat resistance of *L. monocytogenes*. Gaze et al. (1989) recommended that the slowest heating point in a food product should be held at 70°C for 2 min. This time and temperature would result in 12-D reduction of the pathogen in salmon and uncured fish fillets processed via *sous vide*.

METHODS FOR CONTROL OF SPORE FORMERS

The assurance of safety from spore formers is a key factor in the success of *sous-vide* processed food products. The effectiveness of any control measures to guard against spore formers is based on the probability that a spore will germinate, grow out, multiply, and elaborate toxin in the preserved *sous-vide* food product. Although the probability of growth increases as a function of the number of contaminating spores on raw materials, it is often difficult to control the number of spores on raw materials and almost impossible to guarantee their absence. Therefore, the safety of such products may depend not only on the specifications of raw materials but also on combinations of preservation regimes designed to cope with normal levels of spores that will be present in *sous-vide* foods.

The intrinsic properties of food products, primarily pH and sodium chloride content, must be considered when designing adequate thermal processes or when trying to prevent toxin production during extended refrigerated storage. Moreover, it is often desirable to formulate foods and use combinations of these parameters that would allow for reduction of undesirable attributes such as high salt or acidity, as well as to design a reduced thermal process that would destroy the spores without negatively impacting product quality. We know that a majority of *sous-vide* processed foods are produced without any preservative factors or with few hurdles to prevent growth of food-borne pathogenic spores; however, it is very important to build or include factors to prevent spore growth. If intrinsic factors (e.g., low pH and reduced water activity) and extrinsic factors (e.g., temperature and time) are not adequate to guard against the hazards associated with spore formers, heat processing must be sufficient to destroy spores if the particular *sous-vide* food is to be safe. For example, although the heat process and storage temperature may be relied upon in certain well-controlled food service operations, it may be important to include additional hurdles to control growth in relatively less well-controlled retail operations. As a first approach, it is logical to know the minimum requirements for growth and the heat resistance of *C. botulinum* types A, B, E, and F.

The most critical step in the production of *sous-vide* products is the heating process for inactivation of spores of nonproteolytic *C. botulinum* strains. Because these are less heat-resistant spores, it is feasible practically to inactivate these spores by heat. Heat resistance increases with decreasing water activity and decreases with decreasing pH. Fat and protein also have a protective effect and increase spore heat resistance. Juneja et al. (1995a) reported that contaminated turkey should be heated to an internal temperature of 80°C for at least 91.3 min to give a 6-D process for type B spores, with the inclusion of 3% salt in turkey; however, 78.6 min at 80°C was sufficient to achieve a 6-D process (Juneja and Eblen, 1995b). Thus, it is suggested to incorporate low levels of salt while formulating foods and to design a reduced thermal process that ensures safety against nonproteolytic *C. botulinum* type B in *sous-vide* processed foods while maintaining the desirable organoleptic attributes of foods.

Juneja et al. (1995b) assessed and quantified the effects and interactions of temperature, pH, salt, and phosphate levels and found that the thermal inactivation of nonproteolytic *C. botulinum* spores is dependent on all four factors. Thermal resistance of spores can be lowered by combining these intrinsic factors. The following multiple regression equation predicts D-values for any combinations of temperature (70 to 90°C), salt (NaCl; 0.0 to 3.0%), sodium pyrophosphate (0.0 to 0.3%), and pH (5.0 to 6.5) within the range of those tested.

$$\begin{aligned}\text{Log}_e\ \text{D-value} = &-9.9161 + 0.6159(\text{temp}) - 2.8600\,(\text{pH}) - 0.2190\,(\text{salt}) \\ &+ 2.7424\,(\text{phos}) + 0.0240(\text{temp})(\text{pH}) - 0.0041(\text{temp})(\text{salt}) \\ &- 0.0611(\text{temp})(\text{phos}) + 0.0443(\text{pH})(\text{salt}) + 0.2937(\text{pH})(\text{phos}) \\ &- 0.2705(\text{salt})(\text{phos}) - 0.0053(\text{temp})^2 + 0.1074(\text{pH})^2 \\ &+ 0.0564(\text{salt})^2 - 2.7678(\text{phos})^2\end{aligned}$$

TABLE 5.3
Observed and Predicted D-Values at 70 to 90°C of Nonproteolytic *Clostridium botulinum* in Ground Turkey

Temperature (°C)	pH	% NaCl	% Phosphate	D-value Observed (min)	D-value Predicted (min)
70	6.50	0.0	0.00	57.7	66.0
70	6.50	1.5	0.15	40.1	46.5
75	6.25	1.0	0.10	39.1	42.3
75	6.25	1.0	0.20	32.9	38.6
90	5.00	0.0	0.00	5.0	6.3
90	5.00	1.5	0.15	3.1	4.8

Source: Juneja, V.K. et al., *J. Food Safety*, 15:349–364, 1995b.

Additionally, Juneja et al. (1995b) developed confidence intervals to allow microbiologists to predict the variation in the heat resistance of nonproteolytic *C. botulinum* spores. Using this predictive model, food processors should be able to design thermal processes for the production of a safe *sous-vide* food with extended shelf-life without adversely affecting the quality of the product substantially. Representative observed and predicted D-values at 70 to 90°C of nonproteolytic *C. botulinum* in ground turkey, at various pH levels (5.0 to 6.5) supplemented with salt (0.0 to 1.5%, w/v) and sodium pyrophosphate (0.0 to 0.2%, w/v), are given in Table 5.3.

Process time and temperature may not always be sufficient to achieve a 6-$\log_{10}$ reduction in numbers of psychrotrophic *C. botulinum* (Stringer et al., 1997). In work by these authors, spores of nonproteolytic B, E, and F strains were heated in laboratory media for between 0 and 60 min at 90°C and incubated at 5, 10, or 30°C in the absence and presence of lysozyme. When samples were incubated at 5°C for 23 weeks, a process of 1 min at 90°C ensured that a 6-$\log_{10}$ reduction was achieved. However, at 10°C, a process of 60 min at 90°C was required to ensure a 6-$\log_{10}$ reduction in the presence of lysozyme. This study emphasizes the importance of chilled storage in addition to heating temperatures when additional safety factors (hurdles) are not incorporated.

Every effort should be made to extend the lag and generation time of the pathogens in foods. By quantifying the effects and interactions of multiple food formulations, it is possible to provide secondary barriers to toxin production in cases of temperature abuse or failure of other primary preservative techniques. *C. botulinum* growth and subsequent toxin production can be prevented by combining several inhibitory parameters at subinhibitory levels during food formulation to produce a safe *sous-vide* food product. However, commercial potential of this concept of "microbial hurdles" should be fully evaluated because there are very few practical applications to date. This concept of "microbial hurdles" describes pictorially the effects of various intrinsic food composition factors and extrinsic environmental conditions and has been described by Leistner (1985, 2001).

Recently, a number of predictive models have been developed based on multi-factorial design experiments, extensive data collection, and analysis. These models quantify the effects and interactions of intrinsic and extrinsic factors and describe the growth responses of spore formers (Fernandez and Peck, 1997; Graham et al., 1996; Meng and Genigeorgis, 1993). Fernandez and Peck (1997) assessed the effects and interactions of heating temperature (70 to 80°C), heating time (up to 2.5 min), and incubation temperature (5 to 25°C) and developed a predictive model for time to growth of nonproteolytic *C. botulinum*. The heating time–temperature combinations that prevented growth and toxin production during 90 days when the incubation temperature were ≤75°C for 1072 min, 80°C for 230 min, 85°C for 36 min, and 90°C for 10 min. These results were in agreement with recommendations in the temperature range of 80 to 90°C (ECFF, 1996; ACMSF, 1992). None of these heat treatments achieved the reduction of nonproteolytic *C. botulinum* by 6 $\log_{10}$ CFU/g when the incubation temperature after heating was 25 to 30°C.

Meng and Genigeorgis (1993) assessed the interactive effects of NaCl (0 to 2%), sodium lactate (0 to 3%), temperature (4 to 30°C), and spore inoculum (10^2 to 10^4/g) on the length of the lag phase of nonproteolytic *C. botulinum* type B and E spores. The spores were inoculated in cooked turkey and chicken meats without nitrite and stored under vacuum for up to 120 days. In this study, toxin was not detected at 4°C in the presence of sodium lactate (NaL) and an inoculum of 10 spores/3-g sample. The authors developed the following predictive regression model for the lag phase duration as affected by NaCl, sodium lactate, inoculum (I), and temperature (T) of 8 to 30°C and their interactions:

$$\text{Log}\ (1/\text{LP}) = -2.29 \text{ to } 0.123(\text{NaCl}) + 0.22(\text{NaL}) + 0.439(\text{T}) + 0.02(\text{T})(\text{I})$$
$$\text{with } R^2 = 0.945$$

where T equals the square root of temperature.

The study demonstrated that lag phase can be extended to >38 days at <8°C in the presence of 2% NaL and 1% NaCl and an inoculum of 100 spores/g. Increasing the NaCl concentration to 2% extended the lag phase to >55 days. At a mild temperature abuse of 12°C, incorporation of 3% NaL and 2% NaCl was required to prevent toxin production for at least 36 days in turkey meat containing 100 spores (Table 5.4). In a previous study, Genigeorgis et al. (1991) found that turkey meat without added NaL but with 2.2% brine became toxic after 130, 10, and 9 days at 4, 8, and 12°C, respectively. From the preceding model, Meng and Genigeorgis (1993) were able to determine that temperature was the most influential factor on toxigenesis. Storage temperature explained 65% of the variation in results, followed by lactate (21.2%), the interaction of inoculum and temperature (4.9%), and NaCl (3.4%). The R^2 values (percentage of result variation) were used to quantify barriers to toxicity (hurdle concept).

Graham et al. (1996) assessed the effects of pH (5.0 to 7.3), NaCl concentration (0.1 to 5.0%), and temperature (4 to 30°C) on growth of nonproteolytic *C. botulinum* in laboratory media. Growth curves were fitted by the Gompertz (Gibson et al., 1987) and Baranyi models (Baranyi and Roberts, 1994); parameters derived from the curve-fit were modeled. Both models predicted that the optimal conditions for growth were

TABLE 5.4
Influence of Salt and Sodium Lactate Levels on Nonproteolytic *Clostridium botulinum* Time to Toxin Production in Cooked Turkey Meat Inoculated with 100 spores and Incubated at 8 and 12°C for up to 80 days

Temperature (°C)	Salt (%)	Sodium Lactate (%) 0	1.2	2	3
8	0	8	>28	>34	>40
	1	14	>28	>38	>55
	2	>28	>32	>55	>80
12	0	7	10	10	24
	1	7	>18	>22	>36
	2	13	22	26	>36

Source: Meng, J. and Genigeorgis, C.A., *Int. J. Food Microbiol.*, 19:109–122, 1993.

temperatures between 25 and 28°C, pH between 6.6 and 6.7, and a NaCl concentration less than 1%. Growth was not observed at pH less than 5.1 or at 5% NaCl. Experimental and predicted values for doubling time and lag time are shown in Table 5.5. Such predictive models can be useful in defining microbiologically safe operating practices, such as conditions for a critical control point in a hazard analysis critical control program (HACCP), or predicting the growth of a microorganism in a new formulation of a product. Food processors can optimize *sous-vide* product formulation by using these predictive models.

TABLE 5.5
Representative Doubling Time and Lag Time of Nonproteolytic *Clostridium botulinum*: Effect of Temperature, pH, and Sodium Chloride

Temperature (°C)	pH	NaCl (%)	Doubling Time (h) Observed	Doubling Time (h) Predicted	Lag Time (h) Observed	Lag Time (h) Predicted
5.0	6.04	1.0	22	20	246	242
5.0	5.98	0.1	28	20	314	259
5.0	6.12	2.0	32	24	290	267
7.0	5.89	2.0	14	16	180	213
7.9	5.33	1.0	11	23	596	561
8.2	6.81	0.1	4.7	5.1	59	62
9.9	6.52	0.1	4.4	3.1	46	36

Source: Graham, A.F. et al., *Int. J. Food Microbiol.*, 31:69–85, 1996.

Microbiological challenge studies must be conducted to validate the predictive models; these studies involve inoculation of foods with the bacteria of interest and simulating the conditions of any stage from preparation to consumer use. The microflora of foods is then monitored throughout the study to determine the potential safety hazard with the food. It is important that challenge studies be designed specifically for each product.

For *B. cereus* and *C. perfringens,* the most effective control measure is to assure that the rate and extent of cooling to <4°C is rapid to prevent potential spore germination and growth of vegetative cells. If the cooling rate is not rapid, sufficient reheating before consumption would be required to kill large numbers (6 $\log_{10}$ CFU/g) of vegetative cells. As with *C. botulinum*, storage of *sous-vide* food products below 4°C should control the potential hazards associated with *B. cereus* and *C. perfringens* in these foods. The effects and interactions of temperature (12, 19, 28, 37, and 42°C), initial pH (5.5, 6, 6.25, 6.5, and 7), sodium chloride content (0, 0.1, 1.5, 2, and 3%), and sodium pyrophosphate concentration (0, 0.1, 0.15, 0.2, and 0.3) on the growth of a three-strain mixture of *C. perfringens* vegetative cells indicated that the growth kinetics was dependent on the interaction of all four variables, particularly in regard to exponential growth rates and lag phase duration (Juneja et al., 1996). The authors concluded that sodium pyrophosphate can have significant bacteriostatic activity against *C. perfringens* and may provide processed meats with a degree of protection against this microorganism, particularly if employed in conjunction with a combination of acidic pH, high salt concentrations, and adequate refrigeration.

Baker and Griffiths (1993) developed a predictive model for psychrotrophic *B. cereus*. The authors used a response surface analysis to determine the effects and interactions of water activity, pH, temperature, glucose, and starch concentration on the growth and toxin production by the organism. The authors reported that the factors that had the greatest influence on growth and toxin production were water activity and temperature.

Other interventions to prevent surviving spores to grow out in *sous-vide* processed foods may include:

1. Addition of competing microflora. These microorganisms can have antagonistic action on germination, growth, and toxin production of spore formers. Mossel and Struijk (1991) suggested the addition of *Sporolactobacillus inulus* to *sous-vide* foods. The spores of this organism survive pasteurization temperatures and the growth from spores does not occur at temperatures <5°C (Botha and Holzapfel, 1988). At mild temperature-abuse conditions, spores will germinate, grow out, and produce acid, which can warn the consumer that the product is no longer edible. Also, lactic acid bacteria may be incorporated in foods. These bacteria may survive processing and grow when there is temperature abuse, producing acid and bacteriocins and making the product inedible, thus warning the consumer of a possible hazard. However, further research needs to be done in this area.
2. The use of bacteriolytic enzymes. Lysozyme is present in a variety of foods of plant and animal origin and is relatively heat stable, particularly

under acidic conditions. The implications for enhanced germination and outgrowth of spores in *sous-vide* foods during storage must be considered while designing the heat treatment and assessing the safety of such foods.

3. Use of time–temperature indicators (TTIs). Rigorous control of temperature during transportation, distribution, retail storage, or handling before consumption is extremely important. Since temperature abuse is a common occurrence at the retail and consumer levels, producers should not rely on low temperature storage and should use TTIs to track the time and temperature history of the products from production to consumption. Taoukis and Labuza (1989) published a review on TTIs in which they recommended that TTI devices be added to individual food packages to indicate temperature abuse and an increased risk of food poisoning. Also, safeguards must be engineered into TTIs to prevent or indicate removal after placement on a package. Additionally, TTIs should be designed in such a way that expiration, as affected by storage temperature and time, will follow biochemical or physical kinetics slightly faster than the rate of growth of pathogens or toxin production. Skinner and Larkin (1998) compiled data from the published literature and presented a time–temperature curve for *C. botulinum* toxin formation as a function of incubation temperature. The authors suggested using the curve to define the most conservative integrated boundary conditions that TTIs must predict in order to indicate potentially hazardous storage conditions.
4. All establishments within the food chain should have HACCP systems in place to ensure that safe practices are carried out and pathogen controls are properly executed and maintained during the maximum permitted shelf-life of the *sous-vide* foods. The HACCP approach is probably the single most important strategy for controlling the safety of these products. It is a preventive system in which safety is designed into the food formulation and the process by which it is produced. In fact, HACCP is a proactive process that looks critically at all ingredients as well as production, distribution, marketing, preparation, and consumption. Although specific hazards and critical control points will depend on the product formulation, the hazards, in general, can be addressed by good manufacturing practices (GMPs), good quality raw materials or incoming product control (achieved by decontamination of raw materials and ingredients to reduce initial microbial load), adherence to minimum cooking and pasteurization times and temperatures (achieved by addition of heat-sensitizing additives, reduction of pH, etc. for reducing the thermal resistance) and shelf-lives appropriate for the heat treatments, and strict temperature control throughout the distribution chain.

Smith and coworkers (1990) gave a useful description of the application of HACCP to ensure the microbiological safety of *sous-vide* meat and pasta products. They also identified raw-material quality, time–temperature relationships, packaging control, and the use of additional barriers such as pH and reduction of water activity in formulated products as some of the critical control points. Snyder (1995) described the application of HACCP for *sous-vide* products and gave examples of the particular microbiological, chemical, and physical hazards. If the safety of *sous vide* products is to be ensured in the future, then it is critical that all potential hazards for each product be identified and controlled using an HACCP approach.

5. An integral part of preventing hazards associated with spore formers in *sous-vide* foods is education. Workers in food-processing, food-distribution, and food-service establishments should be exposed to continuing training for safe food production, including the consequences of temperature abuse of these food products. Additionally, consumers must be made aware of the potential hazards associated with these products and must receive adequate knowledge regarding their handling and storage.

REGULATIONS

Worldwide, food industry and retail food establishments are required to comply with the published guidelines and recommendations for microbiologically safe production, distribution, and sale of ready-to-eat, refrigerated foods. Also, a code of practice that is advisory rather than prescriptive has been developed in some countries. In comparison with legislation, the codes can be more easily updated to take into account industry practices changing in response to consumer demands.

In North America, the documents are based on the principles and practices of the HACCP system and suggest that processors follow GMP sanitation guidelines, build additional hurdles or barriers into a particular product for additional degrees of safety, store products at refrigerated temperatures, conduct inoculated pack or challenge studies to validate the efficacy of the multiple barriers, and use time–temperature indicators (TTIs) to track the time and temperature history of products from production to consumption (Health and Welfare Canada, 1992; FIOC, 1990; Agriculture Canada, 1990; NACMCF, 1990; Rhodehamel, 1992).

In Europe, recommendations, guidelines, and codes of practice (ACMSF, 1992, 1995; ECFF, 1996; Betts, 1996; Gould, 1996; Martens, 1997) have been developed to ensure the safe production of *sous-vide* foods with respect to preventing growth and toxin production by nonproteolytic *C. botulinum*. Proteolytic *C. botulinum* growth and toxin production is prevented by ensuring that storage of *sous-vide* foods is at temperatures <10°C.

In the U.K., the Advisory Committee on the Microbiological Safety of Food (ACMSF) concluded that safety with respect to nonproteolytic *C. botulinum* could be ensured by one of the following (ACMSF, 1992, 1995):

1. Storage at < 3.3°C
2. Storage at ≤ 5°C and a shelf-life of ≤ 10 days
3. Storage at 5 to 10°C and a shelf-life of ≤ 5 days
4. Storage at chill temperature combined with heat treatment of 90°C for 10 min or equivalent lethality (e.g., 70°C for 1675 min, 75°C for 464 min, 80°C for 129 min, 85°C for 36 min, 90°C for 10 min) (ACMSF, 1992)

5. Storage at chill temperature combined with ≤ pH 5.0 throughout the food
6. Storage at chill temperature combined with a salt concentration of ≥ 3.5% throughout the food
7. Storage at chill temperature combined with ≤ a_w 0.97 throughout the food
8. Storage at chill temperature combined with combinations of heat treatment and other preservative factors that can be shown consistently to prevent growth and toxin production by *C. botulinum*

Some of these recommendations cannot be applied and used for all *sous-vide* foods because most of these foods have a pH ≥ 5.0, a high water activity, and a salt-on-water concentration ≤ 3.5%. To guard against the hazards for these foods, the principal factors controlling microbiological safety and quality are likely to be heat treatment, storage temperature, and shelf-life.

The time and temperature combinations recommended by the European Chilled Food Federation (ECFF, 1996) were 80°C for 270.3 min, 85°C for 51.8 min, 90°C for 10.0 min, 95°C for 3.2 min, and 100°C for 1.0 min. These recommendations were based on a Z-value of 7°C for temperatures less than 90°C, whereas ACMSF calculations were based on a Z-value of 9°C. Nevertheless, it is recommended that a heat treatment or combination process that reduces the number of viable spores of nonproteolytic *C. botulinum* by a safety factor of 6 $\log_{10}$ cycles (a six-decimal process) in *sous-vide* foods be used (ACMSF, 1992; ECFF, 1996). A heat treatment of 90°C for 10 min or equivalent lethality, followed by subsequent chilled storage, was given as reference criteria.

CONCLUDING SUMMARY AND FUTURE RESEARCH DIRECTION

Survival of spore formers and the occurrence of temperature abuse throughout distribution, in retail markets, and in home refrigerators is a challenge for innovative interventions. The safety of *sous-vide* processed foods cannot be considered to rely on only one single chilled-storage factor. Research has assessed and quantified the combination of hurdles to decrease the heat processing requirements and control subsequent germination of surviving spores during storage. Combining hurdle technologies has enormous potential to improve the margin of safety of *sous-vide* foods. Further research employing complex multifactorial experiments and analysis to define and quantify the effects and interactions of additional intrinsic and extrinsic factors, as well as development of "enhanced" predictive models, is needed. These models should be validated by appropriate challenge studies to ensure safety.

Recently, challenge studies were conducted with formulated *sous-vide* products to assure inactivation and control of vegetative and spore-forming food-borne microbial pathogens. In view of continued interest in lowering the heat treatment, it would be logical to define a specific lethality at low temperatures. Development of quantitative risk assessment models based on product composition and formulation, processing, and storage, in conjunction with implementation of HACCP plans and employee training in HACCP principles, should provide an adequate degree of protection against food-borne spore formers and nonspore-forming psychrotrophic pathogens. Also, this approach should result in a higher degree of confidence in product safety than is possible using traditional end sampling approaches to microbiological control. Finally, good manufacturing practices, consumer education, and use of TTIs are advocated to enhance the safety of *sous-vide* products.

REFERENCES

Advisory Committee on the Microbiological Safety of Foods (ACMSF), 1992, Report on vacuum packaging and associated processes, Her Majesty's Stationery Office, London, U.K.

Advisory Committee on the Microbiological Safety of Foods (ACMSF), 1995, Annual Report 1995, Her Majesty's Stationery Office, London, U.K.

Agriculture Canada, 1990, Canadian Code of Recommended Manufacturing Practices for Pasteurized/Modified Atmosphere Packaged/Refrigerated Food, Agri-Food Safety Division, Agriculture Canada, Ottawa, Canada.

Anonymous, 1974, Réglementation des conditions d'hygiène relatives à la préparation, la conservation, la distribution et la vente des plats cuisinés à l' avance, Arrêté du 26 Juin 1974, République Française Ministère de l'Agriculture.

Anonymous, 1988, Prolongation de la durée de vie des plats cuisinés à l' avance, modification du protocole permettant d' obtenir les autorisations, Note de Service DGAL/SVHAIN 881 No. 8106, du 31 Mai 1988, République Française Ministère de l'Agriculture.

Aran, N., 2001, The effect of calcium and sodium lactates on growth from spores of *Bacillus cereus* and *Clostridium perfringens* in a *sous-vide* beef goulash under temperature abuse, *Int. J. Food Microbiol.*, 63:117–123.

Baker, J.M. and Griffiths, M.W., 1993, Predictive modeling of psychrotrophic *Bacillus cereus*, *J. Food Prot.,*, 56:684–688.

Baranyi, J. and Roberts, T.A., 1994, A dynamic approach to predicting bacterial growth in food, *Int. J. Food Microbiol.*, 23:277–294.

Ben Embarek, P.K., Jeppesen, V.F., and Huss, H.H., 1994, Antibacterial potential of *Enterococcus faecium* strains isolated from *sous vide* cooked fish fillets, *Food Microbiol.*, 11:525–536.

Ben Embarek, P.K. and Huss, H.H., 1993, Heat resistance of *Listeria monocytogenes* in vacuum packaged pasteurized fish fillets, *Int. J. Food Microbiol.*, 20:85–95.

Betts, G.D., Ed., 1996, Code of practice for the manufacture of vacuum and modified atmosphere packed chilled foods with particular regard to the risk of botulism, Campden and Chorleywood Food Research Association, Chipping Campden, U.K.

Botha, S.J. and Holzapfel, W.H., 1988, Effect of reduced water activity on vegetative growth of cells and on germination of endospores of *Sporolactobacillus*, *Int. J. Food Microbiol.*, 6:19–24.

Brown, H., 1990, Evaluation of shelf-life for chilled foods, Technical manual no. 28, The Campden Food and Drink Research Association, Chipping Campden, U.K.

Brown, K.L., 1993, Principles of heat preservation, in *Processing and Packaging of Heat Preserved Foods*, Rees, J.A.G. and Bettison, J. Eds., London, U.K.: Blackie Academic & Professional, 33.

Brown, G.D. and Gaze, J.E. 1990, Determination of the growth potential of *Clostridium botulinum* types E and nonproteolytic B in *sous-vide* products at low temperatures, Technical memorandum no. 593, Campden Food and Drink Research Association, Chipping Campden, U.K.

Brown, G.D., Gaze, J.E., and Gaskell, D.E., 1991, Growth of *Clostridium botulinum* nonproteolytic type B and type E in *sous-vide* products stored at 2–15°C, Technical memorandum no. 635, Campden Food and Drink Research Association, Chipping Campden, U.K.

Chavez-Lopez, C. et al., 1997, Evaluation of the safety and prediction of the shelf life of vacuum cooked foods, *Ital. J. Food Sci.*, 9:99–110.

Conner, D.E. et al., 1989, Potential *Clostridium botulinum* hazards associated with extended shelf-life refrigerated foods: a review, *J. Food Safety*, 10:131.

Crandall, A.D., Winkowski, K., and Montville, T.J., 1994, Inability of *Pediocuccus pentosaceus* to inhibit *Clostridium botulinum* in *sous vide* beef with gravy at 4 and 10°C, *J. Food Prot.*, 57:104–107.

Cunningham, F.E., Proctor, V.A., and Goetsch, S.J., 1991, Egg-white lysozyme as a food preservative: an overview, *World's Poultry Sci. J.*, 47:141–163.

Daniels, R.W., 1991, Applying HACCP to new-generation refrigerated foods at retail and beyond, *Food Technol.*, 45(6):122–124.

European Chilled Food Federation (ECFF), 1996, Guidelines for the hygienic manufacture of chilled foods, Helsinki, Finland.

Fernandez, P.S. and Peck, M.W., 1997, Predictive model describing the effect of prolonged heating at 70 to 80°C and incubation at refrigeration temperatures on growth and toxigenesis by non-proteolytic *Clostridium botulinum*, *J. Food Prot.*, 60: 1064–1071.

FIOC (Food Institute of Canada), 1990, The Canadian code of recommended handling practices for chilled food, Ottawa, Ontario, Canada.

Foegeding, P.M. and Berry, E.D., 1997, Cold temperature growth of clinical and food isolates of *Bacillus cereus*, *J. Food Prot.*, 60:1256–1258.

Garnier, J. P., 1990, *Sous vide* and more, in *Symposium Proceedings on Catering for the 90's*, Leatherhead, U.K.: Leatherhead Food and Drink Research Association, 52–57.

Gaze, J. E. et al., 1989, Heat resistance of *Listeria monocytogenes* in homogenates of chicken, beef steaks and carrots, *Food Microbiol.*, 6:251–259.

George, S.M. et al., 1998, Effect of oxygen concentration and redox potential on recovery of sublethally heat-damaged cells of *Escherichia coli* O157:H7, *Salmonella enteritidis* and *Listeria monocytogenes*, *J. Appl. Microbiol.*, 84:903–909.

Genigeorgis, C.A., Meng, J., and Baker, D.A., 1991, Behavior of nonproteolytic *Clostridium botulinum* type B and E spores in cooked turkey and modeling lag phase and probability of toxigenesis. *J. Food Sci.*, 56:373–379.

Gibson, A.M., Bratchell, N., and Roberts, T.A., 1987, The effect of sodium chloride and temperature on the rate and extent of growth of *Clostridium botulinum* type A in pasteurized pork slurry, *J. Appl. Bacteriol.*, 62:479–490.

Gill, C.O. and Reichel, M.P., 1989, Growth of cold-tolerant pathogens *Yersinia enterocolitica*, *Aeromonas hydrophila* and *Listeria monocytogenes* on high pH beef packaged under vacuum or carbon dioxide, *Food Microbiol.*, 6:223–230.

Gould, G.W., 1996, Conclusions of the ECFF botulinum working party, in *Proceedings of 2nd European Symposium on Sous-Vide*, Leuven, Belgium: ALMA, 173–180.

Gould, G.W., 1992, Ecosystem approach to food preservation, *J. Appl. Bacteriol. Symp. Suppl.*, 73:58S-68S.

Graham, A.F., Mason, D.R., and Peck, M.W., 1996, Predictive model of the effect of temperature, pH and sodium chloride on growth from spores of non-proteolytic *Clostridium botulinum*, *Int. J. Food Microbiol.*, 31:69–85.

Hansen, T.B. and Knochel, S. 1996, Thermal inactivation of *Listeria monocytogenes* during rapid and slow heating in *sous vide* cooked beef, *Lett. Appl. Microbiol.*, 22:425–428.

Hansen, T.B. and Knochel, S., 2001, Factors influencing resuscitation and growth of heat injured *Listeria monocytogenes* 13–249 in *sous vide* cooked beef, *Int. J. Food Microbiol.*, 63:135–147.

Harmon, S.M. and Kautter, D.A., 1991, Incidence and growth potential of *Bacillus cereus* in ready-to-serve foods, *J. Food Prot.*, 54:372–374.

Health and Welfare Canada, 1992, Guidelines for the production, distribution, retailing and use of refrigerated prepackaged foods with extended shelf life, Guideline #7, Food Directorate, Health Protection Branch, Health Canada.

Hudson, J.A., Mott, S.J., and Penny, N., 1994, Growth of *Listeria monocytogenes*, *Aeromonas hydrophila* and *Yersinia enterocolitica* on vacuum and saturated carbon dioxide controlled atmosphere packaged sliced roast beef, *J. Food Prot.*, 57:204–208.

Hutton, M.T., Dhehak, P.A., and Hanlin, J.H., 1991, Inhibition of botulinum toxin production by *Pedicoccus acidilacti* in temperature abused refrigerated foods, *J. Food Safety*, 11:255–267.

Hyytia-Trees, E. et al., 2000, Safety evaluation of *sous vide*–processed products with respect to non-proteolytic *Clostridium botulinum* by use of challenge studies and predictive microbiological models, *Appl. Environ. Microbiol.*, 66:223–229.

ICMSF, 1998, Preventing abuse of foods after processing, in *Microbial Ecology of Food Commodities. Microorganisms in Foods*, Vol. 6, London: Blackie Academic and Professional, 579–597.

Johnson, E.A., 1990a, *Clostridium perfringens* food poisoning in *Foodborne Diseases*, Cliver, D.O., Ed., San Diego, CA: Academic Press, Inc., 229–240.

Johnson, E.A., 1990b, *Bacillus cereus* food poisoning, in *Foodborne Diseases*, Cliver, D.O., Ed., San Diego, CA: Academic Press, Inc., 128–134.

Juneja, V.K., Snyder O.P., and Cygnarowicz-Provost, M., 1994, Influence of cooling rate on outgrowth of *Clostridium perfringens* spores in cooked ground beef, *J. Food Prot.*, 57:1063–1067.

Juneja, V.K. and Majka, W.M., 1995a, Outgrowth of *Clostridium perfringens* spores in cook-in-bag beef products, *J. Food Safety*, 15:21–34.

Juneja, V.K. and Eblen, B.S., 1995b, Influence of sodium chloride on thermal inactivation and recovery of non-proteolytic *Clostridium botulinum* type B spores, *J. Food Prot.*, 58:813–816.

Juneja, V.K. et al., 1995a, Thermal resistance of non-proteolytic type B and type E *Clostridium botulinum* spores in phosphate buffer and turkey slurry, *J. Food Prot.*, 58:758–763.

Juneja, V.K. et al., 1995b, Influence of the intrinsic properties of food on thermal inactivation of spores of non-proteolytic *Clostridium botulinum:* development of a predictive model, *J. Food Safety*, 15:349–364.

Juneja, V.K. et al., 1996, Interactive effects of temperature, initial pH, sodium chloride and sodium pyrophosphate on the growth kinetics of *Clostridium perfringens*, *J. Food Prot.,* 59:963–968.

Juneja, V.K. and Marmer, B.S., 1996, Growth of *Clostridium perfringens* from spore inocula in *sous-vide* turkey products, *Int. J. Food Microbiol.,* 32:115–123.

Juneja, V.K. et al., 2001, Growth of *Clostridium perfringens* from spore inocula in cooked cured beef: development of a predictive model, *Innovations Food Sci. Emerging. Technol.*, 2:289–301.

Kim, K., Murano, E.A., and Olson, D.G., 1994, Heating and storage conditions affect survival and recovery of *Listeria monocytogenes* in ground pork, *J. Food Sci.*, 59:30–32, 59.

Knabel, S.J. et al., 1990, Effects of growth temperature and strictly anaerobic recovery on the survival of *Listeria monocytogenes* during pasteurization, *Appl. Environ Microbiol.*, 56:370–376.

Knochel, S., Vangsgaard, R., and Soholm Johansen, L., 1997, Quality changes during storage of *sous vide* cooked green beans (*Phaseolus vulgaris*), *Z Lebensm. Unters. Forsch.*, A, 205:370–374.

Leistner, L., 1985, Hurdle technology applied to meat products of the shelf stable product and intermediate moisture food types, in *Properties of Water in Foods*, Simatos, D. and Multon, J.L., Eds., Dordrecht, The Netherlands: Martinus Nijhoff, 309–329.

Leistner, L., 2001, Hurdle technology, in *Control of Foodborne Microorganisms,* Juneja, V.K. and Sofos, J.N., Eds., New York: Marcel Dekker, Inc., chap. 20, 493–508.

Lund, B.M. and Peck, M.W., 1994, Heat resistance and recovery of spores of nonproteolytic *Clostridium botulinum* in relation to refrigerated, processed foods with an extended shelf-life, *J. Appl. Bacteriol.*, Symp. Suppl., 76:115S–128S.

Maas, M.R., Glass, K.A., and Doyle, M.P., 1989, Sodium lactate delays toxin production by *Clostridium botulinum* in cook-in-bag turkey products, *Appl. Environ. Microbiol.*, 55:2226–2229.

Martens, T., 1997, Harmonization of safety criteria for minimally processed foods, Inventory Report, FAIR Concerted Action CT96–1020.

Meng, J. and Genigeorgis, C.A., 1993, Modeling lag phase of nonproteolytic *Clostridium botulinum* toxigenesis in cooked turkey and chicken breasts as affected by temperature, sodium lactate, sodium chloride and spore inoculum, *Int. J. Food Microbiol.*, 19:109–122.

Meng, J. and Genigeorgis, C.A., 1994, Delayed toxigenesis of *Clostridium botulinum* by sodium lactate in *sous-vide* products, *Lett. Appl. Microbiol.*, 19:20–23.

Mossel, D.A.A. and Struijk, C.B., 1991, Public health implication of refrigerated pasteurized (*sous-vide*) foods, *Int. J. Food Microbiol.*, 13:187–206.

NACMCF (National Advisory Committee on Microbiological Criteria for Foods), 1990, Recommendations for refrigerated foods containing cooked, uncured meat or poultry products that are packed for extended, refrigerated shelf-life and that are ready to eat or prepared with little or no additional heat treatment, Washington, D.C., adopted January 31, 1990.

NFPA (National Food Processors Association), Refrigerated Foods and Microbiological Criteria Committee, 1988, Factors to be considered in establishing good manufacturing practices for the production of refrigerated foods, *Dairy Food Sanit.*, 8:5–7.

Nielsen, H.K., 1991, Novel bacteriolytic enzymes and cyclodextrin glycosyl transferase for the food industry, *Food Technol.*, 45:102–104.

Notermans, S., Dufrenne, J., and Lund, B.M., 1990, Botulism risk of refrigerated, processed foods of extended durability, *J. Food Prot.*, 53:1020–1024.

Peck, M.W., Fairbairn, D.A., and Lund, B.M., 1992, The effect of recovery medium on the estimated heat resistance of spores of non-proteolytic *Clostridium botulinum*, *Lett. Appl. Microbiol.,* 15:146–151.

Peck, M.W. and Fernandez, P.S., 1995, Effect of lysozyme concentration, heating at 90°C, and then incubation at chilled temperatures on growth from spores of non-proteolytic *Clostridium botulinum*, *Lett. Appl. Microbiol.*, 21:50–54.

Proctor, V.A. and Cunningham, F.E., 1988, The chemistry of lysozyme and its use as a food preservative and a pharmaceutical, *CRC Crit. Rev. Food Sci. Nutr.*, 26:359–395.

Quintavala, S. and Campanini, M., 1991, Effect of rising temperature on the heat resistance of *Listeria monocytogenes* in meat emulsion, *Lett. Appl. Microbiol.*, 12:184–187.

Rhodehamel, E.J., 1992, FDA's concerns with *sous-vide* processing, *Food Technol.*, 4673–76.

Scott, V.N., 1989, Interaction of factors to control microbial spoilage of refrigerated foods, *J. Food Prot.*, 52:431–435.

Shamsuzzaman, K. et al., 1992, Microbiological and other characteristics of chicken breast meat following electron beam and *sous vide* treatments, *Int. J. Food Prot.*, 55:528–533.

Shamsuzzaman, K., Lucht, L., and Chuaqui-Offermanns, N., 1995, Effects of combined electron-beam irradiation and *sous-vide* treatments on microbiological and other qualities of chicken breast meat, *J. Food. Prot.*, 58:497–501.

Simpson, M.V. et al., 1995, Challenge studies with *Clostridium botulinum* in a *sous-vide* spaghetti and meat-sauce product, *J. Food. Prot.*, 58:229–234.

Skinner, G.E. and Larkin, J.W., 1998, Conservative prediction of time to *Clostridium botulinum* toxin formation for use with time–temperature indicators to ensure the safety of foods, *J. Food Prot.*, 61:1154–1160.

Smith, J.P. et al., 1990, A hazard analysis critical control point approach (HACCP) to ensure the microbiological safety of *sous-vide* processed meat/pasta product, *Food Microbiol.*, 7:177–198.

Snyder, O.P., 1995, The application of HACCP for MAP and *sous-vide* products, in *Principles of Modified Atmosphere and* Sous-Vide *Products Packaging*, Farber, J.M. and Dodds, K.L., Eds., Lancaster, PA: Technomic Publishing Company, 325–379.

Stephens, P.J., Cole, M.B., and Jones, M.V., 1994, Effect of heating rate on thermal inactivation of *Listeria monocytogenes*, *J. Appl. Bacteriol.*, 77:702–708.

Stillmunkes, A.A. et al., 1993, Microbiological safety of cooked beef roasts treated with lactate, monolaurin or gluconate, *J. Food Sci.*, 58:953–958.

Stringer, S.C., Fairbairn, D.A., and Peck, M.W., 1997, Combining heat treatment and subsequent incubation temperature to prevent growth from spores on non-proteolytic *Clostridium botulinum, J. Appl. Microbiol.*, 82:128–136.

Taoukis, P.S. and Labuza, T.P., 1989, Applicability of time-temperature indicators as shelf-life monitors of food products, *J. Food Sci.*, 54:783–788.

Tolstoy, A., 1991, Practical monitoring of the chill chain, *Int. J. Food Microbiol.*, 13:225–230.

Torrey, G.S. and Marth, E.E., 1977, Temperatures in home refrigerators and mould growth at refrigeration temperatures, *J. Food Prot.,*, 40:393–397.

Turner, B.E. et al., 1996, Control of *Bacillus cereus* spores and spoilage microflora in *sous vide* chicken breast, *J. Food. Sci.*, 61:217–219, 234.

Section II

Pathogen Detection and Assessment

6 HACCP and Regulations Applied to Minimally Processed Foods

O. Peter Snyder, Jr.

CONTENTS

1-58716-041-2/03/$0.00+$1.50

WHAT ARE MINIMALLY PROCESSED FOODS?

There is no regulatory definition for minimally processed foods. An operating definition is:

> Minimally processed foods are foods in which the biological, chemical, and physical hazards are at a tolerable level.

They are not sterile, because they have not been retorted for a 12-D *Clostridium botulinum* kill. These foods have left the farm or raw-food production area but need refrigeration and time controls to minimize pathogen and spoilage growth and to meet customer quality satisfaction. Some minimally processed foods can be frozen to extend shelf-life and then thawed when sold.

The first group of minimally processed foods is those that **need no cook intervention**. The supplier or the person growing and harvesting the food assures that the pathogenic substances are at a safe level. These foods include:

1. Raw beef, lamb, and pork
2. Raw fish and shellfish
3. Raw milk, cheese, and other dairy products
4. Raw fruit and vegetable juices
5. Nuts and grains
6. Eggs

In order to assure safety, the growers and harvesters control the pathogenic substances in the environment so that they can provide a letter of certification that the product is safe to consume without any intervention strategies.

The next group is foods with a **mild disinfection**, i.e., a reduction by 10^{-2} to 10^{-5} of pathogenic substances and chemicals. The pathogens are on the surface and difficult to remove. The common disinfection is washing with or without chemicals in water that is 15°F warmer than the food being washed so that the food items will not suck in contaminated water through the stem ends and become internally contaminated. They can also be blanched in water at 90 to 200°F for 15 to 30 sec. This is highly effective in reducing pathogens to a safe level. This group includes:

1. Cut melon pieces
2. Cut leafy and root vegetables
3. Sprouts
4. Berries of all kinds

The third group of minimally processed foods includes **pasteurized** foods. This means a 10^{-5} to 10^{-7} reduction of pathogens (typically *Salmonella*) on the surface and in the center of the food. Pasteurization can be done by irradiation, heat, pulsed light, ultraviolet light, or high pressure. There are two basic processes for pasteurization.

1. Pack in a plastic or glass container and then process to a pasteurization temperature, typically 160 to 180°F, cool, and sell refrigerated. Examples include pasteurized crab, *sous-vide* products, etc.
2. Process, as in a large kettle, and then pack in a container such as a plastic bag or glass jar. If the food has high enough levels of preservatives, as with tomato sauces and salad dressings, the food can be displayed at room temperature. Otherwise, it needs refrigeration at 40°F. Examples include vacuum pouch meat, poultry, fish, and seafood; vegetables such as potatoes, carrots, beets, and asparagus; fruits such as pears and peaches; salads; sandwiches; pizza; kettle-cooked foods such as soups and casseroles; and eggs in the shell.

The last minimally processed foods group contains foods that are **formulated** to be safe by fermentation, drying, salting, or acidification. These include:

1. Dairy products such as buttermilk, butter, cheese, and yogurt
2. Fermented and dried meats, fish, and sausage
3. Fermented and dried fruits and vegetables, including kimchee and sauerkraut
4. Wine and beer
5. Pasta, as an example of a dried food

SOUS-VIDE AND CHILLED FOODS

A major group of foods now appearing in the market includes *sous-vide* and chilled foods. *Sous-vide* means packaged "under vacuum;" these foods are packaged and then cooked, which retains the natural flavor. Chilled foods are generally cooked first, placed in a container, and then chilled.

Table 6.1 provides a list of some common menu items from a large chilled-foods manufacturer in Europe. These foods can be processed, chilled, and held for about 21 days if minimally pasteurized at 160°F, or, if given a high-temperature pasteurization of 190 to 200°F, the shelf-life will be 45 to 90 days, depending on the chemical degradation of the product.

A major advantage of chilled foods is that they consume half of the energy of frozen foods because they require no freezing or thawing. Hence, for the modern family and restaurants that do not want to take time to thaw food, these foods take approximately half the time to prepare that "conventional" foods do. Although people think of minimally processed foods as being plain chicken or meat, etc., a wide variety of recipes will fit into this category.

THE USDA PROCESS GROUPS

The USDA has specified two basic chilled food process groups: **cook-in-package** (food is cooked in the package, cooled, and distributed) and **cook-then-package**. In the second group, the food is cooked and perhaps cooled, then packaged, or it is

TABLE 6.1
European *Sous-Vide*/Chilled Food

Starters and Soups	**Main Course Fish**	**Sauces — Multiportion**
Tomato soup	Salmon and cod mornay	Bolognaise sauce
Minestrone soup	Eel with green herbs	Béarnaise sauce
Cream of mushroom soup		White wine sauce
Bouillabaisse (fish soup)	**Main Course Poultry**	
Green pea soup	Double chicken fillet, wood mushrooms	**Garnishes**
Terrine of young wild boar		Apples Marechal
Smoked salmon mousse	Chicken fillet/lobster sauce	Pears with cinnamon
	Chicken Dijonnaise	
Pasta		**Potato Dishes — Single Portion**
Tagliatelli carbonara	**Main Course Meat**	
Cannelloni in tomatoes and basil	Lentil bake stew	Hash brown (loose)
Macaroni with ham and cheese	Veal stew in cream sauce	Baked potatoes
	Rabbit with mustard	
Rice/Pulse		**Vegetables — Single Portion**
Couscous (semolina)	**Plain/Roasts/Meats**	Green peas and carrots
Rice with vegetable	Rack of lamb	Puree of carrots (tray)
Basmati rice	Half duck	Vegetable mix for stews and casseroles
	Duck breast (Magret)	

cooked, packaged, and then cooled. The first process, cook-in-package, eliminates the risk of cross-contamination after the food is pasteurized. With the cook-then-package method, the packaging must be done in a sterile room, absolutely free of measurable *Listeria monocytogenes*, with scrupulously clean personnel, in order to avoid post-pasteurization contamination. This can be done, but it complicates production.

Examples of cook-in-package processes are:

1. Assemble–cook/pasteurize–package above 160°F (71.1°C)–chill: e.g., stews, sauces, soups.
2. Assemble–sear (perhaps)–package–cook/pasteurize–chill: e.g., *sous-vide*: rolls and roasts, canned crab, fish, and ham; casseroles: meat, pasta, vegetable, sauce combinations.

Examples of cook-then-package processes are:

1. Cook/pasteurize–chill–assemble ingredient–package: e.g., roast or fried chicken, other roasts, uncured sausages; uncured luncheon meat, diced meat; meat and pasta dinners, sauces, sandwiches, pizza; meat pies, quiches, patties, patés.
2. Assemble with cooked and raw ingredients–package (chill, pasteurize, or serve cold): e.g., chef's salad, chicken salad, sandwiches, pizza with raw ingredients, uncured meats.

TABLE 6.2
Effect of Storage Temperature on the Shelf-Life of Pasteurized Milk

Pasteurized Milk Temperature (°F)	Time to Spoilage
70	12 hours
60	1 day
50	3 days
45	5 days
40	10 days
32	32 days

Source: Farber, J.M. and Dodds, K.L., Eds., *Principles of Modified-Atmosphere and* Sous Vide *Product Packaging*, Technomic Publishing Co., Inc., Lancaster, PA, 1995, 328.

SHELF-LIFE OF PASTEURIZED FOOD

The maximum safe storage temperature for all pasteurized food except fish is 40°F. This controls the growth of *Bacillus cereus*, which is the organism of concern in pasteurized foods. Pasteurized fish is an exception and requires 38°F to control nonproteolytic *C. botulinum.*

Refrigeration systems are designed with an average storage temperature capability of 40°F. During times when not in active use, for instance, at night, a refrigeration unit might get down to 35°F; during times of active use, when the door is frequently opened, the temperature may rise to 45°F. The microorganisms respond to these changes, and the result is roughly the average of the temperature.

Table 6.2 provides information on the storage temperature vs. the shelf-life of pasteurized milk. Using 40°F for 10 days as the norm, 70°F has a spoilage time of 12 h. This is not a "bad" temperature; it means that the milk spoilage bacteria multiply much more quickly at this warmer temperature. If the milk can be at a temperature of 32°F, the storage life is approximately 32 days. This points out the importance of storing chilled foods that would be held a long time at low temperatures.

The freezing point of most protein products (e.g., meat, fish, and poultry) is about 28°F. The freezing point of most soups is about 31°F. Hence, operators need different storage refrigerators if they want maximum shelf-life storage on some products.

Table 6.3 shows the effect of pasteurization time at 158.0°F (70°C) on the shelf-life of liquid whole eggs. If the egg is pasteurized for only 1.3 sec, the time to spoilage is about 8 weeks. If it is pasteurized for 150 sec, the shelf-life is three times longer, or about 24 weeks.

Finally, high-temperature pasteurization is illustrated by the expected shelf-life of pasteurized, canned crab (Table 6.4). If the pasteurization at 185°F is 10 to 15 min, the time to spoilage is about 1.5 months. If pasteurization is more than 40 min, the time to spoilage can be 12 to 36 months. Depending on the times and temper-

TABLE 6.3
Effect of Pasteurization Time at 158.0°F on the Refrigerated Shelf-Life of Liquid Whole Eggs

Pasteurized Time (sec)	Time to Spoilage
1.3	8 weeks
3.8	12 weeks
10.3	16 weeks
40	20 weeks
150	24 weeks

Source: Farber, J.M. and Dodds, K.L., Eds., *Principles of Modified-Atmosphere and* Sous Vide *Product Packaging*, Technomic Publishing Co., Inc., Lancaster, PA, 1995, 328.

TABLE 6.4
Effect of Pasteurization Time at 185°F on the Refrigerated Shelf-Life of Pasteurized, Canned Crab

Pasteurized Time (min)	Time to Spoilage (months)
10 to 15	1.5
15 to 20	2 to 4
20 to 25	4 to 6
25 to 30	6 to 9
30 to 40	9 to 18
>40	12 to 36

Source: Farber, J.M. and Dodds, K.L., Eds., *Principles of Modified-Atmosphere and* Sous Vide *Product Packaging*, Technomic Publishing Co., Inc., Lancaster, PA, 1995, 328.

atures, pasteurized foods can have extended shelf-lives. Also dependent on the contamination of the raw product is how much pasteurization is needed in order to reduce the spoilage microorganisms to such an extent that it takes a very long time for them to grow back and cause off-flavors and sensory changes in the product.

THE RAW FOOD CONTAMINATION PROBLEM

One critical control point in minimally processed chilled food systems is the colonization of raw meat, fish, and poultry with pathogenic organisms. The higher the level of contamination, the more the food must be processed. The contamination problem is controlled by the cleanliness of the farm, whether a land-based farm or a water-based fish or seafood "farm," as well as the farming practices.

FIGURE 6.1 The raw food contamination problem.

As shown in Figure 6.1, the problems begin with raw animal and human waste, which gets into rivers, lakes, and oceans. This water is used to irrigate and wash the fruits and vegetables grown on the farm and is consumed by the animals raised on the farm. On the farm, there is also cross-contamination from rats and mice, birds, and wild animals such as deer and foxes that eat the feed. Cats on the farm carry *Toxoplasma gondii* and contaminate grain and feed.

When fish are pulled from the ocean, or shellfish are harvested, depending on the water temperature, there can be more or less contamination from seawater bacteria such as *Vibrios, Proteus morganii*, or the toxigenic dinoflagellates.

Next, the food goes to a food plant that makes it safe to eat, or the food goes to a food market, where there may be a deli or restaurant that makes the food safe to eat. The food going directly to a restaurant is made safe by the cook or chef. When the home food preparer goes to the market and buys raw ingredients, it is the responsibility of this preparer to make — and keep — the food safe after preparation.

With minimally processed foods, the pathogenic vegetative microorganisms are kept at a safe level by good farming practices or are reduced by some form of process intervention. Toxic chemicals are kept at a safe level by the farmer, grower, and harvester. Physical contaminants are removed by the processing plant or by the cook, e.g., removing rocks from dried beans.

HAZARDS IN THE FOOD SYSTEM

The goal, then, is to process the food as little as possible in order to meet consumer demand for raw-like products. For each hazard to be controlled, one must know the risk. This consists of:

1. Proof that a hazard exists
2. Level at which normally healthy people get ill
3. Likelihood of a given level or size making people ill
4. Likelihood that a person will have a weakened immune system and become ill

There are three classes of hazards.

1. Microbiological — caused by pathogenic microorganisms and their toxins. Examples are bacteria vegetative cells, spores, molds, yeasts, viruses, and parasites. This group of hazards also includes marine animals as sources of toxic compounds, such as fish and shellfish.
2. Chemical — poisonous substances and foods that cause adverse food reactions.
3. Physical — hard foreign objects in the food and functional hazards.

MANAGER CONTROL SYSTEMS

It is essential that managers have effective hazard control programs to minimize risk to their businesses. This is tantamount to showing due diligence in a court of law, if necessary.

A due diligence defense consists of two parts. The first is to take all reasonable precautions. This means setting up a precontrol quality assurance system to ensure that nothing goes wrong and that no defects are in the foods served to customers. HACCP is an important component to due diligence. Knowledgeable oversight and participation of top management are essential. Management must be able to say that the food production operation has been, is, and will be controlled because of its food safety program.

Second, the manager must show due diligence, which means seeing that the system works properly at all levels, from the executive board level down to on-line employees. He or she must show that the system involves careful monitoring and recording at control points. A system that strives for zero deviation using validated, safe process procedures and standards must be in place. There must be precontrol and then a review of all incidents to eliminate the causes of minor deviations in the process.

WHAT FOOD OPERATIONS USE MINIMALLY PROCESSED FOODS?

Figure 6.2 diagrams the retail food system and the users of minimally processed foods. Fundamentally, any retail food outlet, from food stores and markets to fair and camping environments, could be expected to use minimally processed foods. These systems will operate under a management continuous quality improvement program supplied with operating information by employees monitoring their

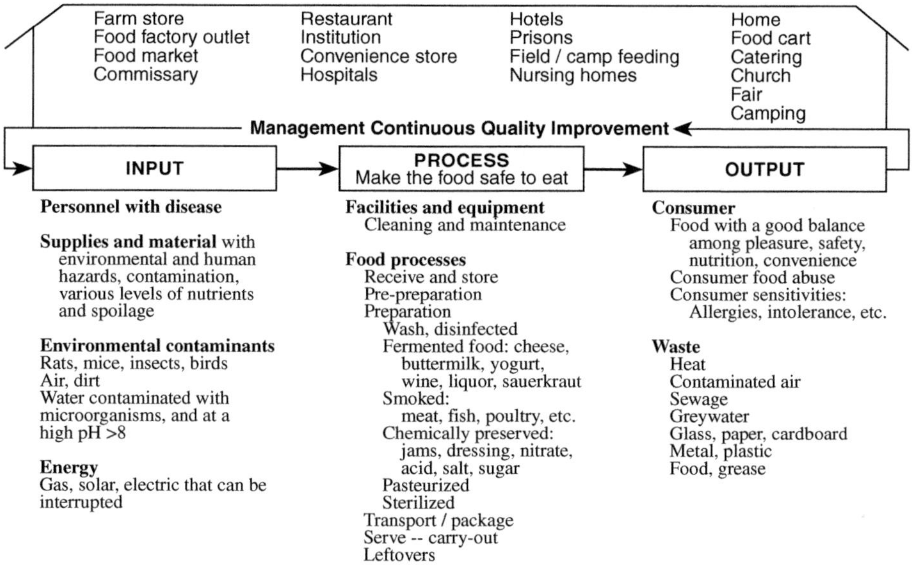

FIGURE 6.2 HACCP TQM minimally processed food system.

processes. A system consists of *input* (*processes* that make the food safe) and an *output* (food that nourishes the consumer and does not cause illness).

The input to this minimally processed food system includes personnel with disease, contaminated supplies and materials, environmental contaminants (mentioned earlier), and a partially reliable energy system. The processes are designed to make the food, with its various levels of contamination, safe. First, one considers the facilities and equipment, which exclude the environmental contaminants. Then, there are the food processes, whereby the food is received and stored, pre-prepared, prepared, transported and served. The final food process is dealing with leftovers.

The output, then, consists of the consumer's food and waste products from the process, to include heat; contaminated air; sewage; greywater; glass, paper, and cardboard; metal and plastic; food waste and grease.

RAW- AND COOKED-FOOD PERFORMANCE STANDARDS

Table 6.5 provides the starting point for the system design. The expected raw food contamination level is shown in column 2, the cooked food contamination level is shown in column 3, and column 4 shows the allowed increase. This table points out that the food must be purchased with cooked-food contamination levels, or the food is prepared in-house to achieve cooked-food pasteurization levels. Once it is prepared, during storage and service, some increase in *Staphylococcus aureus* is expected, which could cross-contaminate the product, as could *C. botulinum*,

TABLE 6.5
Chilled Food, Hazard Thresholds, and Performance Standards

Hazards	Raw Food Expected Contamination	Cooked Food Expected Contamination	Allowed Increase in Prepared Food
Vegetative Pathogens			
Salmonella spp.	<100/g	<1/25 g	none
Vibrio spp.			
Shigella spp.			
E. coli spp.			
Listeria spp.			
Campylobacter spp.			
Post-Cook			
Staphylococcus aureus	1000/g	100/g	10× = 1000/g
Clostridium botulinum	1/10 g	1/10 g	10× = 1/g
Clostridium perfringens	100/g	100/g	10× = 1000/g
Bacillus cereus	100/g	100/g	10× = 1000/g
Toxins and Poisons			
Mold toxin	Below government standards	none	
Chemicals	Below government standards	none	
Physical Hazards			
Rocks, metal, bones	—	<1/16 in.	none

Clostridium perfringens, and *B. cereus*. Through correct system management, the other pathogenic substances will not increase.

MICROBIOLOGICAL CONTROL

Table 6.6 provides the microbiological control information necessary for safe process design. It shows pH levels for growth, minimum water activity, the growth temperatures and rates, and the death rates.

An analysis of the microorganisms that could cause food-borne illness in a minimally processed food system shows that, unless the food is to be eaten raw, the concern is *Salmonella*, which causes many food-borne illnesses each year and is a common contaminant of all kinds of raw foods. Although *L. monocytogenes* and *Yersinia enterocolitica* can be expected to be present, they will be killed, along with the *Salmonella*, if the food is given a 10^5 reduction according to the FDA, or a 10^7 reduction if processed in accordance with USDA regulations. USDA regulations are somewhat more conservative than those of the FDA in the case of cooking meat and poultry products.

After the food is cooked, the organism of primary concern in hot holding is *C. perfringens*. Below 59°F, at which point *C. perfringens* stops multiplying, *B. cereus*

TABLE 6.6
Microbiological Control

Microorganism	pH	a_w	Growth Temp.	Growth Rate	Death Rate
Yersinia enterocolitica	4.6 to 9.0	0.945	29.3 to 111°F (–1.5 to 44°C)	G [41°F (5°C)] = 17 h in raw beef	D [145°F (62.8°C)] = .24 to 0.96 min
Listeria monocytogenes	4.5 to 9.5	0.92	29.3 to 112°F (–1.5 to 44°C)	G [41°F (5°C)] = 24.5 h in corned beef	D [140°F (60°C)] = 2.85 min
Clostridium botulinum Type E, nonproteol. B	5.0 to 9.0	0.97	38 to 113°F (3.3 to 45°C)	*Time for toxin production* [41°F (5°C)] = 426 h (fish)	D [180°F (82.2°C)] = 0.49 to 0.74 min (spores)
Clostridium botulinum Types A and proteol. B	4.6 to 9.0	0.94	50 to 118°F (10 to 47.8°C)	G [68°F (20°C)] = 1.2 h pork slurry (Type A)	D [250°F (121.1°C)] = 0.3 to 0.23 min for A, B (proteolytic spores)
Bacillus cereus	4.3 to 9.0	0.912	39.2 to 122°F (4 to 50°C)	G [46.4°F (8°C)] = 3.6 h in infant formula	D [140°F (60°C)] = 1 min (veg. cells) D [212°F (100°C)] = 2.7 to 3.0 min (spores)
Salmonella spp.	4.1 to 9.0	0.95	41.5 to 114°F (5.5 to 45.6°C)	G [46°F (7.8°C)] = 21.8 h in Chicken á la King	D [140°F (60°C)] = 1.73 min (veg. cells)
Vibrio parahaemolyticus	4.5 to 11.0	0.937	41 to 109.4°F (5.0 to 43°C)	G [98°F (37°C)] = 7.6 h in squid	D [116°F (47°C)] = 6.5 to 48.2 min (dependent on culture media)
Staphylococcus aureus	4.5 to 9.3	0.83	43.8 to 122°F (6.5 to 50°C)	G [46°F (7.8°C)] = 29.5 h in Chicken á la King	D [140°F (60°C)] = 5.2 to 7.8 min (veg. cells)
Clostridium perfringens	5.0 to 9.0	0.95	59 to 127.5°F (15 to 52.3°C)	G [75°F (24°C)] = 1.55 h in turkey roll	D [138°F (59°C)] = 7.2 min (veg. cells)
Campylobacter jejuni	4.9 to 8.0	0.987	90 to 113°F (30 to 45°C)	G [107.6°F (42°C)] = 50 min in egg yolk	D [137°F (58.3°C)] = 12 to 21 sec (veg. cells)
E. coli (pathogenic strains)	4.0 to 9.0	0.95	44.6 to 120.9°F (7.0 to 49.4°C)	G [99°F (37.2°C)] = 30 min in media	D [140°F (60°C)] = 45 sec (veg. cells)

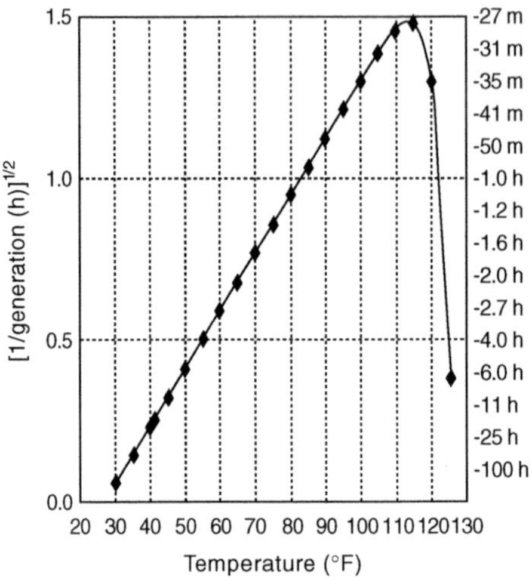

Temp. (°F)	1 Generation (h)	10 Generations
30	293.5	123.8 da
35	45.7	19.3 da
40	18.2	7.5 da
41	15.6	6.5 da
45	9.5	4.0 da
50	5.9	2.4 da
55	4.0	1.7 da
60	2.9	1.2 da
70	1.69	16.9 h
80	1.11	11.1 h
90	0.79	7.9 h
100	0.59	5.9 h
110	0.47	4.7 h
115	0.46	4.6 h
120.0	0.56	5.6 h
125.0	3.1	31.0 h

* m = minutes; h = hours, da = days

FIGURE 6.3 FDA code-based growth of vegetative bacteria 30°F to 125°F.

becomes the organism of choice to control, because it will multiply down to 39.2°F under favorable conditions. If the food is temperature-abused (above 50°F), proteolytic *C. botulinum* types A and B will become a problem. However, in chilled food systems, these abuses are controlled or the food is thrown away. Also, the food is labeled "keep refrigerated."

GROWTH OF VEGETATIVE CELLS IN FOOD

Figure 6.3 provides information concerning the accepted growth of vegetative cells in food. This is based on the FDA Food Code allowance that cooked food can be held at 41°F for 7 days or at any other temperature up to 140°F for 4 h. Actually, as shown previously in Table 6.6, the pathogens multiply over the range of 29.3 to 127°F. The Ratkowsky equation (Ratkowsky et al., 1983) provides a scientific method for predicting growth over the full kinetic range. Using 30°F as the bottom anchor point and 125°F as the upper anchor point, adjusting 41°F for 7 days to 41°F for 6.5 days, and putting 4 h at 115°F, one can obtain an excellent estimate of growth over the real bacterial growth range. When compared to actual bacterial growth, this is equivalent to about 10 multiplications of *L. monocytogenes* at 41°F and 10 multiplications of *Salmonella* or *S. aureus* at 115°F.

THE DESTRUCTION OF *SALMONELLA* IN FOOD

Salmonella destruction in food, shown in Figure 6.4, is based on the work of Goodfellow and Brown (1978) in the table, "Destruction of *Salmonella* in food." The FDA allows a 5D reduction in hamburger and ground meats, but the USDA requires

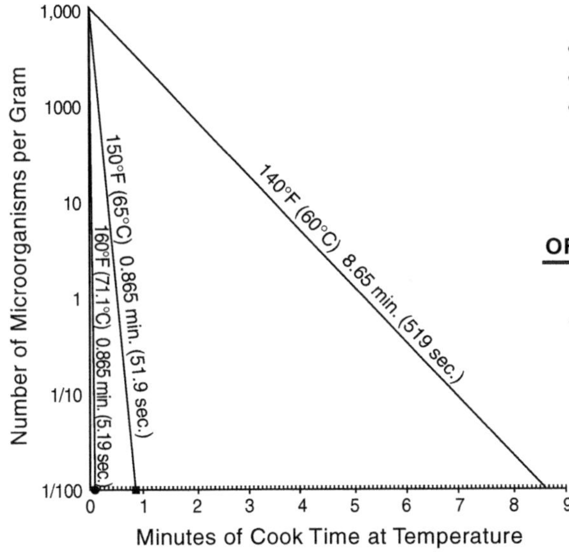

DEATH CONTROLS

- **Time and temperature**
- **Nutrients and acids**
- **Water activity**

DESTRUCTION OF *SALMONELLA* SPP. IN FOOD

Temp. °F (°C)	5 D (min.) (100,000:1)	7 D (min.) (10,000,000:1)
130 (54.4)	86	121
135 (57.2)	27	38
140 (60.0)	8.7	12
145 (62.8)	2.7	3.8
150 (65.6)	52 sec.	72 sec.
155 (68.3)	16 sec.	23 sec.
160 (71.1)	5.2 sec	7.2 sec

FIGURE 6.4 Destruction of *Salmonella* in food.

about a 7D reduction in red meat and poultry. Note that, at 160°F, 7D reduction of *Salmonella* is about 7.2 sec. However, in chilled food shelf-life, the vegetative pathogens are not the issue, but rather, spoilage microorganisms at 100,00 per gram. It can take many minutes at 160°F to get a 6- to 8-week shelf-life, depending on the spoilage bacteria level on the raw food.

WHAT DOES A TYPICAL COOKING PROCESS LOOK LIKE IN CHILLED FOODS?

Figure 6.5 shows the center temperature profile for a pork roast cooked in a water bath at 172°F. It took about 12 h for the roast to reach a center temperature of about 172°F and then, it took about another 12 h to get back down to <40°F. Actually, all of the time under the curve from 130 to 172°F was lethal, when integrated lethality is considered. Hence, this pork roast received many hours of pasteurization, when government regulations require only a few seconds.

PROGRAMS THAT ASSURE THE PRODUCTION OF SAFE FOOD

The answer to the problem of how to produce safe chilled foods is to use Hazard Analysis and Critical Control Point technology to develop a robust food production process in which virtually no defects will be produced. If food is left at room temperature for 2 days, the outgrowth and toxin production of *C. botulinum* in

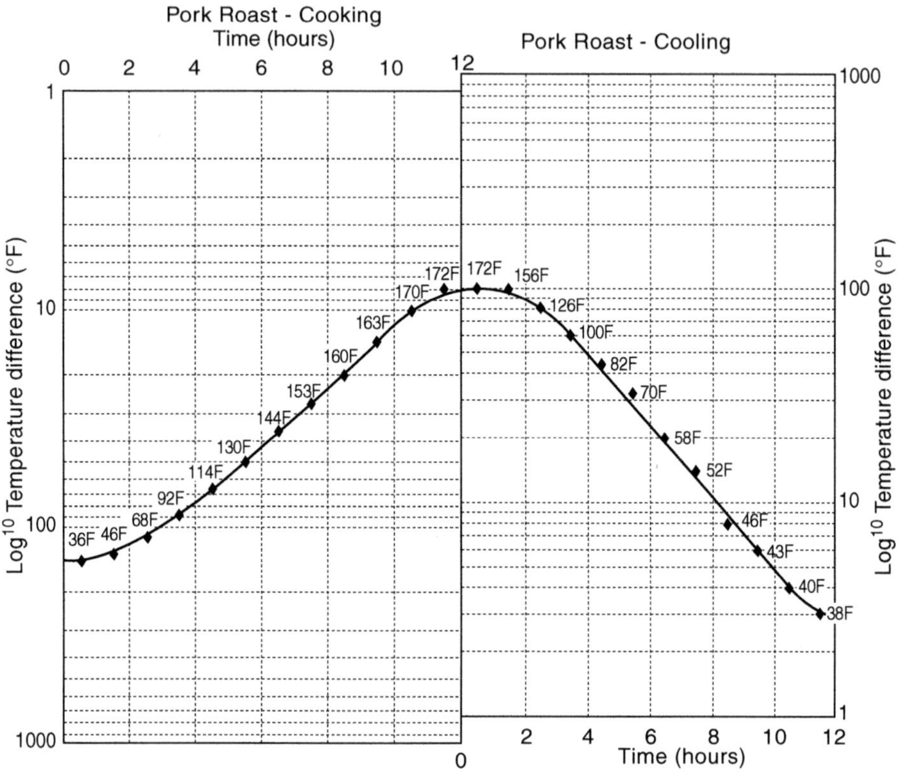

FIGURE 6.5 Water tank cooking and cooling of pork roast.

many of the neutral-acidity, high-water-activity foods can have serious consequences. However, the commercial chilled food industry has had a remarkable history of safety during the past 30 or more years in which processors have been producing ready-to-eat foods for deli foodservice and home use. The most common examples of highly perishable chilled foods are the cooked, sliced products sold in delis. Few preservatives are in these products, and they are hazardous if not refrigerated. Nonetheless, even in the commercial environment an outstanding history of safety exists.

The National Advisory Committee on Microbiological Criteria for Foods (NACMCF) HACCP program begins with the development of the prerequisite programs (NACMCF, 1998). Below is a list of the prerequisite programs as defined by the NACMCF:

- Strong management
- Listing of HACCP management team and responsibilities
- Description of the food and its distribution
- Description of the intended use and consumers of the food
- Facilities constructed and maintained according to sanitary design principles

- Linear product flow and traffic control to minimize cross-contamination from raw to cooked materials
- Supplier controls: GMPs and food safety programs, supplier guarantees, HACCP system verification
- Specifications written for ingredients, products, and packaging
- Production equipment constructed, installed according to sanitary design principles; preventive maintenance calibration schedules established and documented
- Cleaning and sanitizing procedures of equipment and facility written and followed; master sanitation schedule
- Personal hygiene: all employees and visitors follow requirements
- Documented training covering personal hygiene, GMPs, cleaning and sanitation, safety, and their role in HACCP program
- Documented chemical control and segregation procedures; proper use of non-food chemicals (e.g., cleaning chemicals, pesticides, baits)
- Receiving and storage under sanitary and environmental conditions, shipping procedures, temperature, humidity
- Traceability and recall; lot coding, recall system
- Pest control programs in place
- QA procedures
- Process and recipes
- Product formulation
- Labeling
- Glass control

NACMCF follows this with the development of flow charts and the HACCPs for each food process. In retail, it is essential that the processes be grouped because a large kitchen can have 300 or 400 different recipes. Actually, the recipes can be grouped simply by whether the food is cooked to pasteurization, the spores are controlled, and there is prevention toxin production by *S. aureus* after the food is cooked, where there are no more competitive microorganisms.

An HACCP process is diagrammed according to these steps in Table 6.7.

TABLE 6.7
NACMCF HACCP Process

Process Step	CCP	Chemical, Physical, Biological Hazards	Critical Limits	Monitoring Procedures/ Frequency/ Person(s) Responsible	Corrective Action(s)/ Person(s) Responsible	Verification Procedures/ Person(s) Responsible	HACCP Records
1.	Yes or no	1. 2. 3. etc.					

System Risk / HACCP Management

HACCP →	Risk Analysis / Failure Mode Effect Analysis →	If Necessary, Risk Reduction
• Description of consumer • System description • Process procedures • Hazards; amount, size, etc. • Validated controls; technical, management, consumer risk communications	• Possible process control failures • Monitoring and probability that the process deviation will be seen and corrected • Effect of failure / severity • Risk calculation	• Cost of the risk is not acceptable • Action to change; change control • New risk that is acceptable

FIGURE 6.6 System risk/HACCP management.

RISK ANALYSIS

In an HACCP plan for the production of a safe group of food products, there will never be zero risk. Even though all prerequisite programs are in place, employees are trained, and all ingredients are supplier certified, mistakes will happen. The best that management can do is to reduce the chance of a mistake by performing a risk analysis. Figure 6.6 lays out the three steps for risk analysis.

First, one performs the HACCP with a description of the consumer or customer, the system, procedures, hazards, and validated controls for technical management and consumer risk communications.

Next, one looks at the risk analysis, the failure-mode effect analysis, and ways that the controls might fail. When the chance of failure of control is unacceptable, management takes action to reduce the risk of failure. Eventually, it is necessary to have insurance against a highly likely adverse event.

The food safety program is then documented in the retail food HACCP TQM (Total Quality Management) manual. The following sections of a retail manual are based on controlling the prerequisites and the processes:

1. Management — Good Manufacturing Practices
2. Organization and Personnel
3. System Description
4. *Reserved* (for special use)
5. Supplier Policies, Procedures, and Standards
6. Food Production Policies, Procedures, and Standards
7. Sanitation Guidelines
8. Maintenance Guidelines
9. Pest Control Schedule and Instructions
10. HACCP TQM Employee Training Program and Record
11. Self-inspection, Continuous Quality Improvement
12. Food Safety Program Verification and Capability Certification

Figure 6.7, showing the unit as a food process system, illustrates the input, process, and output of a typical food system.

There are eight control elements to the food process system:

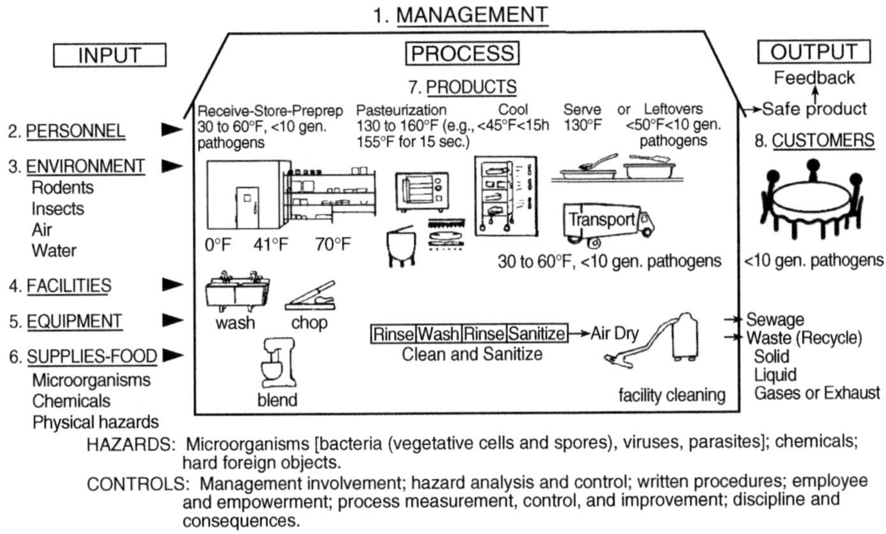

FIGURE 6.7 The unit as a food process sytem.

1. **Management** oversees that operational controls validated as capable of very low food safety risk are used by everyone on a daily basis. Management also provides the money necessary to keep the equipment and facilities in adequate operating condition.

Input

2. **Personnel** work in the facility; this includes personal hygiene.
3. **Environment** means control of the environment outside the building, including rodents, insects, contaminated air, and safe water and sewer systems.
4. **Facilities** include walls, floors, and ceilings, which keep the contaminated environment outside and allow the food to be produced in a semisterile processing plant.
5. **Equipment** must be designed to be easily cleanable. Floor drains are included in this category, as are equipment corners and niches, which must be scrubbed regularly in order to eliminate accumulation of spoilage microorganisms and prevent the growth of *L. monocytogenes*.
6. **Supplies and foods**: The supplier certifies the levels of pathogenic substances on the food. If the food is to be processed without any pathogenic substance control, the supplier must certify that the incoming produce is safe to eat without any hazard control procedure because the supplier did the hazard control.

Process

7. **Products**: The food is stored, pasteurized, and cooled. The cook reduces the pathogenic substances to a safe level, based on the sensitivity of the customer being served.

Output

8. If the **customer** removes the food from the facility, labels on the take-out containers warn the customer not to abuse the food; otherwise, spores can outgrow and multiply, thus making the food toxic and harming the customer.

PROCESS PERFORMANCE STANDARDS FOR PATHOGEN CONTROL

There are only a few common steps to all processes that must be controlled. The process performance standards are as follow and are used to develop HACCP-qualified and validated food processes:

1. Fingertip washing: reduce *Shigella* spp. by 10^{-5}.
2. Food contact surface cross-contamination: reduce *Campylobacter jejuni* by 10^{-5}.
3. Receiving and storage: *L. monocytogenes* control to <10 generations multiplication.
4. Prepreparation: wash fruits and vegetables to be eaten without pasteurization to reduce *L. monocytogenes* by 10^{-2}/g (10^{-5} if surface heat pasteurization is used).
5. Preparation:
   a. Control growth of *S. aureus* by heating from 50 to 130°F <15 h.
   b. Pasteurize to get a 10^{-5} or 10^{-7} *Salmonella* spp. reduction.
   c. Hold hot food >130°F, Rh >90% to prevent outgrowth of *C. perfringens.*
   d. Cool food from 130 to 45°F in <15 hours to control growth of *C. perfringens* to <1 log.
   e. Mix cold food <50°F to prevent toxin production by *S. aureus* from hands.
   f. Store cold food <40°F to prevent growth and toxin production by *B. cereus*.
   g. Acid control *Salmonella* <pH 4.1 and *C. botulinum* <pH 4.6.
   h. A_w control 0.86 (*S. aureus*) or 0.92 (*C. botulinum*).
6. Take-out hot food: eat cooked food that is at 95°F in <4 h, or begin refrigerator cooling <2 in. deep in <2 h to control *C. perfringens* in the food.

Figure 6.8 shows a basic layout of the HACCP worksheet for the analysis of each step in a food process to determine the risk priority of a step. The risk analysis

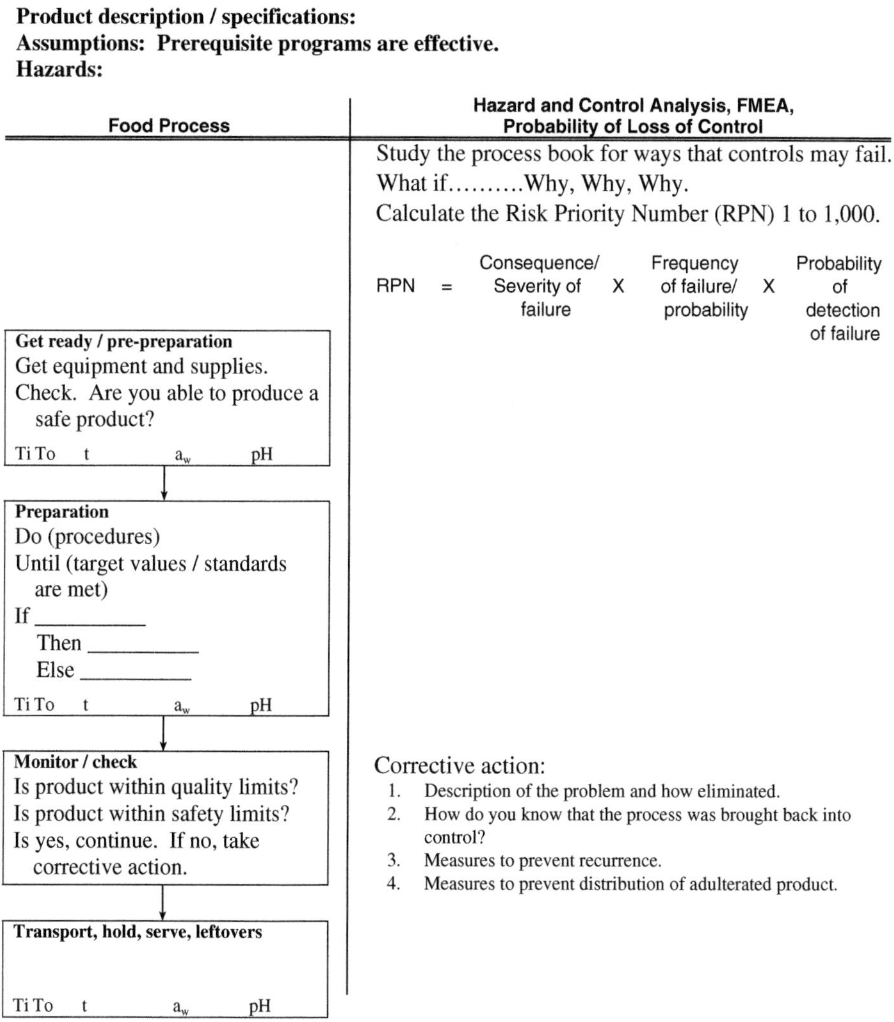

FIGURE 6.8 HACCP-qualified/validated food processes.

determines the probability of the failure of the hazard control program and personnel needed to detect and control the hazard when a hazard needs to be controlled. The analysis can be done for groups of food as well as for individual foods.

PROCESS QUALIFICATION

What is involved in process qualification (i.e., validating that the process is capable of producing a safe product and determining the probability of accomplishing this)? A qualified process is one that has demonstrated in operation that all necessary procedures, training, documentation, measurement, controls, and checks and balances are in place to ensure that the process can produce uniform quality output, even under stress conditions. Until one can predict the performance of the process, one does not have adequate control. To predict through the hazard analysis, one must know the key process variables, have control of them, and be able to produce products on a repeatable basis. When there is a problem, the root cause of the problem is removed; this is documented on a corrective action sheet.

WHAT IS A "BETTER" PROCESS?

A better process is one that is more stable. This means that the C_{pk} (Process capability index) = >1. The C_{pk} is calculated as the upper specification limit minus the average divided by 3 standard deviations of the process variable. Under these conditions, the common and special causes are known, and they are reduced, as is waste. Production is greater, because fewer mistakes are made, and customer satisfaction is better.

$$C_{pk} = \frac{USL\text{-}\bar{X}}{3\sigma}$$

EXAMPLES OF AN HACCP'D PROCESS

Figure 6.9 shows an HACCP process worksheet for sushi rice. Note that the flow chart depicts the exact handling of the food, including times and temperatures. On the right-hand side of the paper, each step is analyzed in accordance with the requirements of the NACMCF's HACCP criteria.

These flow charts are not intended to be read by the cook, so the flow chart must be turned into a narrative flow chart — in other words, the recipe. The recipe shown in Figure 6.10 for sushi rice illustrates a simple way to include validated process controls such as times and temperatures in the recipe at critical steps and produce a document to be used by the cook to assure the production of safe, standardized product.

EMPLOYEE TRAINING

This entire HACCP TQM program is dependent on employee behavioral management. There are three components to employee behavioral control. If management wants specific **behavior**, it must think in terms of the **antecedents** of the behavior as well as the **consequences**. Figure 6.11 shows this process.

Antecedents include validated HACCP procedures that are written down and taught to mastery. Supervisors provide constructive comments as a consequence of the employee behaviors; these consequences are positive, certain, and soon. They do this by complimenting and promoting employees who do their work according to the HACCP plan. The results are HACCP-based behaviors, whereby hazards are controlled. Employees monitor their own correct food production habits and record

Process: Preparation of the acidified rice that is used as the core ingredient of sushi menu items.

Assumption: Good manufacturing practices are effective in the operation.

Hazard: Traditionally, the sushi is not refrigerated. If the tuna species is chosen carefully, it will be free of parasites. If the fish have been frozen, this changes the spoilage bacteria so that they will not convert histadine to histamine. The risk that remains is the possible multiplication of *Bacillus cereus* multiplication in the rice. Lowering the pH of the rice to a level at which there is no *B. cereus* multiplication (<4.9 pH) can control this possible hazard in sushi rice as it "sits out" at room temperature for 24 hours. (If the rice is refrigerated, and the tuna is kept below 40°F, *B. cereus* will not multiply and pH control is not necessary. However, this is not sushi.) The actual pH of the rice in most recipes is 3.9 to 4.2.

Process Step, Procedure, and Control	**Hazard and Control Analysis: a. hazard identification; b. critical limit; c. employee monitoring procedure / frequency and person; d. verification-who, when, how**
Pre-preparation	
1. Assemble all ingredients and utensils. **O** Check that everything is ready.	Check that the tuna or other fish is fresh and has a very low APC count (for example, <10,000 CFU / g) to control the possible production of histamine. Use frozen fish to control the risk of parasites since it is very difficult to buy parasite-free fish. If appropriate, check for ciguatoxin. Check that the correct amount of acid (rice vinegar) has been measured (4.47% of the recipe total weight).
2. Wash rice in colander until water runs clean. **O** Drain rice. (If rice is enriched, do not wash it since that will remove the added B vitamins.)	Inspect the rice for rocks and remove if any are present..
Preparation	
3. Place drained rice in pan or rice cooker. Add **O** water (70°F). Cover container and bring rice and water to a boil (212°F). Ti 70°F To >200°F t 10 m.	This pasteurization-cook will reduce all vegetative pathogens to a safe level. Spores of *C. botulinum, B. cereus* and *C. perfringens* will be activated.
4. Reduce heat to simmering temperature **O** (190°F). Continue to cook until rice is done. Ti >200°F To 190°F t 20 m.	The *Clostridia* and *Bacillus* spores survive.
5. While rice is cooking, combine the vinegar, **O** sugar, and salt in small stainless steel bowl or pan. Heat the vinegar mixture until sugar has dissolved. Set aside Ti 70°F To 150°F t 5 m.	This must be done in a stainless steel container, or other type of container that does not react with the acid.
6. After rice is done, empty the pan of rice into a **O** *hangiri* (small shallow container). Spread rice evenly over bottom of the pan with a *shamoji* (a large wooden or stainless steel spoon). Let cool Ti 200°F To <120°F t 5 m.	The time is too short for any risk.
7. CCP Run spatula through the rice (~80°F) **C** using right and left slicing motions to separate grains of rice. At the same time, slowly add the vinegar mixture (~80°F). Fan the rice as the vinegar mixture is being added. Ti 120°F To <80°F t 10 m.	**CCP** The vinegar, sugar, and salt mixture will reduce the pH of the rice to<4.4 pH. This controls the outgrowth of *B. cereus*, which will not multiply at<4.9 pH. At pH of <4.6, the risk (if there is one) for *C. botulinum* is controlled. CCP-pH is 4.4 or less
8. Check the pH of the rice. It must be <4.6. The **T** target pH is 4.3 (± 0.3pH) **No**	This is the monitoring step to assure that the hazard control standard has been met.
Storage Yes	
9. Store at room temperature (70 to 80°F.) Use **S** within 24 hours. Ti 80°F To 80°F t <24 h.	The rice will spoil safe, because of airborne yeast and mold that get into the rice during mixing. The foodborne illness hazards are controlled.
Leftovers - >24 hour none. Discard any leftover sushi rice.	

FIGURE 6.9 Sushi rice HACCP.

Recipe Name: **Sushi Rice**		Portion size (vol.):	Number of casings:
Recipe #:		Number of portions:	Preparation time:
Production style:		Final yield (AS):	To be prepared by:
		Final Yield:	Supervisor:
Written by:	Date:2/99	SA/QA by: P. Snyder Date:3/99	

Gp. #	Ingred. #	Ingredients and Specifications	Weight %	Weight	
I	1	Rice, short grain, 3 1/2 cups	38.14	682.5 g.	24.1 oz.
	2	Water, 4 cups	52.53	940.0 g.	33.2 oz.
II	2	Rice vinegar*, 5 tablespoons plus 1 teaspoon	4.47	80.0 g.	2.8 oz.
	3	Sugar, 5 tablespoons	3.63	65.0 g.	2.3 oz.
	4	Salt, 4 teaspoons	1.23	22.0 g.	0.8 oz.
		Total weight	100.00	1789.5 g.	63.2 oz.

* Nakano Rice Vinegar (4.2% acetic acid)

Pre-preparation

1. Assemble all ingredients and equipment
2. Wash rice in colander until water runs clean. Drain thoroughly. (If enriched rice is used, do not wash the rice because washing removes enrichment B-vitamin and mineral mixture.)

Preparation

3. Place drained rice in pan or rice cooker. Add water. Cover container with close fitting lid and bring the water containing the rice to boil (212°F).
4. Reduce heat to a simmering temperature (190°F) and continue to cook for 15 to 20 minutes (until all the water has been absorbed).
5. Remove from heat. Take off the lid and spread a clean, white cloth or paper towel over the top of the pot. Replace the lid and let stand for 10 to 15 minutes. (The towel absorbs any excess moisture in the rice.)
6. While the rice is cooking, combine the vinegar, sugar and salt in a small stainless steel bowl or pan. Heat the mixture until the sugar has dissolved (150°F), stirring constantly. Remove from heat. Set aside.
7. Empty the rice into a *hangiri* (nonmetallic shallow container) and spread the rice evenly over the bottom with a *shamoji* or large wooden spoon (or stainless steel spoon). Let cool at room temperature. (As an alternative, the rice can be spread on a stainless steel pan and cooled to 80°F in about 30 minutes.)
8. CCP Run a spatula through the rice (~80°F) using right and left slicing motions to separate the grains. At the same time, slowly add the vinegar mixture (~80°F). (You may not need all of it. Avoid using too much or the rice will become mushy.)
9. The rice should be fanned as the vinegar mixture is added. A helper may be required for this step.
10. Check the pH of the rice mixture. It must be less than 4.4. The expected pH is about 4.3.
11. The fanning and mixing take about 10 minutes [until the rice reaches room temperature (75°F)].
12. Do not refrigerate the rice, but keep it covered with a clean, white cloth or paper towel, at room temperature (75°F) until it is ready to be used.
13. Sushi lasts just one day. It should not be used as a leftover. (There is no hazard, but the sushi rice will spoil due to yeast and mold growth.)

Process step #	Start food ctr. temp., °F	Thickest food dimension (in.)	Container size H x W x L (in.)	Cover Yes/No	Temp. on/ around food	End food ctr. temp., °F	Process step time, hr./min.

FIGURE 6.10 Quality-assured HACCP recipe procedure for sushi rice.

the results so that there can be review by the HACCP team and continual improvement of the process. A typical task performance statement includes:

1. What must be done to which standard
2. Where
3. Why
4. Who will do it
5. When which situation exists
6. How to do it and check it

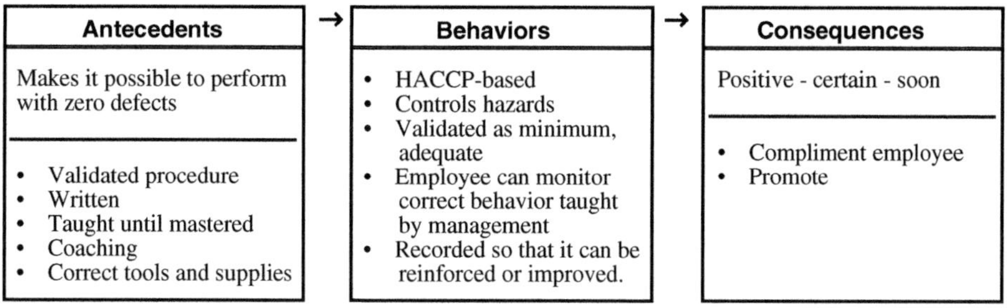

FIGURE 6.11 Employee behavioral management.

SUMMARY

This chapter has described opportunities available to the retail food industry to please the customer with minimally processed foods. It has provided the technical base for control of the pathogens and times and temperatures necessary to ensure a long shelf-life, if desired.

Pasteurization is invariably easier to accomplish than reducing the higher levels of spoilage microorganisms to achieve a long shelf-life. The spoilage microorganisms are at least as difficult to inactivate as the vegetative pathogen cells, and there will be more of them. Hence, refrigerated foods "fail safe" if temperature is abused, because the higher levels of spoilage microorganisms spoil the food first.

At the same time, one must realize that raw food will always be contaminated with pathogens. The water and dirt on farms are not sterile. Both land and water on farms are contaminated by wild birds, insects, and other animals that surround them.

An occasional government inspection is not the answer to a zero-defect food safety program. Management must know the hazards in food processes and must specify controls. Then, management must train employees to perform the processes according to validated recipe standards and check that the employees are doing their work according to these standards. When employees cannot accomplish a task according to how they have been trained, they must ask for supervisory help. Employees must never perform a process their way simply because they think that the hazard control program is wrong.

The real controller in the kitchen is the cook or chef. Cooks look at every batch of food produced, and they know if the food was produced according to the policies, procedures, and standards manual.

The rules are not complicated, but a scientifically validated system is required in case of a lawsuit in order to show due diligence.

REFERENCES

Farber, J.M. and Dodds, K.L., Eds., 1995, *Principles of Modified-Atmosphere and Sous Vide Product Packaging*, Lancaster, PA: Technomic Publishing Co. Inc., 328.

Goodfellow, S.J. and Brown, W.L., 1978, Fate of *Salmonella* inoculated into beef for cooking, *J. Food Prot.*, 41(8):598–605.

National Advisory Committee on Microbiological Criteria for Foods (NACMCF), 1998, Hazard analysis and critical control point principles and application guidelines, *J. Food Prot.*, 61(9):1246–1259.

Ratkowsky, D.A. et al., 1983, Model for bacterial culture growth rate throughout the entire biokinetic temperature range, *J. Bacteriol.*, 154(3):1222–1226.

7 Rapid Methods for Microbial Detection in Minimally Processed Foods

Karl R. Matthews

CONTENTS

INTRODUCTION

During the past four decades significant advances have been made in the identification of microorganisms. Emphasis has been directed toward the development of methods that permit rapid detection and identification of the target microorganism in a specific sample. Initially, the medical microbiology community led development of these methods; however, food microbiologists have made great strides in recent years. This is the result of heightened concern by the consumer, government agencies, and food processors with respect to the safety of the food supply and food-borne pathogens.

The number of cases, deaths, and costs attributed to food-borne illness each year is staggering, resulting in questioning the safety of the food supply. Many food-borne pathogens are recalcitrant to traditional methods of control such as low pH. Food-borne pathogens traditionally tested for by food microbiologists, *Staphylococcus aureus*, *Salmonella*, and *Clostridium botulinum*, have been supplanted by the "emerging" pathogens of the last decade: *Escherichia coli* O157:H7, *Campylobacter*,

1-58716-041-2/03/$0.00+$1.50

and *Listeria monocytogenes*. Many factors have changed the complexity of microorganisms associated with food, including a global marketplace, changes in processing practices, and consumer preferences.

Testing foods for pathogenic and spoilage bacteria is the cornerstone to ensuring a safe and wholesome food supply. Although cost effective and sensitive, conventional methods are often time consuming to perform and require several days before results are known. Many products that are minimally processed inherently have a short shelf-life, restricting the use of conventional testing methods. Rapid methods for isolation, detection, characterization, identification, and enumeration based on microbiological, biochemical, immunological, molecular, and serological methods are available (Fung, 1995).

"Rapid" is a general term and may relate to seconds, minutes, hours, or days and encompass the number of samples that can be analyzed within a set time frame. Although some restraints to the compatibility of food sample and rapid method used for analysis exist, most rapid methods developed are amenable for testing minimally processed foods. This holds true because most rapid methods rely on culturing methods to recover injured cells and amplify the number of target cells (Feng, 1993).

This chapter examines rapid methods, showcasing the breadth and scientific principle of the methods used for detection of target organisms in food samples. Discussion will focus on commercially available methods and methods that, although not yet available, show promise for the future.

IMMUNOLOGICAL METHODS

Many rapid methods available today are based on immunoassay technology. These assays utilize polyclonal antibodies or monoclonal antibodies. Polyclonal antisera contain antibodies to an array of cellular targets, whereas monoclonal antisera are directed to a specific target. Detection limits of many immunoassays are such that an enrichment step is required for detection of the target. Typically, enzyme-linked immunosorbent assays (ELISA) have detection limits of 10^3 to 10^5 colony forming units (CFU)/ml. The ELISA-based assays typically involve a solid phase (microtitre plate or polystyrene dip stick) coated with an antibody specific to the antigen (e.g., target pathogen).The sample is introduced to the solid phase and incubated for a period to allow binding of the antigen to the target antibody. Unbound materials are washed away and an enzyme-labeled antibody, or conjugate, with specificity to the antigen is then added. Following an incubation period, the unbound conjugate is washed away and a chromogen is added to enact a color change that can be detected by colorimetric or fluorometric techniques (de Boer and Beumer, 1999; Feng, 1993). ELISA-based systems include the Assurance EIA system (BioControl), TECRA system (Tecra Diagnostics), and the EHEC Tek (Organon Teknika). Fully automated ELISA-based systems are available and include the Vitek Immuno Diagnostic Assay System (VIDAS, bioMerieux).

Development of rapid assays for the detection of molds has not received the same attention as development of methods for the detection and identification of pathogenic bacteria. ELISA-based assays that detect the water-soluble extracellular secretions of fungi, the exoantigens, can detect fungi at the genus or species level. Heat-stable polysaccharides are generally more specific for one or more genus of fungi (Li et al., 2000); ELISA assays are available for mycotoxin detection (Pestka et al., 1995). As technology advances, perhaps other methods, such as PCR, fluorescence, and biological sensors, will become available for detection and identification of molds.

Aside from ELISA-based assays, immunodiffusion, agglutination, and immunocapture methods are also available. The Salmonella 1–2 test from Biocontrol is based on immunodiffusion and the formation of an antigen–antibody complex that forms a visual three-dimensional ImmunoBand. The VIP tests (BioControl) and similar tests are self-contained units and based on the lateral flow of target antigen through the system. Another system, based on migration of the sample along a chromatographic strip, is the Reveal test. During migration the sample passes through a reaction zone containing colloidal gold-labeled anti-O157 antibodies, which bind the antigen and form a target–antibody complex. Migration continues until the complex reaches the reaction zone that is an immobilized line of secondary antibody, producing a dark line of colloidal gold particles (Power et al., 2000). The Reveal (Neogen Corp.), SafePath (SafePath Laboratories LLC), and PATH-STIK (LUMAC) have been used to screen beef, fecal, and environmental samples and have demonstrated a broad range of sensitivity and specificity, based on the sample tested (Brinkman et al., 1995; Heuvelink et al., 1998; Power et al., 2000). Latex agglutination tests are available for most pathogens; in these tests, latex beads are coated with specific antibodies that, upon contact with specific antigens, form a visible precipitate.

A fast and convenient method for separating a target organism from a sample is to use immunocapture methods. Perhaps the most popular of these methods is immunomagnetic separation (IMS). In IMS, samples are mixed with beads coated with antibodies specific for the target organism. Target organisms in the sample bind to the immunomagnetic beads and the complex is isolated from the sample using a magnetic field. The beads can be plated on medium, used to inoculate broth, or used directly in PCR (polymerase chain reaction) assays (Gooding and Choudary, 1997; Mansfield and Forsythe, 1993; Mitchell et al., 1994; Wright et al., 1994).

Representative systems using IMS include the anti-*E. coli* from Dynal and screen/verify assays for *Salmonella* and *Salmonella* Enteritidis from VICAM. IMS separation is well suited for isolation of target organisms from complex food samples. Recovery of *Listeria* from environmental samples by IMS reduced test time to 24 h and improved sensitivity (Mitchell et al., 1994). Coupled with PCR, IMS has been used to detect *Salmonella* in fecal samples and *E. coli* O157:H7 in raw milk and ice cream (Chen and Griffiths, 2001; Gooding and Choudary, 1997; Widjojoatmodjo et al., 1992; Wright et al., 1994). A modification of IMS is magnetic capture hybridization polymerase chain reaction (MCH-PCR) (Chen et al., 1998). In this procedure, magnetic beads are used to capture DNA that is then used in PCR. The MCH-PCR assay was modified to detect *Salmonella* and shiga toxin producing *E. coli* (Chen and Griffiths, 2001). IMS is a malleable method and has been coupled with ELISA detection procedures (Mansfield and Forsythe, 2000) and use of fluorescent antibodies and laser scanning instruments that count fluorescent cells (Robinson, 1998).

NUCLEIC ACID-BASED METHODS

A plethora of nucleic acid-based typing methods is available for the differentiation and identification of food-borne pathogens. These include, but are not limited to, plasmid typing, ribotyping, PCR, randomly amplified polymorphic DNA (RAPD), and restriction fragment length polymorphism. Some of these methods have been partially or fully automated and systems are available commercially. The key advantages to these systems are that DNA can be isolated from all bacteria and that discriminatory power is greater than with phenotypic methods (Farber, 1996).

One of the most popular nucleic acid based methods is the PCR, a method based on amplification of a specific segment of cellular DNA. There are three essential steps in the PCR reaction: first, dsDNA (double-stranded DNA) is denatured into single strands, followed by addition of primers (specific short pieces of DNA complementary to a region on the target DNA) and annealing. This is an extremely important step because the annealing temperature will determine stringency of the reaction or how specific the attachment will be. Finally, extension of the primers is accomplished using a thermal stable DNA polymerase. From a single target DNA, one ends up with approximately a million-fold increase in product sequences that can be visualized as a band on an ethidium bromide-stained agarose gel.

The BAX System from Qualicon is a commercially available PCR detection system for *Listeria monocytogenes*, *Salmonella*, and *E. coli* O157:H7 (Maciorowski et al., 2000; Stewart and Gendel, 1998). The test simplifies PCR by including all reagents required for the PCR — primers, enzyme, deoxyribnucleosides — in a single lyophilized tablet. The BAX system for *E. coli* O157:H7 outperformed conventional and immunodiffusion methods for detection of the target pathogen in ground beef (Johnson et al., 1998). The authors reported that the BAX for screening *E. coli* O157:H7 assay had a detection rate of 96.5%, compared to 39% for the best cultural method and 71.5% for a commercially available immunodiffusion method. The efficacy of the BAX for screening *Salmonella* was tested using feed samples artificially contaminated with 1200 CFU/10 g of feed and agreed with conventional plating results for 16 of 18 samples and 13 of 18 samples spiked with 40 CFU/10 g of feed (Maciorowski et al., 2000).

The TaqMan test from PE Applied Biosystem is another commercially available PCR test that uses a 5′ nuclease assay in an automated PCR amplification and detection system. The system was highly effective in detecting *Salmonella* in contaminated food samples (Fach et al., 1999). A PCR-based system has been developed for detection of pathogenic protozoan parasites *Cyclospora cayetanensis*, *Cryptosporidium parvum*, and microsporidia (Orlandi and Lampel, 2000). The method is a filter-based protocol for isolation of the protozoan and preparation of DNA templates for use in PCR. The assay can be adapted for detection of parasites from a wide variety of foods.

Alternatives to conventional PCR-based systems are the NASBA (nucleic acid sequence-based amplification) system and RT-PCR (reverse transcription-PCR). In RT-PCR, the enzyme RT (reverse transcriptase) converts RNA into DNA molecules, which then serve as the targets for PCR amplification (de Boer and Beumer, 1999). Although kits are not available commercially, RT-PCR has been used successfully for detection of *L. monocytogenes* (Klein and Juneja, 1997) and *E. coli* O157:H7 (Berry, 2000) in various samples. Following a 2-h enrichment incubation, RT-PCR detected *L. monocytogenes* in a cooked meat sample originally inoculated with approximately 0.3 CFU/g (Klein and Juneja, 1997). RT-PCR overcomes a decided disadvantage of PCR because the assay is based on detection of RNA that has a relatively short half-life. In PCR, DNA from live and dead cells can potentially be amplified, whereas RT-PCR, when performed properly, specifically detects presence of live cells.

NASBA is similar to RT-PCR in that RNA is the initial template in which RNA is amplified below the melting point temperature for DNA. Problems in RT-PCR associated with false positives that result from contaminating DNA are reduced in NASBA. Utilization of mRNA rather than rRNA would make the system a means for detection of live cells in a sample. The NucliSens® Basic Kit (Organon Teknika) is commercially available for conducting NASBA analysis, which has been used for detection of viable *Salmonella, Campylobacter*, and *L. monocytogenes* (Simpkins et al., 2000; Uyttendaele et al., 1994).

Ribotyping is a very reproducible method with excellent discriminatory power that can be useful to the food industry, particularly in food processing applications for determining sources of contamination. The basics of Ribotyping include isolation of bacterial chromosomal DNA and cleaving with a restriction enzyme to obtain small pieces of DNA. The DNA fragments are separated by electrophoresis and transferred electrophorectically to a nitrocellulose membrane. Membranes are probed with labeled rRNA, highlighting the fragments of chromosomal DNA containing a ribosomal gene and thereby creating a fingerprint pattern. The patterns are unique and can be used to differentiate isolates at the strain level. An automated method is available from Dupont in which the whole process is broken down into seven basic steps. Results are obtained within 16 h. Patterns can be stored to create a unique database associated with a production facility or process line. The Riboprinter has been used to study strains of *C. botulinum*, *Salmonella*, and *L. monocytogenes* associated with various foods and cases of food-borne illness (Gendel and Ulaszek, 2000; Oscar, 1998; Skinner et al., 2000).

Plasmid typing and pulsed-field gel electrophoresis (PFGE) are methods with advantages and disadvantages. Plasmid typing is easy to perform and relatively quick; however, plasmids are unstable and some organisms contain few or no plasmids. PFGE is highly reproducible and discriminatory but time consuming and tedious. Moreover, highly trained laboratory personnel are required to conduct the procedure and interpret the results (Johnson et al., 1995).

MODIFICATION OF TRADITIONAL METHODS

Traditional methods are based on separation of bacteria from the sample matrix, dilution of sample in diluent, and plating onto differential media or into media to determine a specific biochemical response. A variety of methods is available to speed up the process of sample preparation and eliminate or reduce the need for media preparation. Samples must be prepared before they can effectively be plated and permit determination of microbial load. Traditionally, samples were prepared by addition of sample and diluent to sterile blender cups. The Stomacher is now used routinely to massage samples in sterile bags, eliminating the need to sterilize and use blender cups for sample preparation. During massaging, the microorganisms are dislodged from the sample and the initial dilution is made in the bag. To aid in sample dilution, several companies make automated diluters that deliver the appropriate amount of diluent to make the first dilution.

In the same vein, automated plating systems eliminate the need to make dilutions by distribution of liquid sample onto the surface of a rotating plate. This permits counting over a 1000-fold dilution on one plate. Counting colonies on a plate can be a time consuming process. Colony counting systems make use of line scanners and specialized software and thus eliminate the need for hand counting. Moreover, images of plates can be stored for future viewing and analysis or printed.

A number of methods are available that eliminate the need to prepare agar-poured plates. The Petrifilm system (3M), hydrophobic grid membrane filter (HGMF; ISO-GRID system, QA Life Sciences), and the SimPlate (IDEXX Laboratories) are examples of alternatives to agar-poured plates. The Petrifilm system consists of rehydratable nutrients embedded into a series of films coupled with a water-soluble gelling agent. A liquid sample (1 ml) is placed onto the center of the film, permitting rehydration of the media to support growth of microorganisms. After an appropriate incubation period, colonies can be counted directly on the plate. Petrifilm products include *E. coli* O157:H7; *E. coli* aerobic plate counts; and yeast, mold, and coliform systems (Russell, 2000). A newly improved product, Petrifilm 2000, provides results of coliforms in food, including frozen green beans and bakery product, in less than 12 h (Priego et al., 2000). The Petrifilm 2000 method was compared to the widely used violet red bile agar (VRBA) for enumeration of coliforms in food. The results provided by Petrifilm 2000 presented a close correlation ($r = 0.86$) to VRBA and greater sensitivity.

The SimPlate device looks similar to a petri dish containing numerous wells. The system for total plate count uses a specialized media containing dark blue resazurin that comes as a presterilized powder and is reconstituted with sterile water before use. The plate is incubated for 24 h and results are read. Growing bacteria reduce the dark blue resazurin to a pink resorufin or a colorless dihydroresorufin. Positive wells are counted and compared to a most probable number (MPN) chart to determine the MPN of the sample. Regardless of whether the sample analyzed was raw or ready-to-eat, the system performed well (Smith and Townsend, 1999). Results after 24 h of incubation were highly correlated, ranging from 0.94 to 0.98, in side-by-side comparisons against standard plate count agar. The system performed well for testing foods including fresh vegetables, beef ravioli, pasta salad, pesto, various juices, and ice cream.

The HGMF uses a unique system that confines colony growth to a set of 1600 grid cells. Homogenized samples are prefiltered to trap food particles larger than 5 μm and then the sample is filtered through the membrane filter, trapping microorganisms within the grid cells. The inoculated HGMF is placed onto the appropriate media and, after a suitable incubation period, colonies are counted. HGMF grids are available for aerobic plate count, yeast and mold counts, *Salmonella*, Coliform–*E. coli*, and *E. coli* O157:H7. The HGMF for *E. coli* O157:H7 was shown to recover more injured cells and more total cells from apple cider inoculated with *E. coli* O157:H7 and subjected to freeze–thaw conditions than MPN and spread plating (Sage and Ingham, 1998). Sharpe and Parrington (1998) found that testing foods and carcass rinse solution using the HGMF system for *E. coli* was simple and provided accurate results.

In recent years, a number of media have been developed permitting the detection, enumeration, and identification on a single media and thus eliminating the need for subculture and further biochemical tests. The number of media for detection of specific bacteria is staggering; for example, media for detection of *E. coli* O157:H7 include SD-39 (QA Life Sciences), BCM®O157:H7(+) (Biosynth Biochemica and Synthetica), and Fluorocult *E. coli* O157:H7 agar (EM Science). Many selective media contain chromogenic or fluorogenic substrates that yield brightly colored or fluorescent products when reacting with specific bacterial enzymes or bacterial metabolites. MacConkey sorbitol agar (MSA) is routinely used as a selective differential plating medium. Other media reported as adequate differential or selective plating media for *E. coli* O157:H7 include phenol red sorbitol agar containing 4-methylumbelliferyl-β-d-glucuronide, Levine eosin methylene blue agar, and MSA containing 5-bromo-4-chloro-3-indoxy, 1-β-d-glucuronic acid cyclohexylammonium salt (Ahmed and Conner, 1995; Blackburn and McCarthy, 2000). Media containing the substrates 5-bromo-4-chloro-3-indoxy, 1-β-d-glucuronic acid and 4-methylumbelliferyl-β-d-glucuronide are used for the detection of β-d-glucuronidase activity. This enzyme is present in 94 to 96% of *E. coli* strains but not in *E. coli* O157:H7 strains, thereby providing differential capability.

LUMINESCENCE

Rapid assays based on chemical reactions and biological reactions are useful in determining the cleanliness of food contact surfaces and presence of a target microorganism in a food. Bioluminescent-based methods can provide a rapid estimate of total bacterial load by making use of the ubiquitous presence of ATP in all living cells and its reaction with the luciferase enzyme complex found in fireflies. The production of light can be correlated to microbial number; this forms the basis of the ATP bioluminescence assay of microbial biomass. For chemiluminescence assay, microbial iron protoporphyrin IX (heme) is used as an indicator of bacterial load. Basically, the assay relies on the oxidation of luminescent chemicals, such as luminol, resulting in the emission of light (Pietrzak and Denes, 1996).

The Uni-lite system by Biotrace, Inc. (Plainsboro, NJ) consists of a portable luminometer reading unit and tests swabs with prepackaged reagents. The user swabs the desired surface, activates the swab by placing it in an enzyme solution, and then inserts the swab into a chamber in the luminometer to obtain a measurement. The system provides results within 2 min. Similar systems to measure ATP are the Lightning bioluminescence system (IDEXX Laboratories, Inc., Westbrook, ME) and the Lumac Hygiene Monitoring kit (Integrated BioSolutions, Inc., Monmouth Junction, NJ).

The direct epifluorescence filter technique (DEFT) can be performed in as little as 10 min, but the assay can yield varying results because acridine orange staining of cells may be variable. The assay has been recently modified, and now antibodies specific for a given microorganism are used rather than staining cells with acridine orange. The technique involves membrane filtration of a food sample to collect microbial cells associated with the sample on a membrane surface, fluorescent antibody staining of the target microorganism, and epifluorescence microscopy. The antibody–DEFT assay has been used successfully for determination and enumeration of *L. monocytogenes* in fresh vegetables and *E. coli* O157:H7 in apple juice (Tortorello et al., 1997, 1998).

A FEW MORE RAPID METHODS

Methods based on flow cytometry, impedance, nanotechnology, and bacterial ice nucleation detection offer alternatives to the methods discussed previously (deBoer and Beumer, 1999; Irwin et al., 2000; Peng and Shelef, 1999; Shelef and Eden, 1996; St John et al., 1998). The use of nanotechnology is an exciting approach to the detection of food-borne pathogens in a sample. The system makes use of a silicon chip specific for a target microorganism by stamping an antibody grating pattern on the silicon surface. The antibody grating alone produces insignificant optical diffraction, but an optical phase change occurs upon immunocapture of cells, producing a diffraction pattern. The diffraction intensity increases in proportion to the cell density bound on the surface (St. John et al., 1998).

Impedance analysis of food samples is commonly used to estimate total bacterial counts and to screen large numbers of samples; it is based on resistance in an electric circuit to the flow of alternating current corresponding to the actual resistance to a direct current. Impedance works for detection of microorganisms because, as they grow, they metabolize substrates of low conductivity into products of high conductivity and thereby decrease impedance. Using impedance microbiology, low levels of *S. aureus* (10^1 CFU/ml) on poultry could be detected in less than 24 h (Glassmoyer and Russell, 2000). Several automated systems are commercially available that are based on impedance, such as the Bactometer from bioMerieux and the Malthus System V from Malthus Diagnostics (Shelef and Eden, 1996).

SAMPLE PROCESSING

A plethora of rapid methods is available for the detection and identification of microorganisms associated with food. Specific methods mentioned in this chapter are generally acceptable for detection of a target organism in a wide range of food products including ground beef, lettuce, milk, apple juice, and yogurt. In this chapter methods used for processing a sample have been discussed; however, most rapid methods, particularly ones not based on modification of a traditional method, include enrichment, pre-enrichment, selective enrichment, or both, to facilitate detection of the target organism. Enrichment increases the likelihood of detecting low levels of a target microorganism in a food sample because many rapid tests have detection limits ranging from 10^2 to 10^5 CFU/gram or milliliter. Through enrichment, numbers of a target microorganism are amplified in order for the assay to detect its presence.

Usually, enrichment incubation is of short duration, ranging from 3 to 12 h, but in some instances 24 h are required. Media used for enrichment are dependent on the assay and the target microorganism. When considering whether to use an assay enrichment, incubation should be considered because this may significantly impact time required to obtain a result.

BASIC REQUIREMENTS OF RAPID METHODS

Before rapid methods can be developed, basic information concerning the target organism is required. Once that information is obtained, unique features can be identified and used for rapid detection. Rapid systems must have certain attributes or meet specific requirements before qualifying as a rapid method (Fung, 1995; de Boer and Beumer, 1999). The accuracy of a new method is paramount; basically, this refers to the sensitivity or ability of the method to detect low numbers of the target organism and specificity or ability to differentiate the target organism from background in the sample. A failure to detect the target pathogen when it is present upon culture is considered a false negative result; a culture negative result with a positive test result is considered a false positive result. The following equations are used to calculate sensitivity and specificity rates (Beumer et al., 1991):

$$\text{Sensitivity} = \frac{\text{Number of true positives (p)}}{\text{p + number of false negatives}} \times 100$$

$$\text{Specificity} = \frac{\text{Number of true negatives (n)}}{\text{n + number of false positives}} \times 100$$

Clearly, the method must be as sensitive as possible and the detection limit as low as possible. In many instances, the demand is less than 1 cell per 25 g of food for microorganisms that cause disease.

Since the method is supposed to be rapid, the speed at which an accurate result is obtained must be considered. Ideally, accurate results of a rapid test would be obtained almost instantaneously but certainly within a typical 8-h workday. However, as indicated previously, the detection limit on many tests is 10^2 to 10^5 organisms per gram or milliliter; therefore, enrichment from 4 to 24 h is required prior to running the assay. A second component of this equation concerns the number of samples that can be processed at one time. Although many rapid systems are designed to perform many tests for a single sample, the microtiter plate format has become very popular and can handle 96 samples at one time.

The three most important criteria when selecting a test are accuracy, speed, and cost. When considering costs, the purchase of the test, reagents, supplies, instruments and equipment, training of personnel, and maintenance and service of equipment must be considered. The volume of testing will dictate whether single-unit diagnostic kits are used or large instruments designed for high-volume testing are purchased.

The test should be simple to perform and required equipment easy to operate. This facilitates use by a range of laboratory workers, regardless of degree of technical expertise. The test should give good performance in the matrices to be tested. Consideration should be given to interference from sample components, resident microbial flora, or other debris that may interfere with the accuracy of the test. The acceptability of the test by industry or government agencies may be a deciding factor as to whether a particular test can be used. The Bacteriological Analytical Manual contains procedures used by the Food and Drug Administration and recognized as standard for comparison of methods. The AOAC (Association of Official Analytical Chemists) International is a recognized body that provides test kit manufacturers with a third-party validation of product performance claims. The "Performance Testing" program is a four-step process that, upon successful completion, permits the manufacturer to indicate that the test is an official AOAC-approved method.

FUTURE DEVELOPMENTS

Development of new rapid methods is in part contingent on the collaboration of multidisciplinary teams with chemists, immunologists, physicists, microbiologists, and food technologists. Real-time detection of pathogens for monitoring in HACCP programs and ensuring the safety of minimally processed food that may have a short shelf-life is an exciting endeavor.

Nanotechnology can be exploited and can permit the development of silicone chips imprinted with specific DNA sequences for detection and identification of a target microorganism. Similarly, chips imprinted with specific antibodies to target microorganism are a possibility. Development in PCR technology has resulted in "real-time" PCR in which amplified product is nearly simultaneously detected spectrophotometrically. Linking methods such as IMS, PCR, and ELISA permits accurate rapid detection or identification of low levels of a pathogen in a food matrix. Completely automated systems are available that isolate plasmids, prepare DNA for PCR, perform restriction digests of DNA, transfer nucleic acids to membranes, and store profiles. Epidemiologists can use such systems to trace outbreaks or follow production practices.

The greatest challenge may rest in development of techniques that separate target microorganisms from a food matrix prior to testing. Combining methods is one approach; however, problems still arise with interference from the food components. The fouling of sensors for in-line detection of microorganisms is a distinct problem associated with sensor technology. However, given the advances in the last decade with respect to development of rapid microbiological methods for detection and identification of food-borne microorganisms, this problem, too, can be overcome.

REFERENCES

Ahmed, N.M. and Conner, D.E., 1995, Evaluation of various media for recovery of thermally-injured *Escherichia coli* O157:H7, *J. Food Prot.*, 58:357–360.

Berry, E.D., 2000, Development and demonstration of RNA isolation and RT-PCR procedures to detect *Escherichia coli* O157:H7 gene expression on beef carcass surfaces, *Lett. Appl. Microbiol.*, 31:265–269.

Beumer, R.R., Brinkman, E., and Rombouts, F.M., 1991, Enzyme-linked immunoassays for the detection of *Salmonella* spp.: a comparison with other methods, *Int. J. Food Microbiol.*, 12:363–374.

Blackburn, C.W. and McCarthy, J.D., 2000, Modifications to methods for the enumeration and detection of injured *Escherichia coli* O157:H7 in foods, *Int. J. Food Microbiol.*, 55:285–290.

Brinkman, E. et al., 1995, Evaluation of a new dip-stick test for the rapid detection of *Salmonella* in food, *J. Food Prot.*, 58:1023–1027.

Chen, J. and Griffiths, M.W., 2001, Detection of *Salmonella* and simultaneous detection of *Salmonella* and shiga-like toxin-producing *Escherichia coli* using the magnetic capture hybridization polymerase chain reaction, *Lett. Appl. Microbiol.*, 32:7–11.

Chen, J., Johnson, R., and Griffiths, M.W., 1998, Detection of verotoxigenic *Escherichia coli* by magnetic capture hybridization PCR, *Appl. Environ. Microbiol.*, 1:147–152.

deBoer, E. and Beumer, R.R., 1999, Methodology for detection and typing of foodborne microorganisms, *Int. J. Food Microbiol.*, 50:119–130.

Fach, P. et al., 1999, Evaluation of a polymerase chain reaction–based test for detecting *Salmonella* spp. In food samples: Probelia *Salmonella* spp., *J. Food Prot.*, 62:1387–1393.

Farber, J.M., 1996, An introduction to the hows and whys of molecular typing, *J. Food Prot.*, 59:1091–1101.

Feng, P., 1993, Rapid methods for the detection of *Salmonella* in foods, *J. Food Drug Ann.*, 1:119–131.

Fung, D.Y.C., 1995, What's needed in rapid detection of foodborne pathogens, *Food Technol.*, 49:64–67.

Gendel, S.M. and Ulaszek, J., 2000, Ribotype analysis of strain distribution in *Listeria monocytogenes, J. Food Prot.*, 63:179–185.

Glassmoyer, K.E. and Russell, S.M., 2000, Evaluation of a selective broth for detection of *Staphylococcus aureus* using impedance microbiology, *J. Food Prot.*, 64:44–50.

Gooding, C.M. and Choudary, P., 1997, Rapid and sensitive immunomagnetic separation-polymerase chain reaction method for the detection of *Escherichia coli* O157:H7 in raw milk and ice-cream, *J. Dairy Res.*, 64:87–93.

Heuvelink, A.E. et al., 1998, Isolation and characterization of verocytotoxin-producing *Escherichia coli* O157:H7 strain from Dutch cattle and sheep, *J. Clin. Microbiol.*, 36:878–882.

Irwin, P. et al., 2000, Minimum detectable level of salmonellae using a binomial-based bacterial ice nucleation detection assay, *J. AOAC Int.*, 83:1087–1095.

Johnson, J.L., Brooke, C.L., and Fritschel, S.J., 1998, Comparison of the BAX for screening/*E. coli* O157:H7 method with conventional methods of detection of extremely low levels of *Escherichia coli* O157:H7 in ground beef, *Appl. Environ. Microbiol.*, 64:4390–4395.

Johnson, J.M. et al., 1995, Use of pulse-field gel electrophoresis for epidemiological study of *Escherichia coli* O157:H7 during a food-borne outbreak, *Appl. Environ. Microbiol.*, 61:2806–2808.

Klein, P.G. and Juneja, V.K., 1997, Sensitive detection of viable *Listeria monocytogenes* by reverse transcription-PCR, *Appl. Environ. Microbiol.*, 63:4441–4448.

Li, S., Marquardt, R.R., and Abramson, D., 2000, Immunochemical detection of molds: a review, *J. Food Prot.*, 63:281–291.

Maciorowski, K.G., Pillai, S.D., and Ricke, S.C., 2000, Efficacy of a commercial polymerase chain reaction–based assay for detection of *Salmonella* spp. in animal feeds, *J. Appl. Microbiol.*, 89:710–718.

Mansfield, L.P. and Forsythe, S.J., 1993, Immunomagnetic separation as an alternative to enrichment broths for *Salmonella* detection, *Lett. Appl. Microbiol.*, 16:122–125.

Mansfield, L.P. and Forsythe, S.J., 2000, The detection of *Salmonella* using a combined immunomagnetic separation and ELISA end-detection procedure, *Lett. Appl. Microbiol.*, 31:279–283.

Mitchell, B.A. et al., 1994, Use of immunomagnetic capture on beads to recover *Listeria* from environmental samples, *J. Food Prot.*, 57:743–745.

Orlandi, P.A. and Lampel, K.E., 2000, Extraction-free, filter-based template preparation for rapid and sensitive PCR detection of pathogenic parasitic protozoa, *J. Clin. Microbiol.*, 38:2271–2277.

Oscar, T.P., 1998, Identification and characterization of *Salmonella* isolates by automated ribotyping, *J. Food Prot.*, 61:519–524.

Peng, H. and Shelef, L.A., 1999, Automated rapid screening of foods for the presence of salmonellae, *J. Food Prot.*, 62:1341–1345.

Pestka, J.J., Abouzied, M.N., and Sutikno, 1995, Immunological assays for mycotoxin detection, *Food Technol.*, 49:120–123.

Pietrzak, E.M. and Denes, A.S., 1996, Comparison of luminol chemiluminescence with ATP bioluminescence for the estimation of total bacterial load in pure cultures, *J. Rapid Methods Autom. Microbiol.*, 4:207–218.

Power, C.A. et al., 2000, Evaluation of the Reveal and Safepath rapid *Escherichia coli* O157:H7 detection tests for use on bovine feces and carcasses, *J. Food Prot.*, 63:860–866.

Priego, R., Medina, L.M., and Jordano, R., 2000, Evaluation of Petrifilm series 2000 as a possible rapid method to count coliforms in foods, *J. Food Prot.*, 63:1137–1140.

Robinson, K., 1998, Markers and magnets speed up search for *E. coli, Biophometrics Int.*, 3:60–61.

Russell, S.M., 2000, Comparison of the traditional three-tube most probable number method with the Petrifilm, Simplate, BioSys Optical, and Bactometer conductance methods for enumerating *Escherichia coli* from chicken carcasses and ground beef, *J. Food Prot.*, 63:1179–1183.

Sage, J.R. and Ingham, S.C., 1998, Evaluating survival of *Escherichia coli* O157:H7 in frozen and thawed apple cider: potential use of a hydrophobic grid membrane filter-SD-39 agar method, *J. Food Prot.*, 61:490–484.

Sharpe, A.N. and Parrington, L.J., 1998, Membrane filter method based on FC-5-bromo-4-chlor-3-indolyl-β-D-glucuronide medium facilitates enumeration of *Escherichia coli* in foods and poultry carcass rinses, *J. Food Prot.*, 61:360–364.

Shelef, L.A. and Eden, G., 1996, Optical instrument rapidly detects and enumerates microorganisms in food, *Food Technol.*, 50:82–85.

Simpkins, S.A. et al., 2000, An RNA transcription-based amplification technique (NASBA) for the detection of viable *Salmonella enterica, J. Appl. Microbiol.*, 30:75–79.

Skinner, G.E. et al., 2000, Differentiation between types and strains of *Clostridium botulinum* by ribotyping, *J. Food Prot.*, 63:1347–1352.

Smith, C.F. and Townsend, D.E., 1999, A new medium for determining the total plate count in food, *J. Food Prot.*, 62:1404–1410.

Stewart, D. and Gendel, S.M., 1998, Specificity of the BAX polymerase chain reaction system for detection of the foodborne pathogen *Listeria monocytogenes*, *J. AOAC Int.*, 81:817–822.

St. John, P.M. et al., 1998, Diffraction-based cell detection using a microcontact printed grating, *Ann. Chem.*, 70:1108–1111.

Tortorello, M.L., Reineke, K.F., and Stewart, D.S., 1997, Comparison of antibody-direct epifluorescent filter technique with the most probable number procedure for rapid enumeration of *Listeria* in fresh vegetables, *J. AOAC Int.*, 80:1208–1214.

Tortorello, M.L. et al., 1998, Comparison of methods for determining the presence of *Escherichia coli* O157:H7 in apple juice, *J. Food Prot.*, 61:1425–1430.

Uyttendaele, M. et al., 1994, Identification of *Campylobacter jejuni, Campylobacter coli,* and *Campylobacter lari* by the nucleic acid amplification system NASBA®, *J. Appl. Bacteriol.*, 77:694–701.

Widjojoatmodjo, M.N. et al., 1992, The magnetic immunopolymerase chain reaction assay for direct detection of salmonellae in fecal samples, *Appl. Environ. Microbiol.*, 59:1342–1346.

Wright, D.J., Chapman, P.A., and Siddons, C.A., 1994, Immunomagnetic separation as a sensitive method for isolating *Escherichia coli* O157 from food samples, *Epidemiol. Infect.*, 113:31–39.

8 Quantitative Risk Assessment of Minimally Processed Foods

Siobain Duffy, Yuhuan Chen, and Donald W. Schaffner

CONTENTS

INTRODUCTION

Microbial quantitative risk assessment (QRA) is an emerging technique that allows the risk posed by food-borne disease agents to be estimated through mathematical modeling with probability distribution functions (PDFs). Several factors need to be considered when constructing a QRA: the rate of contamination of a food with a specific pathogenic microorganism, the growth and inactivation of that organism through processing of the food, the exposure of the consumer to the bacterium, and the consumer's response to the dose of bacteria ingested (Buchanan, 1997; Schaffner, 1999). Risk assessors combine PDFs and other information to construct Monte Carlo computer simulations to estimate the food-borne illness caused by a given pathogen in specific food products. Although risk assessment has only begun to be applied to food pathogens within the past decade, regulatory agencies have predicted that QRAs

1-58716-041-2/03/$0.00+$1.50

will be important tools for creating science-based regulations for the food industry in the future (Buchanan et al., 1998; Pargas, 1998; Klapwijk et al., 2000).

Risk assessments are created in four steps:

1. **Hazard identification** involves choosing the most problematic food pathogen (bacterium, virus, or parasite) for a given food product.
2. **Exposure assessment** requires mathematical modeling of the concentration and prevalence of the chosen pathogen in the raw food and the changes to those values as a result of processing. The output represents the number of bacteria to which a consumer may be exposed. In the creation of an exposure assessment, portions of the food-processing chain requiring future research by food microbiologists may be identified (Cassin et al., 1998b).
3. **Dose-response** models are then created based on epidemiological data, feeding trials, or previously constructed probability distribution descriptions of the dose response for the selected pathogen.
4. In **risk characterization**, the computer simulation combines exposure and dose-response assessments to characterize the risk posed to the public by the chosen pathogen in the specific food.

The outcome of a QRA is a mathematical statement of risk, usually in the format of a cumulative probability distribution, showing the likelihood and range of the resulting risk.

Risk assessment is the first of three parts of risk analysis; the subsequent steps are risk communication and risk management. The responsibility for risk management falls to regulatory agencies, which enact legislation to protect the health of the public, and to food industries, which aim to produce safe products. These two sectors control the processing and preservation of foods and verify the effectiveness of these methods. Risk assessment not only allows risk managers to predict how safe a food is but also identifies which processing steps or factors in food production are most crucial to the overall safety of a product (Cassin et al., 1998a). Risk communication is the dialog between affected stakeholders regarding a particular risk (i.e., food industry professionals, regulatory agencies, and consumers) and is an important part of the risk analysis process (Pargas, 1998). QRA can ease risk communication by making the basis for risk management decisions more transparent (Morales and McDowell, 1998).

The rising popularity of minimally processed foods (MPFs) (Odumeru et al., 1997; Francis et al., 1999), which necessarily eschew a harsh bactericidal processing step, means that QRA may become important in assuring the safety of these products. The low-acid canning industry, for example, has little need for QRA because the 12-D reduction in *Clostridium botulinum* has historically proved to be very safe. The fresh-tasting, higher-quality MPFs (Sloan, 1995) require a more accurate prediction of the possibility of food-borne illness. Emerging food pathogens such as *Escherichia coli* O157:H7 also present an increased risk in MPF because they have low infectious doses (Buchanan and Doyle, 1997). Risk assessment can combine the effects of a series of smaller intervention steps and predict the final risk of pathogens in MPFs in a scientific manner. For example, the FDA has proposed that juice producers demonstrate a 5-log reduction in pathogenic bacteria in fresh juices to assure that the beverages are safe to drink (Anonymous, 2001). A QRA for *E. coli* O157:H7 in apple cider created in the authors' lab (Duffy et al., 1999) can help fresh juice producers achieve that goal utilizing HACCP without the need to pasteurize their juices. QRA could help prevent escalating rates of food-borne illness while the consumption of minimally processed foods increases.

Unfortunately, most published QRAs have focused on foods that have a terminal thermal processing step and are more common vehicles for food-borne illness, such as egg products and hamburger meat (Whiting and Buchanan, 1997; Cassin et al., 1998a). In the future, more risk assessments will need to be developed for MPFs, as demand for these foods increases. The most important concepts in QRA are presented in this chapter and illustrated with data from QRA developed in the authors' laboratory, published risk assessments, and the literature on minimally processed foods.

VARIABILITY AND UNCERTAINTY

The topics of variability and uncertainty have been a constant source of debate in the field of risk assessment. The definitions of these words are often in flux, as is the boundary between them, and whether or not a risk assessor can separate the two. Variability, as defined by Cox (cited by Vose, 2000), is "a phenomenon of the physical world to be measured, analyzed, and where appropriate explained," which applies to food-borne pathogens.

Uncertainty, on the other hand, can be defined as a risk assessor's lack of understanding or information concerning a pathogen. For example, the density of *E. coli* O157:H7 on dropped apples in a typical orchard is unknown. Uncertainty about a pathogen or its responses to processing and preservation factors calls for further research, whereas variability in data obtained by research can merely be quantified. Gathering more data can reduce uncertainty, while the magnitude of variability associated with a pathogen or process (e.g., a pathogen's responses to processing) is not reducible through more data collection and may be altered only through system modification, i.e., through changing the physical process applied to the pathogen or mutation of the pathogen itself (Cassin et al., 1998a; Vose, 2000). One can argue that the distribution used to describe variability conveys a sense of uncertainty in this case, not due to the lack of knowledge but rather to the invincible property of chance. Making uncertainties evident in a microbial QRA model continues to be a tough challenge for risk assessors (Jolis et al., 1999; Vose, 2000). Variability, on the other hand, can be quantified using probability distribution functions.

PROBABILITY DISTRIBUTION FUNCTIONS

One of the most common PDFs is the normal distribution; many microbiological processes are normally or lognormally distributed. In several situations encountered by the authors' laboratory, normal-like distributions such as the logistic have been very useful and widely employed in QRAs. Many PDFs have been associated with specific biological or physical meanings (Vose, 1996). For example, the Poisson distribution is commonly used to describe the number of microorganisms in a food (Jarvis, 1989; Marks and Coleman, 1998). Five PDFs are shown here with their applications in QRA.

In many experiments, if a researcher performs enough replicates, the results tend to follow a normal distribution. The authors' lab found this to be true when conducting a series of experiments to determine the efficacy of handwashing on reducing cross-contamination in a kitchen setting (Chen et al., 2001). Samples from at least 30 different participants were collected to determine the statistical distribution of cross-contamination rates, which were found to be highly variable. The distribution of the logarithm of bacterial transfer rates appears approximately normal for handwashing efficacy as well as for transfer rates distributions between chicken-to-hand, hand-to-lettuce, cutting board-to-lettuce, and hand and water faucet handles. A higher mean indicates a higher average transfer rate, and a higher standard deviation points to a greater degree of variability.

Figure 8.1 shows the handwashing histogram data and the normal distribution obtained by fitting the data using BestFit (Palisade Corp., Newfield, NY). For the six transfer rates examined in the study, a normal distribution was judged to be a good fit by the Kolmogorov-Smirnov statistic (Chen et al., 2001), though it was not always the best-fitting distribution. Although other, more complicated, three- and four-parameter distributions such as the Beta distribution could more accurately describe the individual cross-contamination rates, normal distributions were chosen to represent the data because of their adequate goodness-of-fit and statistical convenience.

The logistic distribution can look very much like a normal distribution except for a stronger peaking tendency than the bell curve. It is often used to describe

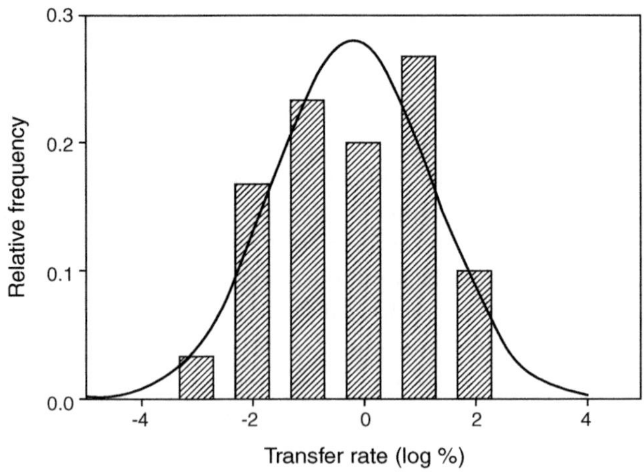

FIGURE 8.1 Change in *Enterobacter aerogenes* B199A populations on hands from handwashing. The bars represent the experimental data from 30 volunteers, and the curve is a normal distribution fitted to the frequency data. Adapted from Chen et al., *J. Food Prot.*, 64(1):72–80, 2001.

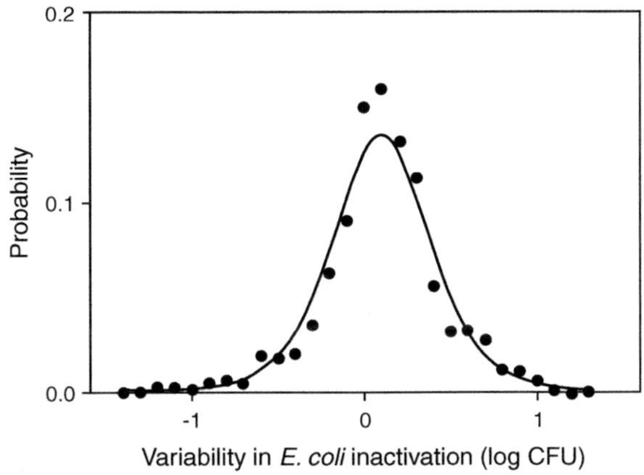

FIGURE 8.2 Variability around the ultraviolet light–induced mean log CFU reduction of *Escherichia coli* ATCC 25922. The solid circles represent a histogram of the variability in the data set, and the solid line is the logistic (0.0997, 0.18) distribution. Adapted from Duffy et al., *J. Food Prot.*, 63(11):1587–1590, 2000.

subsets of data sets (Anonymous, 1996), and the authors have found this to be true in the lab, as well. When analyzing Dr. Randy Worobo's data on UV-inactivation of *E. coli* ATCC 25922, the variability around the mean centered very strongly on 0 — the mean itself (Duffy et al., 2000). Though the variability ranged from –1 to 1 log CFU different from the mean, the logistic distribution in Figure 8.2 shows the strong peak around 0, leaving the mean unchanged.

The Weibull distribution is a PDF with wide applicability (Weibull, 1951), including several recent examples in food. Several research groups have shown that the Weibull distribution is useful in describing the nonlinear inactivation kinetics of spore forming microbes in food, including *Clostridium botulinum* (Peleg and Cole, 2000) and *Bacillus cereus* (Fernandez et al., 1999). A variant of this distribution (called Weibull-gamma, because it incorporates elements of the gamma distribution as well) has also been successfully used to model human dose response to food-borne pathogens, including *Listeria monocytogenes* (Farber et al., 1996; Lindqvist and Westoo, 2000), *Shigella*, *Campylobacter*, and *Salmonella* (Holcomb et al., 1999).

When few data points are available, fitting a more complicated PDF may not be sensible. The distributions commonly used to model sparse data sets are the triangular and the beta-pert. The triangular distribution is defined by three parameters: a minimum value, a most likely value, and a maximum value, which can be estimated by experts in the absence of data. The beta-pert distribution is similar in intent but "tails" near the minimum and maximum values, thus reducing the effects of the very low chance values occurring in the QRA. Some QRAs rely heavily on these distributions because very few, if any, quantitative data are in the literature on emerging pathogens in a specific food product (Baker et al., 1998; Fazil, 1999).

The advantages of triangular and beta-pert distributions are that they are intuitive and easy for nonmodelers to alter as more data become available. However, these

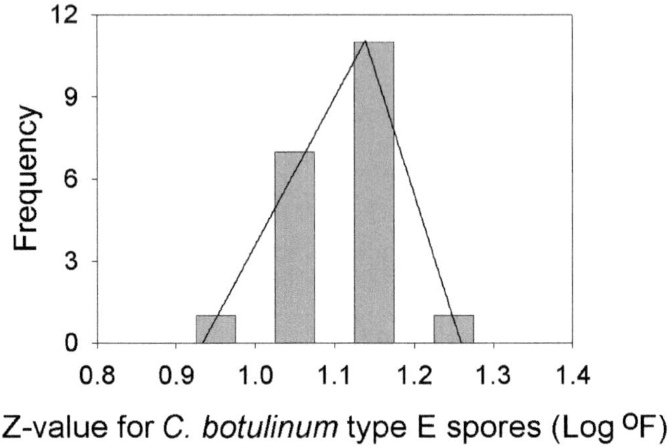

FIGURE 8.3 Thermal resistance characteristics (Z-values) of various strains of *C. botulinum* type E spores in different fish and fishery products. Bars represent frequency of each Z-value (log °F) reported in the literature and the solid line is the triangular distribution (0.93, 1.14, 1.26).

distributions have significant limitations. They are solely empirical distributions and do not offer any apparent mechanistic insight into a process, and they are not able to reflect bimodal and other trends that may appear as the data set grows larger. The triangular and beta-pert have been used commonly in microbial QRA with the idea that further research will refine the choice of distribution to a more biologically relevant PDF (Fazil, 1999). When trying to fit a distribution to a sparse data set, the triangular distribution will often be a highly statistically significant fit. If the data are best reflected with this simple PDF, then there is little reason not to use that distribution. This is illustrated in Figure 8.3, in which a triangular distribution is the best statistical fit, as determined by BestFit (RMS error 0.0029), closely followed by beta (RMS error 0.0044) and logistic (RMS error 0.02) distributions. The Z-value data were reported in literature studies in which various strains of *C. botulinum* type E spores were inoculated into whitefish (Crisley et al., 1968; Pace et al., 1972), crabmeat (Lynt et al., 1977, 1983), surimi (Rhodehamel et al., 1991), and other seafood products (Bohrer et al., 1973; Licciardello, 1983; Betts and Gaze, 1995) to determine the thermal resistance characteristics of the spores.

MODELING VARIABILITY IN QRA

The quality of a quantitative risk assessment is dependent upon how well it models and describes variability. Every process, whether biological or physical, is variable and exhibits a range of responses. Even thermal processing of low-acid canned foods does not always result exactly in a 12-log reduction in *C. botulinum* spores (Charm, 1966). Sometimes the reduction is greater and sometimes it is lower — because of the laws of physics. Because food always contains far fewer than 10^{12} *C. botulinum* spores (Dodds, 1993), the regulations governing the low-acid canned food industry are considered fail-safe. For reasons previously mentioned, however, variability is of great importance to minimally processed foods. Risk assessors incorporate at least three major sources of variability into their QRAs: method variability, biological variability, and process variability.

Method Variability

Method variability is an inevitable component of any literature-based risk assessment. Different laboratories using differing protocols and media will publish differing results for the same experiments. This could partially account for the variability in the handwashing efficacy of antimicrobial soaps shown in Figure 8.4, based on data reported in the literature (Montville et al., 2001). Individual hands' responses to the soaping process may be the primary source for the variability. Figure 8.4 shows data collected for antimicrobial soaps fitted to a beta distribution. Soaping resulted in various degrees of log CFU change in the bacterial count on hands. The bars represent the number of times each reduction was observed in the literature, with different shades representing soaps containing different antimicrobial agents. Visual inspection of these data indicates that combining them into a single distribution would be appropriate to represent variability (Montville et al., 2001).

In some cases, errors in experimental design can be found in published studies once further research shows that these earlier protocols were inadequate. The authors' lab encountered this problem while constructing a handwashing risk assessment (Montville et al., 2001). When trying to model the effects of antimicrobial

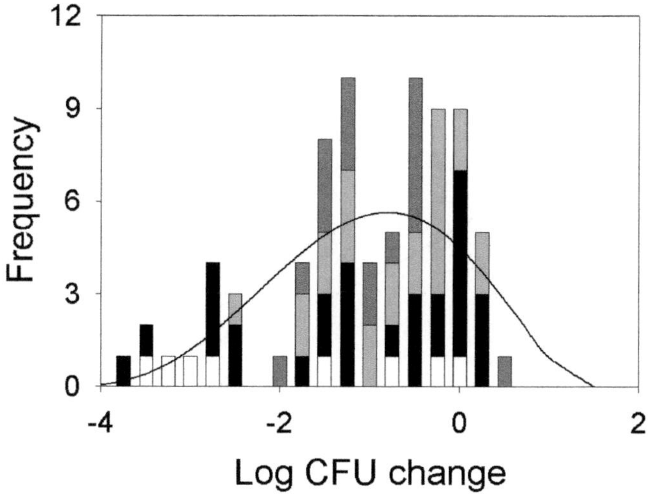

FIGURE 8.4 Efficacy of antimicrobial soaps based on literature data. Bars represent frequency of log change in bacterial count on the hand: soap containing triclosan (light gray), PCMX (black), iodophor (white), and other antimicrobial compounds (dark gray). Solid line represents the beta distribution fit to the combined data (bars). Adapted from Montville et al., *Int. J. Food Microbiol.*, 73(2–3):305–313, 2002.

soap on bacterial counts on hands, soaps with chlorohexadine gluconate (CHG) were clearly more effective than other kinds fof soaps. However, further discussion (Vashon, 1999, personal communication) indicated that this study contained a substantial flaw: the researchers had not quenched the CHG properly and active CHG was carried along with the bacteria onto agar plates, potentially inhibiting growth of cells that had survived the handwashing procedure. By using different PDFs, the effect of CHG soaps was coded separately from those of regular soaps and antimicrobial soaps in the handwashing QRA.

Biological Variability

The greatest source of variability may be biological variability. A risk assessment for *L. monocytogenes* has to be flexible enough to reflect the differing virulence and resistance of individual strains of the pathogen (Begot et al., 1997). At low inoculum sizes (i.e., low levels of contamination), microbes behave differently than they do in large populations (Kaprelyants and Kell, 1996). Inoculum size significantly affects the lag time for *C. botulinum* growth (Zhao et al., 2000). Figure 8.5 shows that decreasing spore inoculum size increases the lag time observed under various experimental conditions. Similarly, the influence of inoculum size on the mean as well as the variability of time-to-spoilage (TTS) was observed for *Bacillus stearothermophilus* spores in culture media; experimental data and computer simulations show maximum TTS variability at 1 CFU/ml and minimum variability at 500 CFU/ml (Llaudes, 1999).

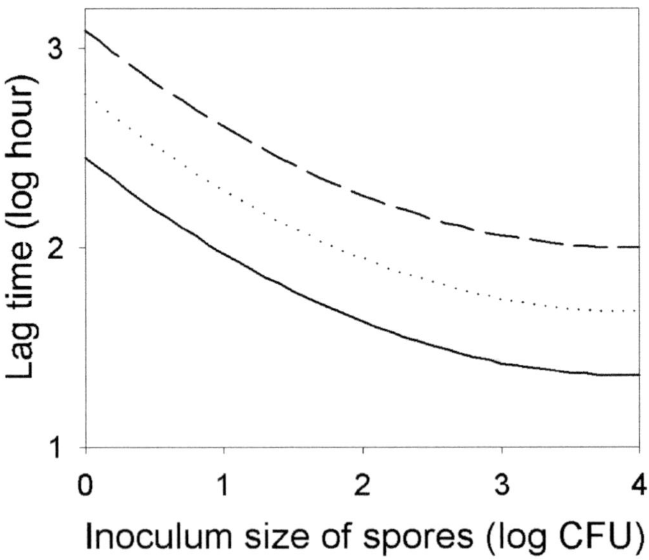

FIGURE 8.5 Effect of inoculum size on *C. botulinum* lag time in culture media at pH 6.0 with varying salt concentrations: 0% NaCl (solid line), 2% NaCl (dotted line), and 4% NaCl (dashed line). Adapted from Zhao et al., *J. Food Sci.*, 65(8):1369–1375, 2001.

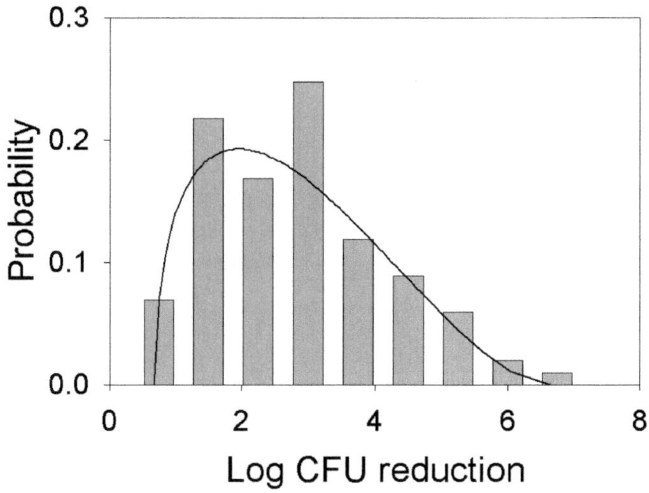

FIGURE 8.6 The pressure resistance of 101 strains of *Listeria monocytogenes*. The bars represent the data from Smelt and Hellemons, in *Fresh Novel Foods by High Pressure*, Autio, K., ed., Helsinki: Technical Research Center of Finland, 27–38,1998, and the solid line is the beta distribution (1.4275, 2.4601, 0.71215, 6.5053) fit to the data set.

Another example of inherent biological variability is the resistance of bacteria to processing. The sensitivity of 101 strains of *L. monocytogenes* to identical pressure treatments is shown in Figure 8.6 (Smelt et al., 1998). A wide range of pressure resistance is evidenced (ranging from less than a 1-log reduction in *L. monocytogenes* to greater than a 6-log reduction), and the distribution of results can be fit to a beta or Weibull distribution with an excellent fit (RMS error ~ 0.0015). The beta distribution (1.4275, 2.5601) bounded at 0.71215 and 6.5053 is shown as the solid line superimposed onto the data set in Figure 8.6. Unfortunately, similar data sets do not exist for most food-borne pathogens and bactericidal processes. Hopefully, increased awareness of QRA and its importance will encourage researchers to produce similar quality data for other bacteria.

The complexity of biological variability is illustrated in the fact that the most difficult data to obtain are adequate dose-response curves in humans. Besides other considerations such as issues regarding the extrapolation of animal data to describe human susceptibility (Haas and Thayyar-Madabusi, 1999), a dose-response curve must incorporate the variability of the pathogen (virulence, acid sensitivity), the properties of the food matrix (protective effect of food components), and the variability of the consumer (gastric pH, immune system strength) (Buchanan et al., 1998).

Process Variability

Process variability can be understood well through experimentation, but other barriers exist for risk assessors in this respect. Just as the variability associated with thermal inactivation of *C. botulinum* in canned foods has not been reported, virtually no published data show a range of efficacy of a given processing step. In

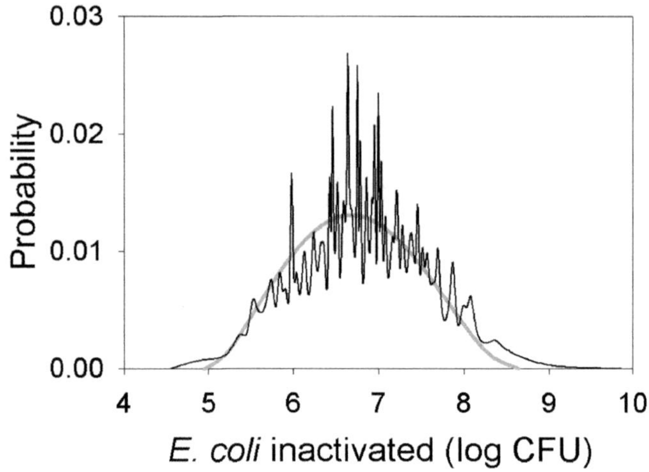

FIGURE 8.7 The simulation of the efficacy of ultraviolet light irradiation on *Escherichia coli* ATCC 29522 in apple cider. The gray, smooth line is the result of using the mean efficacy of all UV tubes in the simulation, and the jagged, black line combines the mean and variability in the simulation. Adapted from Duffy et al., *J. Food Prot.*, 63(11):1587–1590, 2000.

the authors' lab, nonthermal ultraviolet (UV) light inactivation of *E. coli* in apple cider was modeled through access to raw data from Dr. Randy Worobo's validation studies of the *CiderSure* UV equipment, which has gained popularity in the cider industry (Duffy et al., 2000). The Worobo lab tested each UV light tube three times to ensure that, each time, it achieved more than a 5-log CFU reduction of *E. coli* ATCC 25922 (a nonpathogenic surrogate with similar UV sensitivity to *E. coli* O157:H7) in apple cider.

Through analysis, the authors were able to separate the variability of the process from the mean log CFU reduction of each tube and modeled the two in separate probability distributions. Figure 8.7 shows the possible log CFU reductions of *E. coli* predicted by using the mean reductions of the UV unit as a dotted line; the simulation of combining the mean reduction with the variability distribution (shown in Figure 8.2) is shown as the solid line. Figure 8.7 shows that the range of possible log CFU inactivation due to UV light increased by less than 1 log CFU on either side if the variability of the process was included in the simulation (solid line). Although all of the tubes used in this analysis had never caused a lower than 5-log CFU reduction of *E. coli* in the trial tests, these simulations show a small failure rate of 0.1% when considering the variability around the mean.

The authors' modeling of UV inactivation is one of the best-described processes in their QRA for *E. coli* O157:H7 in apple cider (Duffy et al., 1999) and shows another benefit of the risk assessment approach to food processing: it is always better to know the variability associated with a process. A highly reliable process that causes a lower average inactivation of bacteria may assure greater food safety than a widely variable process with a higher mean inactivation of bacteria. It is difficult to assess the risk posed by a particular pathogen in a given food without knowledge of the variability of inactivation processes.

The issues of biological and method variability were combined in a component of the authors' QRA for *E. coli* O157:H7 in apple cider (Duffy et al., 1999, 2001). Seven different laboratories have published studies with widely varying results concerning the survival of *E. coli* O157:H7 in cider, using a total of nine strains in apple ciders that were locally available to each laboratory (Duffy et al., 2001). A variety of agars were used for enumeration, including selective sorbitol McConkey agar (SMAC), which is known to produce lower counts of *E. coli* O157:H7 from acidic environments vs. plating on recovery media (Silk et al., 1997). In addition to the variability caused by the use of different agars, each strain of *E. coli* O157:H7 behaves differently in cider — the behavior of strains 43889 and 43895 differed greatly in one paper (Garland-Miller and Kaspar, 1994) — and a single strain may respond differently to cider made from varying apple cultivars (Dingman, 2000).

This biological variability of strain response to cider and the bactericidal or bacteristatic properties of different apple ciders should be incorporated into a QRA because a range of acid resistance in the *E. coli* O157:H7 that contaminates apple cider, and in the combination of apple cultivars used in a batch of cider, may exist. Method-to-method variability described above is less desirable in this risk assessment. Since several papers on survival of *E. coli* O157:H7 were published before it was known that acid-stressed *E. coli* did not survive or grow as well on selective agars as on nonselective agars, a substantial portion of this research consistently underestimates the amount of surviving *E. coli* O157:H7. This range of results was incorporated into the QRA because no conversion factors are available to accurately transform counts of acid-stressed *E. coli* O157:H7 on SMAC agar to the higher counts observed on recovery media. The resulting distribution contains a great deal of variability and uncertainty.

PREDICTIVE FOOD MICROBIOLOGY AND QRA

Quantitative microbial risk assessment has been shaped by previous research in predictive microbiology and quantitative chemical risk assessment. The EPA and private organizations have been advancing QRA from deterministic models to stochastic Monte Carlo risk assessments for chemical hazards over the last 30 years (Washburn et al., 1998; Suter, 1999). Although these chemical models inspired microbial QRA, they are less dynamic than microbial risk assessments because of bacteria's ability to proliferate as well as adapt and develop resistance to inactivation processes (Buchanan et al., 1998). On the other hand, microbial QRAs do not have to account for cumulative effects of exposure over time — a single dose of bacteria can cause illness. Nevertheless, the advances achieved by chemical QRA present food microbiologists with a useful toolkit to assess and manage risk of food-borne illness.

Predictive food microbiology has been primarily interested in creating kinetic models to describe the growth and inactivation of microbes. A stochastic QRA approach requires PDFs to be more effective than a simple deterministic risk estimate. Nonetheless, predictive models are still useful and included as part of QRA (van Gerwen and Zwietering, 1998), and many predictive modelers are developing probability distribution models as a complement to kinetic modeling of bacterial growth and inactivation (Zhao et al., 2000).

Though PDFs are the most useful mathematical descriptions for data in a QRA, when there are too few data points, traditional polynomial regression can summarize tabular data in a form more easily coded into a QRA. In the apple cider QRA, the authors chose to model the effect of a short-term freeze–thaw cycle with a polynomial model. Uljas and Ingham at the University of Wisconsin studied short-term freezing as a means of causing a 5-log CFU reduction in apple cider. They have studied the effects of holding cider inoculated with *E. coli* O157:H7 at different pH values for 12 h or fewer at various temperatures and then subjecting them to a 24-h freeze and finally a thaw (Uljas and Ingham, 1999). Their results were published in a table, giving one value for each set of conditions and no variability (i.e., a standard deviation) around that value. Apparently, no other labs had performed comparable experiments, so Uljas and Ingham's data could not be combined with data from other papers in order to assess variability. The authors performed a polynomial regression on this sparse data set with SAS (Jandel Corp, Cary, NC) and fit a model with an R^2 of 0.8914. The response-surface curve shown in Figure 8.8 shows the agreement of the polynomial model with the data set but also shows the lack of variability in this description; for any given cider pH, hold time, and hold temperature, only one value is produced by their equation.

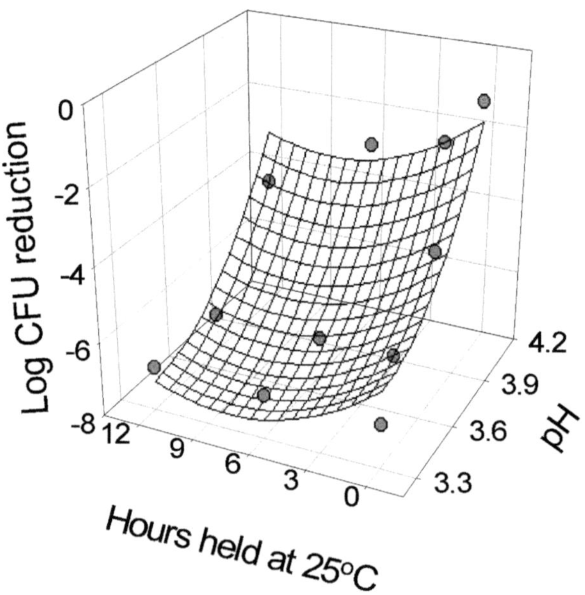

FIGURE 8.8 The polynomial regression model for short-term freeze–thaw reduction of *Escherichia coli* O157:H7 in apple cider held at 25°C prior to freezing (mesh lines). The circles represent the 25°C data from Uljas and Ingham, *Appl. Environ. Microbiol.*, 65(5):1924–1929, 1999.

QRA SOFTWARE

Computer simulation is the factor that has allowed QRA to progress from deterministic point estimates of risk to producing distributions to describe risk. The software programs created to run these simulations were developed primarily for business, not science. The first programs used for microbial QRA in food were Crystal Ball (Decisioneering, London, U.K.) and @Risk (Palisade Corp., Newfield, NY). Both are "add-ins" to spreadsheet programs like Microsoft Excel (Redmond, WA) and Lotus 1–2–3 (Cambridge, MA). Recently, stand-alone simulation programs such as Analytica (Lumina, Los Gatos, CA) have been written and are gaining popularity with microbial food risk assessors. Analytica uses a flowchart-like interface more user friendly than that of the spreadsheet programs and is consequently more understandable to those less familiar with QRA.

However, an advantage of @Risk over Analytica is the integration of @Risk with the software program BestFit, also made by Palisade Corp. BestFit is a distribution-fitting program that ranks the ability of 20 continuous and several discrete PDFs to describe a given data set. Because Palisade Corp. makes both programs, the distribution parameters are the same and in the same order in BestFit and @Risk. The conversion of BestFit parameters to meet the requirements of other programs, such as Analytica, may require that Analytica be programmed to recognize a new distribution or rearranging the order of the parameters. BestFit is commonly used to create PDFs in microbial QRA, though the best statistical fit to a given data set may not always be the most reflective of the data. There is no substitute for good judgment on the part of the risk assessor in choosing an appropriate probability distribution function.

The simulation software mentioned allows linkage into a model for all PDFs describing the concentration and prevalence of a given pathogen in the raw product, the distributions for postharvest processing, storage, and other interventions, as well as dose-response PDFs. The software can then simulate thousands of iterations of the food process using different values chosen from each PDF in each iteration. Monte Carlo simulation is often used with Latin Hypercube sampling, which forces the simulation program to sample from the tails of the distribution more often than the program would otherwise do, in order to define the tails and range of results more quickly (Vose, 2000).

The software can also select the factors that have the greatest impact on the final risk, a procedure known as "sensitivity analysis," by observing how much one distribution's tails impact the result of the simulation when all other factors are held at their median values (Cassin et al., 1998b). A sensitivity analysis, performed to determine the correlation between 15 factors (input variables) on the risk outcome (probability of illness caused by *E. coli* O157:H7) resulting from ground beef hamburger consumption, revealed that the risk was most sensitive to the level of the pathogen carried by the cow, followed by host susceptibility (Cassin et al., 1998a). In the authors' handwashing QRA (Montville et al., 2001), sensitivity analysis revealed that the primary factors influencing the final bacterial counts on the hand were sanitizer use, soap use, and hand-drying method.

Sensitivity analysis can be invaluable when creating HACCP plans for food industries. The QRA ranks the importance of bactericidal processes and other factors for help in identifying critical control points (Buchanan and Whiting, 1998). Risk managers can modify a QRA to assess the effects of additional processing to inactivate microorganisms in the food before expensive new equipment is purchased or new regulations are passed. The QRA provides managers with a new tool to make decisions more firmly based on science (Morales and McDowell, 1998).

Though this stochastic simulation is considered preferable to predicting risk by multiplying 95th-percentile values for all the factors in the farm to fork risk assessment (Zwietering and van Gerwen, 2000), Monte Carlo risk assessment is not without its detractors. Most notably, it has been suggested that Monte Carlo simulation overestimates the uncertainty and variability in the QRA and causes an overly wide range of results (Cox, 1999). QRA is an emerging technique, not only in food microbiology but also as a risk assessment tool in other fields. Currently, most microbial QRAs are empirical, though more mechanistic models could be developed as research improves to better assess the risks of food-borne illness (Buchanan et al., 2000). Future improvements in simulation software or statistical techniques may supercede the techniques presented here.

SPECIAL CONSIDERATIONS

One of the legacies of food microbiology is the notion that multiple preservation factors can act synergistically to inhibit microbial growth. This is the same understanding that drives hurdle technology, which is an important concept for MPF (Leistner and Gorris, 1995). It is possible to model the effects of multiple factors in a QRA through predictive models and through algebraic links between nodes. In a series of food processing steps, for example, the concentration and prevalence of pathogens after each processing step (i.e., washing) get passed in the program to the next PDF for a different process (i.e., sanitizing). When the processing is carried out in a series, it is difficult to discern what effect, if any, a previous bactericidal step has on another processing step. The simplest assumption is to add the effects of the processing steps together in the QRA.

On the other hand, a sanitizing rinse could weaken surviving bacteria and cause them to be more susceptible to subsequent heating, making the effects synergistic. It could also be possible that a washing step would remove a majority of the bacteria that would have perished from the following sanitizing rinse, thus reducing the apparent efficacy of the sanitizer. If a single laboratory examined the entire processing of a product, then the data would be optimal for a QRA. However, when constructing a literature-based QRA, different processing steps of the whole process are often studied separately, with different labs looking at different steps. This proves problematic for the risk assessor.

If a washing step removes 1.5 log CFU of a pathogen from a fruit in one study, and a sanitizing rinse inactivates 2 log CFU of a pathogen from the same fruit in another study that did not account for the effect of prior washing, how can a risk assessor be assured that the two steps cause a combined 3.5 log CFU reduction in the pathogenic bacteria? Because of the lack of farm-to-fork studies that follow the levels of contamination of a food through every step of processing, most QRAs will be left with the caution that they may well be overestimating the efficacy of processing and therefore the safety of the food. Validation of QRA models thus becomes important. QRA was developed as an alternative to the "compounding conservatism" of multiplicative risk estimates (i.e., multiplying 95th percentiles for all factors), but is QRA "compounding liberalism"? Future development and research in QRA will tell.

Finally, it cannot be overstressed that a risk assessment is only as good as the data collected, the distributions chosen to fit the data, and assumptions made to compensate for lack of data (Petersen, 2000; Zwietering and van Gerwen, 2000). The majority of the food microbiology literature is concerned with average bacterial counts and presence or absence data — not with modeling or describing the range of results. Because MPFs are becoming more popular and, presumably, research interest in MPF will also increase, there is a unique opportunity for microbiologists to extend their research or perhaps just improve their reporting of results to accommodate the requirements of QRA. HACCP has been and will continue to be very important to minimally processed foods (Willocx et al., 1994), and increased quantitative research for QRA is the logical step to making HACCP plans more science based in the future (Schaffner, 1999). As risk assessors construct QRA for MPFs, they will find and illustrate the most critical data gaps; MPF researchers should respond to those needs to help assure food safety. Besides the development and validation of QRA for bacterial pathogens in MPFs, opportunity exists for other food-borne disease agents such as viruses (Gerba et al., 1996) and parasites that present significant risk for MPF products due to cross-contamination from food handlers.

CONCLUSION

Quantitative microbial risk assessment is a developing and emerging technique in food safety microbiology. It is especially useful for predicting food-borne illness associated with foods that are not heavily thermally processed and, thus, pose the need to assure their safety through HACCP. The benefits of a QRA range far beyond the resulting prediction of food-borne illness; QRA can draw attention to aspects of food processing that require further research and also determine which steps in processing are truly critical to the safety of the food. QRA has become important in regulatory decisions made by USDA, FDA, and international agencies (Buchanan et al., 1998; Klapwijk et al., 2000) and will aid future decisions regarding the safety of minimally processed foods in the age of emerging food pathogens.

ACKNOWLEDGMENTS

The authors would like to thank Kristin M. Jackson and Rebecca Montville for their contributions to this manuscript and valuable editorial assistance. The authors would also like to thank the New Jersey Agricultural Experiment Station for funding the Food Risk Analysis Initiative.

REFERENCES

Anonymous, 1996, *BestFit User's Guide*, Newfield, NY: Palisade Corporation, 2–18.

Anonymous, 2001, Hazard analysis and critical control point (HACCP); procedures for the safe and sanitary processing and importing of juice; final rule, *Federal Register*, 66(13):6138–6202.

Baker, A.R. et al., 1998, *Salmonella enteritidis* risk assessment — shell eggs and egg products, http://www.fsis.usda.gov/ophs/risk/index.htm.

Begot, C., Lebert, I., and Lebert, A., 1997, Variability of the response of 66 *Listeria monocytogenes* and *Listeria innocua* strains to different growth conditions, *Food Microbiol.*, 14(5):403–412.

Betts, G.D. and Gaze, J.E., 1995, Growth and heat resistance of psychrotrophic *Clostridium botulinum* in relation to *sous vide* products. *Food Control*, 6:57–63.

Bohrer, C.W., Denny, C.B., and Yao, M.G., 1973, Thermal destruction of type E *Clostridium botulinum*, Final report on RF 4603, National Canners Association Research Foundation, Washington, D.C.

Buchanan, R.L., 1997, Principles of risk assessment for illnesses caused by foodborne biological agents, *J. Food Pro.*, 60(11):1417–1419.

Buchanan, R.L. and Doyle, M.P., 1997, Foodborne disease significance of *Escherichia coli* O157:H7 and other enterohemorrhagic *E. coli*. *Food Technol.*, 51(10):69–76.

Buchanan, R.L. et al., 1998, Potential application of risk assessment techniques to microbiological issues related to international trade in food and food products, *J. Food Prot.*, 61(8):1075–1086.

Buchanan, R.L., Smith, J.L., and Long, W., 2000, Microbial risk assessment: dose-response relations and risk characterization, *Int. J. Food Microbiol.*, 58(3):159–172.

Buchanan, R.L. and Whiting, R.C., 1998, Risk assessment: a means for linking HACCP plans and public health, *J. Food Prot.*, 61(11):1531–1534.

Cassin, M.H. et al., 1998a, Quantitative risk assessment for *Escherichia coli* O157:H7 in ground beef hamburgers, *Int. J. Food Microbiol.*, 41(1):21–44.

Cassin, M.H., Paoli, G.M., and Lammerding, A.M., 1998b, Simulation modeling for microbial risk assessment, *J. Food Prot.*, 61(11):1560–1566.

Charm, S.E., 1966, On the margin of safety in canned foods, *Food Technol.*, 20:97.

Chen, Y. et al., 2001, Quantification and variability analysis of bacterial cross contamination rates in common foodservice tasks, *J. Food Prot.*, 64(1):72–80.

Cox, L.A., 1999, Internal dose, uncertainty analysis, and complexity of risk models, *Environ. Int.*, 25(6–7):841–852.

Crisley, F.D. et al., 1968, Thermal resistance of spores of five strains of *Clostridium botulinum* type E in ground whitefish chubs, *J. Food Sci.*, 33:411–416.

Dingman, D.W., 2000, Growth of *Escherichia coli* O157:H7 in bruised apple (*Malus domestica*) tissue as influenced by cultivar, date of harvest, and source, *Appl. Environ. Microbiol.*, 66(3):1077–1083.

Dodds, K.L., 1993, *Clostridium botulinum* in foods, in *Clostridium botulinum Ecology and Control in Foods*, Hauschild, A.H.W. and Dodds, K.L., Eds., New York: Marcel Dekker, Inc., 53–68.

Duffy, S. et al., 2000, Analysis and modeling of the variability associated with UV inactivation of *Escherichia coli* in apple cider, *J. Food Prot.*, 63(11):1587–1590.

Duffy, S. and Schaffner, D.W., 1999, A quantitative assessment of the risk of *Escherichia coli* O157:H7 in apple cider, IAMFES Annual Meeting, August 1–4, Dearborn, MI.

Duffy, S. and Schaffner, D.W., 2001, Modeling the survival of *Escherichia coli* O157:H7 in apple cider using probability distribution functions for quantitative risk assessment, *J. Food Prot.*, 64(5):599–605.

Farber, J.M., Ross, W.H., and Harwig, J., 1996, Health risk assessment of *Listeria monocytogenes* in Canada, *Int. J. Food Microbiol.*, 30(1–2):145–156.

Fazil, A.M., 1999, A quantitative risk assessment model for *C. jejuni* in chicken, IAMFES Annual Meeting, August 1–4, Dearborn, MI.

Fernandez, A. et al., 1999, Application of a frequency distribution model to describe the thermal inactivation of two strains of *Bacillus cereus*, *Trends Food Sci. Technol.*, 10(4–5):158–162.

Francis, G.A., Thomas, C., and O'Beirne, D., 1999, The microbiological safety of minimally processed vegetables, *Int. J. Food Sci. Technol.*, 34(1):1–22.

Garland-Miller, L. and Kaspar, C.W., 1994, *Escherichia coli* O157:H7 acid tolerance and survival in apple cider, *J. Food Prot.*, 57(6):460–464.

Gerba, C.P. et al., 1996, Waterborne rotavirus: a risk assessment, *Water Res.*, 30(12):2929–2940.

Haas, C.N. and Thayyar-Madabusi, A., 1999, Development and validation of dose-response relationship for *Listeria monocytogenes*, *Quant. Microbiol.*, 1:89–102.

Holcomb, D.L. et al., 1999, Comparison of six dose-response models for use with food-borne pathogens, *Risk Anal.*, 19(6):1091–1100.

Jarvis, B., 1989, *Statistical Aspects of the Microbiological Analysis of Foods*, New York, NY: Elsevier.

Jolis, D., Pitt, P., and Hirano, R., 1999, Risk assessment for *Cryptosporidium parvum* in reclaimed water, *Water Res.*, 33(13):3051–3055.

Kaprelyants, A.S. and Kell, D.B., 1996, Do bacteria need to communicate with each other for growth?, *Trends Microbiol.*, 4(6):237–242.

Klapwijk, P.M., Jouve, J.L., and Stringer, M.F., 2000, Microbiological risk assessment in Europe: the next decade, *Int. J. Food Microbiol.*, 58(3):223–230.

Leistner, L. and Gorris, L.G.M., 1995, Food preservation by hurdle technology, *Trends Food Sci. Technol.*, 6(2):41–46.

Licciardello, J.J., 1983, Botulism and heat-processed seafoods, *Mar. Fish. Re.*, 45(2):1–7.

Lindqvist, R. and Westoo, A., 2000, Quantitative risk assessment for *Listeria monocytogenes* in smoked or gravad salmon and rainbow trout in Sweden, *Int. J. Food Microbiol.*, 58(3):181–196.

Llaudes, M., 1999, Simulation of the effect of inoculum size on spoilage of a model system by *Bacillus stearothermophilus*, M.S. Thesis, Rutgers, The State University of New Jersey, New Brunswick.

Lynt, R.K., Kautter, D.A., and Solomon, H.M., 1983, Effect of delayed germination by heat-damaged spores on estimates of heat resistance of *Clostridium botulinum* type E and F, *J. Food Sci.*, 48:226–229.

Lynt, R.K. et al., 1977, Thermal death time of *Clostridium botulinum* type E in meat of the blue crab, *J. Food Sci.*, 42:1022–1025.

Marks, H. and Coleman, M., 1998, Estimating distributions of numbers of organisms in food products, *J. Food Prot.*, 61(11):1535–1540.

Montville, R., Chen, Y., and Schaffner, D.W., 2002, Risk assessment of handwashing efficacy using literature and experimental data, *Int. J.Food Microbiol.*, 73(2–3):305–313.

Morales, R.A. and McDowell, R.M., 1998, Risk assessment and economic analysis for managing risks to human health from pathogenic microorganisms in the food supply, *J. Food Prot.*, 61(11):1567–1570.

Odumeru, J.A. et al., 1997, assessment of the microbiological quality of ready-to-use vegetables for health-care food services, *J. Food Prot.*, 60(8):954–960.

Pace, P.J., Krumbiegel, E.R., and Wisniewski, H.J., 1972, Interrelationship of heat and relative humidity in the destruction of *Clostridium botulinum* type E spores on whitefish chubs, *Appl. Microbiol.*, 23:750–757.

Pargas, N., 1998, Risk assessment increasingly important in USDA decision process, USDA economist reports, *Food Chem. News*, 40(17):22–24.

Peleg, M. and Cole, M.B., 2000, Estimating the survival of *Clostridium botulinum* spores during heat treatments, *J. Food Prot.*, 63(2):190–195.

Petersen, B.J., 2000, Probabilistic modeling: theory and practice, *Food Additives Contamination*, 17(7):591–599.

Rhodehamel, E.J. et al., 1991 incidence and heat resistance of *Clostridium botulinum* type E spores in menhaden surimi, *J. Food Sci.*, 56:1562–1563.

Schaffner, D.W., 1999, Understanding food safety risks through quantitative risk assessment, *J. Assoc. Food Drug Of.*, 63(1):8–15.

Silk, T.M., Ryser, E.T., and Donnelly, C.W., 1997, Comparison of methods for determining coliform and *Escherichia coli* levels in apple cider, *J. Food Prot.*, 60(11):1302–1305.

Sloan, A.E., 1995, Fresh cut gets fresh, *Food Technol.*, 49(5:):38–40.

Smelt, J.P.P.M. and Hellemons, J.C., 1998, High pressure treatment in relation to quantitative risk assessment, in *Fresh Novel Foods by High Pressure*, Autio, K., Ed., Helsinki: Technical Research Center of Finland, 27–38.

Suter, G.W., 1999, Developing conceptual models for complex ecological risk assessments, *Hum. Ecol. Risk Assessment*, 5(2):375–396.

Uljas, H.E. and Ingham, S.C., 1999, Combinations of intervention treatments resulting in 5-log unit reductions in numbers of *Escherichia coli* O157:H7 and *Salmonella typhimurium* DT104 organisms in apple cider, *Appl. Environ. Microbiol.*, 65(5):1924–1929.

van Gerwen, S.J.C. and Zwietering, M.H., 1998, Growth and inactivation models to be used in quantitative risk assessments, *J. Food Prot.*, 61(11):1541–1549.

Vose, D., 1996, *Quantitative Risk Analysis: A Guide to Monte Carlo Simulation Modelling*, Chichester, U.K.: John Wiley & Sons.

Vose, D., 2000, *Risk Analysis: A Quantitative Guide*, Chichester, U.K.: John Wiley & Sons.

Washburn, S.T., Kleiman, C.F., and Arsnow, D.E., 1998, Applying USEPA risk assessment guidance in the 90s, *Hum. Ecol. Risk Assessment*, 4(3):763–774.

Weibull, W., 1951, A statistical distribution function of wide applicability. *J. Appl. Mech.*, 187:293–297.

Whiting, R.C. and Buchanan, R.L., 1997, Development of a quantitative risk assessment model for *Salmonella enteritidis* in pasteurized liquid eggs, *Int. J. Food Microbiol.*, 36(2–3):111–125.

Willocx, F., Tobback, P., and Hendrickx, M., 1994, Microbial safety assurance of minimally processed vegetables by implementation of the hazard analysis critical control point (HACCP) system, *Acta Aliment.*, 23:221–238.

Zhao, L., Montville, T.J., and Schaffner, D.W., 2000, Inoculum size of *Clostridium botulinum* 56A spores influence time-to-detection and percent growth-positive sample, *J. Food Sci.*, accepted.

Zwietering, M.H. and van Gerwen, S.J.C., 2000, Sensitivity analysis in quantitative microbial risk assessment, *Int. J. Food Microbiol.*, 58(3):213–221.

Section III

Current and Future Innovations

9 Microbial Safety during Nonthermal Preservation of Foods

Gaurav Tewari

CONTENTS

NONTHERMAL PROCESSING

Preservation techniques such as salting, pickling, and smoking originated in various ancient civilizations as far back as 3000 B.C. Currently, most liquid foods are preserved commercially by ultra-high temperature (UHT) or high-temperature short-time (HTST) processes. Although heating inactivates enzymes and microorganisms, the organoleptic and nutritional properties of the food suffer because of protein denaturation and the loss of vitamins and volatile flavors. Thus, extending the shelf-life of food by heat treatment is not only energy intensive but in most cases also adversely affects the flavor, chemical composition, and nutritional quality of the preserved food. There is a great need for a nonthermal method for inactivating microorganisms that is economical, compact, energy efficient, safe, and socially and environmentally acceptable and does not adversely affect nutrition, texture,

1-58716-041-2/03/$0.00+$1.50

and flavor of the treated food. The consumer demand for and awareness of quality foods is also increasing. The following sections detail nonthermal food preservation, with major emphasis on technologies that have been recently commercialized or are close to being commercialized, such as pulsed electric field and hydrostatic high-pressure processing.

IRRADIATION

Food irradiation is not new; however, it is not used widely due to nontechnical reasons. The reader is referred to the chapters on food irradiation for a comprehensive analysis not covered in this chapter. The use of radiation for processing foods has been approved by FAO/WHO for more than 50 types of foods since 1981 (Mertens and Knorr, 1992). This technology uses ionizing radiation for inactivation of microorganisms and pests (Thayer, 1990, 1994). Cobalt-60, x-rays, or accelerated electrons are some commonly used sources to produce ionizing radiation. A number of researchers have studied the changes induced in foods after irradiation and have uniformly agreed that radiation-related changes are minimal (Thayer, 1990; Becker, 1983; Thayer et al., 1991). Vegetative organisms could be easily destroyed by irradiation; however, additional data are needed on the effect of ionizing radiation on viruses and spore-forming microorganisms, especially pathogens.

Sterilization of packaging materials during aseptic packaging has been done by ionizing radiation for many years. However, public perception has been the main reason for lack of acceptance of irradiated foods. Irradiation has a tremendous amount of potential for use, not only from a public health safety point of view but also from consumers' demand for no nutrient degradation as a result of any pathogen reduction process. Ionizing radiation could be more suitable for reducing pathogenic vegetative organisms such as *Salmonella*, *Listeria monocyotegenes*, and *Escherichia coli* O157:H7 in meat and poultry products. However, more work is required to find its suitability for inactivating spore formers and viruses.

In addition, there is debate about the availability of suitable polymers for irradiated packages, because currently only a few polymers are approved in the U.S. Code of Federal Regulation (CFR) Title 21§149.5 for use with irradiated foods (Tewari and Jayas, 1999a). Among these, only EVA (ethylene vinyl acetate) can receive electronic beam treatment. Most polymers listed in this section were approved in the 1960s, and it is very difficult to make a contemporary multilayer structure with currently approved polymers. As per radiation chemistry, electron beam, gamma, and x-ray treatments should have the same effect on polymer packaging materials; however, they are treated differently in the CFR because initial requests only mentioned gamma treatment (Tewari et al., 1999a). Therefore, coming up with a multilayer film for irradiation is another challenging area.

PULSED ENERGY PROCESSING

Pulsed energy processing is based on the concept of applying any energy (electric, magnetic, or light) that has been stored for a long amount of time, in a very short amount of time. This results in huge power generation and causes microbial death. Pulsed energy has been applied in three forms to food, namely, pulsed electric field (PEF), pulsed light (PL), and pulsed magnetic field (PMF).

Pulsed Electric Field (PEF)

Recently, researchers and the food industry in North America, Japan, and Europe have shown a great interest in the pulsed electrical field (PEF). Due to its nonthermal and chemically free nature, this process has a future; however, more systematic research work is needed to make this process commercially feasible. Mittal (1993) reviewed earlier work on this area when proper equipment for the technology was the main problem. Castro et al. (1993) reviewed the application of PEF, mainly based on the work conducted at Washington State University and PurePulse®, San Diego. Knorr et al. (1994) provided a general review, including microbial decay, plant cell rupture, and juice extraction. Zhang et al. (1992) reviewed engineering aspects, and Qin et al. (1995) emphasized equipment development and shelf-life studies. Recently, Ho and Mittal (2000) provided a comprehensive review on this technology.

Pulsed power refers to the general technology of accumulating energy on a relatively long time scale (pulse charging, slow systems) and then compressing that energy in time and space to deliver large power pulses (pulse discharging, high-speed systems) to a desired load. Orientation and assembly of some high voltage pulse generators and treatment chambers for batch processing are summarized by Ho and Mittal (2000). A typical batch treatment system consists of a high-voltage pulse generator and a treatment chamber. Other auxiliary devices may also be used for degassing, vacuuming, preheating, and cooling the treatment medium. A unique low-energy pulsed electric field treatment system, developed by Ho et al. (1995), consists of a 30-kV direct current (DC) high-voltage pulse generator and a circular treatment chamber. Figure 9.1 shows the block diagram of the design unit.

The 110-V alternating current (AC) is raised in voltage through a high capacitor through a 6 Mμ resistor. The generation of high-voltage pulses (exponential decay)

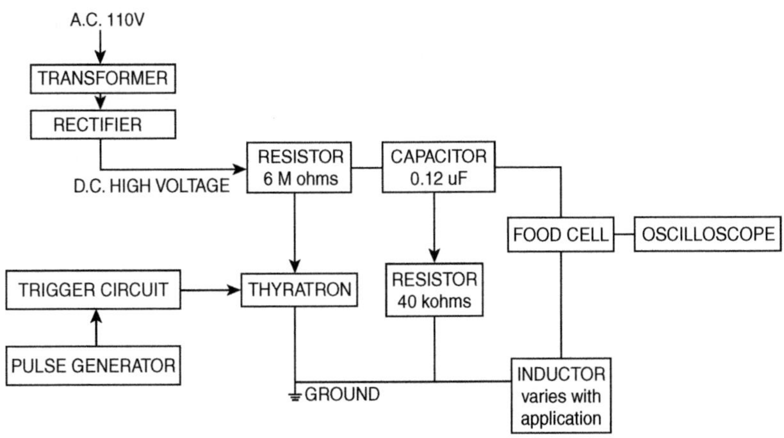

FIGURE 9.1 Low-energy pulsed electric field treatment system.

relies on the discharge of the 0.12°F capacitor through the thyratron. The batch unit can generate short duration pulses (2-µsec width) with electric field strength up to 100 kV/cm. The circular treatment chamber (25.0-cm diameter) has two circular and parallel stainless steel electrodes (16.5-cm diameter). The insulation, Delrin, was constructed to have close physical contact with the electrodes. The distance between the electrodes can be adjusted by inserting Delrin circular plates (14.5-cm diameter) with thickness 0.3, 0.6, or 0.9 cm. Thus, the process volume can be varied between 49.5, 99.1, and 148.6 mL, respectively.

The effects of these pulses on microbial inactivation in an aqueous solution were studied under different operating conditions (electric field strength, pulse period, and number of pulses applied) and different fluid properties (electrical conductivity, density, and rheological characteristics). Electric field strength at 10 kV/cm for 10 pulses (2-sec pulse period and 2-µsec pulse width) was found to deliver significant microbial inactivation (Ho et al., 1995). *P. fluorescens* in various aqueous solutions were reduced in population by more than 6 log cycles. However, the critical electric field strength was proved to be affected by the nature of the pulse waveform across the treatment chamber.

Most batch treatment chambers consist of two parallel electrodes spaced apart to create the treatment volume. The electrodes are usually inserted in insulation material such as Teflon or Plexiglas for safety and, in some cases, provide an enclosed environment for the fluid under treatment. Configuration of the chamber and electrodes should minimize sparking or dielectric breakdown (the breakdown of the microbial cell membrane, not the food matrix, is desired). This can be achieved by developing a uniform electric field in the chamber, using a smooth electrode surface (bead-blasting), using round-edged electrodes (especially at the contact area of electrodes and insulation), and eliminating impurities such as air in the food.

The process parameters used for batch processing had a very wide range: DC voltage — 2.5 to 43 kV; electric field strength — 0.6 to 100 kV/cm; electrode distance — 3 to 77 mm; pulse width — 1 µs to 10 msec; pulse frequency — 0.2 to 50 Hz; number of applied pulses — 1 to 120; and process volume — 0.5 mL to 1.6 L (Ho and Mittal, 2000). Properties of the suspending media (electrical conductivity, pH, and compositions), process temperature, and microbiological conditions were sometimes not reported. In most studies, microorganisms (usually pure cultures of commonly found food spoilage or pathogenic microbes) were grown in nutrient media to a desired level and then washed and inoculated in the selected suspension fluid (usually buffer solutions or conventional food products) for pulse treatments. Studies on real or natural-state food products, e.g., raw milk or fresh pressed juice, are limited. The use of various pulse shapes and their energy levels has a large influence on the results, so it is difficult to compare various reported studies. Based on all the available studies, the microbial reduction rate was found to range from a moderate one to three log cycles to a significant six to nine log cycles and seemed to be a function of various process parameters, conditions, and procedures.

Continuous systems utilize the same operating principles, but the concept of flow dynamics must be implemented in the design process so that the food product can receive the necessary treatment conditions. Because the majority of researchers have been using small-size, batch mode operations, only a few studies describe the concept of a continuous system, and even fewer studies report experimental studies on them. For most research work, the high-voltage pulse generator employed was conceptually the same as the batch operation.

Researchers from Washington State University (Qin et al., 1994, 1995) described two chambers for continuous processing. The first one was a parallel plate chamber based on their batch chamber. The sample would flow through the horizontal test chamber in a series of U-shaped channels. Electrode gap was 0.51 or 0.95 cm, giving a volume of 8 or 20 mL. The operation parameters were a 35- to 70-kV/cm field, 2- to 15-μsec pulse width, 1-Hz pulse frequency, and a flow rate of 600 to 1200 mL/min. Circulation of water around the circular stainless steel electrodes through jackets was implemented for cooling purposes. The electric field strength and flow profile may be difficult to monitor with the use of U-shaped channels. The second chamber was a coaxial treatment chamber. The operation parameters were a 50- to 80-kV/cm field, 2- to 6-mm electrode gap along the chamber, 2- to 15-μsec pulse width, 1-Hz pulse frequency, and a flow rate of 2 to 10 L/min. Cooling jackets were attached to both electrodes. Researchers from Ohio State University also reported studies on continuous treatments using a so-called "co-field flow" model treatment chamber, where the flow direction of the liquid medium was parallel to the electric field (Qiu et al., 1998). Log cycle reductions of 3 to 4.2 were reported for naturally occurring microbes in orange juice.

Like batch operations, the microbial reduction rate under a continuous treatment process seemed to be a function of various process parameters, conditions, and procedures. Although most researchers are still focusing on the treatment chamber design, it seems that no information is available to describe mathematically the application of pulsed power in a continuous processing mode or the relationships between batch and continuous processing. Also, studies on real, natural-state food products, such as raw milk or fresh-pressed juice, are very limited.

The optimum field strength for maximum microbial control may be reliant on various components such as pulse shape, width, frequency, voltage, applied and electrode separation, configuration and orientation of the treatment chamber, physical properties of the fluid medium, and the type of microorganisms. At this point, most of the studies have been conducted using small-scale, batch mode, treatment systems with artificial or simulated treatment media and microorganisms. The process parameters, conditions, and procedures vary to such an extent that the information available is difficult to compare and standardize. Furthermore, only a few studies have treated real food products under a pilot-scale, continuous-flow operation. As a result, the actual number and the type of food products that can profit from pulsed power treatment are yet to be identified.

A PEF system consists of a set of electrodes, a pulse generator, a capacitor, a switch, and a small chamber for the food to be processed. The pulse generator charges the capacitor, which is then discharged. The resulting high electric field pulse is applied across the food sample, and the lethal effect of the pulse is a result of the induction of an electric potential across the cell membrane, causing a charge separation in the cell (Mertens and Knorr, 1992; Knorr et al., 1994). When the electrical potential exceeds the critical value by a large amount, the change in the cell membrane is irreversible, which leads to cell death. The critical electrical potential for vegetative cells is approximately 15 kV/cm; this would not inactivate spores (Mertens and Knorr, 1992). The major advantage of this technique is that the product temperature increase is minimal. However, no commercial success of this technology has been reported so far.

Pulsed Light (PL)

Pulsed light (PL) is produced using technologies that multiply power many-fold. In this process, power is magnified by storing energy in an energy storage capacitor over relatively long times (fractions of a second) and releasing this stored energy to do work in much shorter times (millionths or thousandths of a second) (Dunn et al., 1995). The result is very high power during the cycle and expenditure of only modest average power consumption. A PL system uses a pulsed power system to drive an inert gas lamp that produces intense broadband white light pulses. The intense flashes of light have a duration of 10^{-6} to 10^{-1}/sec and an energy density approximately 20,000 times that of sunlight on Earth's surface. This technology is very effective at the surface of food, in transparent foods, and on packaging materials.

Comparison of antimicrobial effects obtained using pulsed light with that of nonpulsed or continuous-wave conventional UV sources shows a significantly higher killing effect for pulsed light. Only one to a few pulsed light flashes kill high levels of all exposed organisms; several to tens of flashes can be applied per second, and high kill levels can be obtained in a fraction of a second (Dunn et al., 1995). Dunn and colleagues also showed that PL is effective against vegetative cells, spores, and molds. The most resistant microorganism and spore former were *Aspergillus niger* and *Bacillus pumilus*, respectively. Microbiological data on spore-forming pathogens is needed before this technology can be used commercially for low-acid foods. Until then, it can be used for acid, acidified, and refrigerated foods. Again, no commercial success of this technology has yet been reported.

Pulsed Magnetic Field (PMF)

Similar to PEF, alternate magnetic field pulses (PMF) can be used to inactivate microorganisms (Mertens and Knorr, 1992). Magnetic fields are usually generated by supplying current to coils. The inactivation of microorganisms requires magnetic field strengths of 5 to 50 T, which can be generated using superconducting coils, coils that produce DC fields, and coils that are energized by the discharge of energy stored in a capacitor (Pothakumary et al., 1995a,b,c). The technological advantages of PMF include minimal thermal denaturation of nutritional and organoleptic properties, reduced energy requirement for sufficient processing, and potential treatment of foods inside a flexible film package. Additional work is required to correlate the inactivation of microorganisms in food to PMF strength, PMF to the denaturation of nutritional characteristics of food, and the energy efficiency of PMF to the extended shelf-life of food. Little information is available on spore-forming pathogens' inactivation using PMF, which makes this technology several years away for possible commercialization for shelf-stable low-acid foods. However, there seems to be a potential for producing foods with minimal nutritional loss using PMF.

TABLE 9.1
Commercial Products Using High-Pressure Processing (HPP)

Product	Company
Orange juice	UltiFruit®, Pernod Richard Company, France
Acidified avocado purée and salsa	Avomex Company, TX, U.S.
Sliced ham (cured-cooked and raw-cooked)	Espuna Company, Spain
Oysters, clams, mussels, scallops, shrimps, crabs, and crab meat	Motivatt Sea Foods Inc., Houma, LA, U.S.
Processed meats, certain vegetable products, jams, jellies, fresh fruit cuts, and desserts	Tewari Fresher Foods, San Antonio, TX, U.S.

HIGH-PRESSURE PROCESSING (HPP)

Among nonthermal processing technologies, high-pressure processing (HPP) is the one that has gained the most attention since late 1980s. HPP is gaining in popularity with food processors not only because of its food preservation capability but also because of its potential to achieve interesting functional effects (Leadley and Williams, 1997) High-pressure processing has had application for years in industries that process or use ceramics, carbon graphite, diamond, steel/alloy, and plastics. Hite et al. (1914) were the first researchers who reported the effects of HPP on food microorganisms by subjecting milk to pressures of 650 MPa and obtaining a reduction in the viable numbers of microbes. For the last 15 years, the use of HPP has been explored extensively in the food industry and related research institutions due to increased consumer demand for improved nutritional and sensory characteristics of food without loss of "fresh" taste. In recent years, HPP has been extensively used in Japan and a variety of food products such as jams and fruit juices have been processed (Cheftel, 1995). Examples of commercial pressurized products in Europe and U.S. are:

1. Orange juice by UltiFruit®, Pernod Richard Company, France
2. Acidified avocado purée (guacamole) by Avomex Company in U.S. (Texas/Mexico) and several types of prepared food products by Tewari Fresher Foods, San Antonio, Texas
3. Sliced ham (cured-cooked and raw-cooked) by Espuna Company, Spain

Pressurized fruit preparation from yogurt should be on the market soon. A few commercialized products using HPP are presented in Table 9.1. The capability and limitations of HPP have recently been reviewed and studied extensively by Tewari et al. (1999b).

Mechanisms of HPP

Any phenomenon in equilibrium (chemical reaction, phase transition, change in molecular configuration), accompanied by a decrease in volume, can be enhanced by pressure (Le Chatelier's Principle). Thus, HPP affects any phenomenon in food

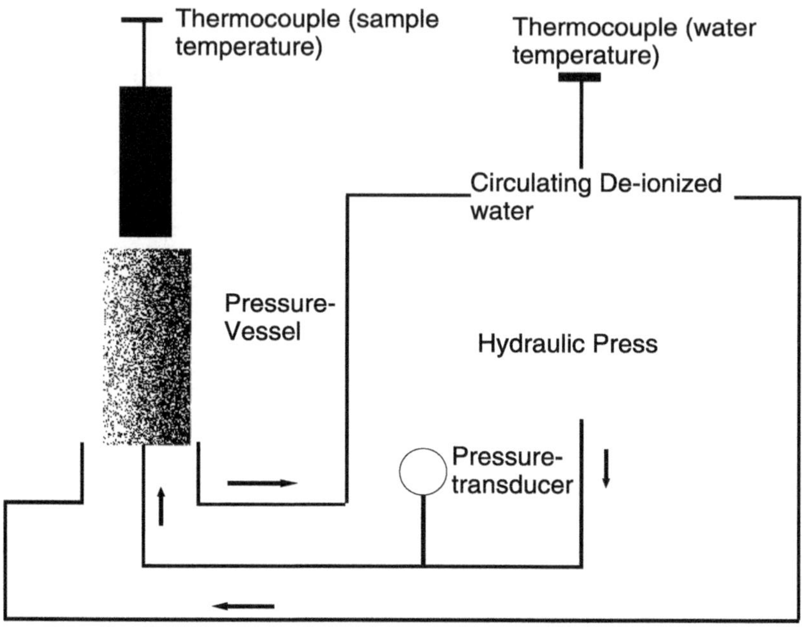

FIGURE 9.2 Schematic of high-pressure food processor (batch-type).

systems where a volume change is involved and favors phenomena that result in a volume decrease. The HPP affects noncovalent bonds (hydrogen, ionic, and hydrophobic bonds) substantially because some noncovalent bonds are very sensitive to pressure, which means that low molecular-weight food components (responsible for nutritional and sensory characteristics) are not affected, whereas high molecular-weight components (whose tertiary structure is important for functionality determination) are sensitive. Some specific covalent bonds are also modified by pressure.

The other principles that govern HPP are the *Isostatic Principle*, which implies that the transmittance of pressure is uniform and instantaneous (independent of size and geometry of food), and the *Microscopic Ordering Principle*, which implies that, at constant temperature, an increase in pressure increases the degree of ordering of the molecules of a substance (Heremans, 1992). Another interesting rule concerns the small energy needed to compress a solid or liquid to 500 MPa as compared to heating to 100°C, because compressibility is small. HPP offers several advantages: reduced process times; minimal heat-penetration and heat-damage problems; good retention of freshness, flavor, texture, and color; no vitamin C loss; multiple changes in ice-phase forms resulting in pressure-shift freezing; and minimized functionality alterations compared with traditional thermal processing. A schematic of a high-pressure food processor is given in Figure 9.2.

Microbial Inactivation during HPP

Application of HPP as a method for microbial inactivation has stimulated considerable interest within the food industry. The effectiveness of HPP on microbial inactivation must be studied in great detail to ensure the safety of food treated in this manner. Currently, research in this area has concentrated mainly on the effect of HPP on spores and vegetative cells of different pathogenic bacterial species. Detectable effects of HPP on microbial cells include an increase in the permeability of cell membranes and possible inhibition of enzymes vital for survival and reproduction of the bacterial cells (Farr, 1990). To design appropriate processing conditions for HPP of food materials, it is essential to know the precise tolerance levels of different microbial species to HPP and the mechanisms by which those tolerance levels can be minimized. Knowledge of critical factors that affect the baroresistance of different bacterial species will help in the development of more effective and accurate high-pressure processors.

Inappropriate use of a variety of parameters like pressure range, processing temperature, initial temperature of sample, holding time, and packaging type may adversely affect the outcome of HPP. Thus, a thorough understanding of the effect of a variation in critical factors on the intracellular changes undergone by pressure-treated microbial species is essential for documentation of safe HPP. The physicochemical environment can adversely change the resistance of a bacterial species to pressure. In most cases, the effect of HPP on Gram-positive bacteria is less pronounced than on Gram-negative species. Factors such as water activity and pH also influence the extent to which foods need to be treated to eliminate pathogenic microorganisms.

Hoover et al. (1989) reported that most bacteria are baroduric; i.e., they are capable of enduring high pressures but grow well at atmospheric pressures. Hauben et al. (1997) studied HPP resistance development in *E. coli* MG1655 mutants. Three barotolerant mutants (LMM1010, LMM1020, and LMM1030) were isolated and pressure treated. Mutants showed 40 to 85% survival at 200 MPa for 15 min (ambient temperature) and 0.5 to 1.5% survival at 800 MPa for 15 min (ambient temperature). In contrast, survival of the parent strain (MG1655) decreased from 15% at 220 MPa to 8% at 700 MPa. It should be noted that pressure sensitivity of the mutants increased from 10 to 50°C, as opposed to the parent strain, which showed minimum sensitivity at 40°C. This research indicated that the development of high levels of barotolerance should be properly understood in order to predict the safety of HPP. Similar studies are needed to document the barotolerance of other potentially pathogenic bacterial species.

Maggi et al. (1994) studied the effects of HPP on the heat resistance of fungi in apricot nectar and distilled water. They reported complete inactivation of *T. flavus* ascospores at 900 MPa for 20 min at 20°C and a two-log cycle reduction in *N. fischeri*, but no effect was seen on *B. fulva* and *B. nivea* populations. In contrast, preheating apricot nectar (50°C) followed by pressure treatment of 800 MPa for 1 to 4 min resulted in complete inactivation of all four species. Also, lower pressure resistance was observed in distilled water samples.

Factors Affecting Microbial Inactivation

Several theories on the effect of HPP on bacterial species have been proposed over the years to explain the mechanism behind microbial inactivation and to better optimize HPP of foods. The processing conditions (initial sample temperature, circulating water temperature, pressurizing medium, holding time) under which high pressures are applied significantly influence the level of inactivation as well as the overall effect on the nutritional and sensory characteristics of food.

Ye et al. (1996) studied the pressure tolerance of *Saccharomyces cerevisiae*, *E. coli*, and *Staphylococcus epidermidis* in various media (agar, broth, apple jam, and juice) by subjecting the inocula to a pressure of 300 MPa at 5 to 25°C for 1 to 20 min. They reported that media pH played a very important role in the destruction of microbes; *S. epidermidis* was inhibited >90% at 300 MPa in 11.2 min at pH 7.2 and in 4.8 min at pH 4.0. Variations in the pH and water activity of foods can result in different levels of lethality to a particular bacterium for the same high-pressure processing parameters. Studies conducted by Timson and Short (1965) on *B. subtilis* showed that the pressure resistance of the bacterium (when subjected to 483 MPa for 30 min) was decreased as the pH in milk medium was lowered or raised from a pH value of 8. This value is not a constant for all microorganisms, and the survival of *B. subtilis* at a specific pH can vary with the pressure and temperature of treatment.

The type of culture medium used for growing the microbial species can also have a significant impact on the pressure and heat resistance of any microorganism. In general, the richer the growth medium is, the better the baroresistance of the microorganisms. This is thought to be because of the increased availability of essential nutrients and amino acids to the stressed cell. It must be kept in mind that the parameters governing pressure tolerance are not constant for every bacterial species; they vary from one bacterium to another and may also be different for a single species grown under different conditions or in different growth media. It is very likely that the application of pressure affects a multitude of functions in a cell, thus interacting to retard or even kill the cell (Hoover et al., 1989). Therefore, studies related to the inactivation of bacterial species during HPP should specifically describe and document the processing conditions under which the inactivation took place. Also, specific information should be given about the variation in sample temperature during HPP.

Hayakawa et al. (1994) compared the pressure resistance of spores of six *Bacillus* strains. The spores were cultivated on nutrient agar and suspended in cold sterile distilled water, with the filtrate heated for 30 min at 80°C to destroy any vegetative cells. Spores of these six strains were then treated under pressures ranging from 196 to 981 MPa at 5 to 10°C for holding times of 20 to 120 min. It was found that *B. stearothermophilus* IAM 12043, *B. subtilis* IAM 12118, and *B. licheniformis* IAM13417 had the most resistance to pressure, but *B. coagulans* was actually activated when treated with high pressures. There was no actual correlation between pressure and heat resistance, although they chose correctly a number of spore-forming bacteria varying widely in heat resistance. This could be due to applications of very low pressure in most cases. However, variables including initial and final spore levels and dilution levels were not specified, which may help explain the results better. It should also be noted that in most cases the reference heat treatment is much more inactivating than the pressure treatment to be compared; therefore, direct comparison between heat resistance and pressure resistance is often not fair.

Patterson et al. (1995) studied the sensitivity of vegetative pathogens (*Yersinia enterocolitica* 11174, *Salmonella typhimurium* NCTC 74, *Salmonella enteritidis*, *Staphylococcus aureus* NCTC 10652, *L. monocytogenes*, and *E. coli* O157:H7) in buffer (pH 7.0), UHT milk, and poultry meat to high pressures up to 700 MPa at 20°C. A 10^5 reduction in numbers was obtained in all cases when pressures in the range of 275 to 700 MPa for 15 min were applied at 20°C. Different strains of *L. monocytogenes* and *E. coli* O157:H7 showed significant variation in pressure resistance; these were further used to examine the effect of substrates on pressure sensitivity and indicated that substrate affected the baroresistance of the microorganisms significantly.

The pH of a food material plays a very important role in determining the extent to which HPP affects the microorganisms under study. Several studies have documented and analyzed changes in heat resistance of organisms grown under different pH conditions, but there have not been many studies on the pressure resistance of spores at different pH values.

Roberts and Hoover (1996) investigated the effects of changes in pH values combined with a variety of other factors on the pressure resistance of *B. coagulans* ATCC 7050. They reported an increase in the effectiveness of pressurization as the pH of the buffer was lowered. A decrease of an additional 1.5 log was observed as the pH was decreased from 7.0 to 4.0. On the basis of these results, it is highly probable that, other factors remaining constant, a neutral pH value is most conducive to high-pressure resistance in the cells. Earlier, Timson and Short (1965) observed that at high pressure the spores were most resistant at neutral pH and at low pressure the spores were most sensitive to neutral pH. That is perfectly in line with other findings that a pressure between 50 and 200 MPa enhances germination followed by kill and direct killing at high pressures above 900 MPa.

In contrast, Sale et al. (1970) reported that the inactivation of bacterial spores by pressurization was maximum when the buffer was at a pH near neutral and was lowest at extreme values of pH. The change in pH was thought to affect membrane ATPase and intracellular functions of the spore, thereby destabilizing the microorganisms (MacDonald, 1992). This effect of pH on the pressure resistance of any microorganism is accentuated by other factors like the addition of salts, temperature conditions, and general process parameters. A better understanding of the exact process by which the variations in the pH affect the stability of the spore is still needed to give a better overall picture of the effect of high pressures. Therefore, studies to examine the effect of pH variation on the inactivation of different types of spores by HPP need to be done.

The water activity (a_w) of cells also affects the in-pressure resistance. It is reported that the lower the a_w is, the higher the pressure resistance of cells. Palou et al. (1996) studied the combined effect of HPP and a_w on *Zygosaccharomyces bailii* inhibition. They reported complete inhibition of yeast at a_w >0.98 and an increase in the surviving fraction with a decrease in a_w. They concluded that addition of sucrose (to decrease a_w) acts as a baroprotective layer, preventing inhibition of yeast even at high pressures. Such a mechanism can be utilized to prevent inhibition of favorable microbes during HPP. More work is also needed in this area.

COMBINED PROCESSES: PRESSURE AND TEMPERATURE

Alpas et al. (1998) studied the interaction of pressure, time, and temperature on the viability of *Listeria innocua* strain CWD47 in peptone solution by subjecting samples to pressures in the range of 138 to 345 MPa, temperatures in the range of 25 to 50°C, and exposure times of 5 to 15 min. They showed that the combination of 345 MPa, 50°C, and 9.1 min could reduce the microbial population by 7 log, with a Z-value (temperature change in °C causing a tenfold reduction in D-value; here it stands for pressure in MPa needed to reduce D-value by tenfold) of 173 MPa. Such studies resulting in a description of Z-values for a specific microbial species are aimed toward a standardization of HPP parameters, which will facilitate commercialization of HPP.

Maggi et al. (1995) studied the use of HPP for inactivation of *Clostridium pasteurianum* spores isolated from peeled tomato and inoculated into tomato serum. They reported that high-pressure treatments of 900 MPa for 5 min at 60°C completely destroyed the spores (which was not obtained at temperatures <60°C). Pressures of 700 or 800 MPa for 5 min at 60°C resulted in D-values of 2.4 and 3.4 min, respectively. They also studied spore inactivation using pressure pretreatment followed by heat treatment and found that a pretreatment of 300 MPa or 500 MPa for 1 min at 60°C reduced heat resistance of spores by one-third and one-half, respectively. Their documented D-values for *C. pasteurianum* may be used for developing a database for pressure–temperature–time requirements for complete destruction of different types of spores.

Roberts and Hoover (1996) studied the inactivation of *Clostridium sporogenes* PA3679 and *B. subtilis* 168 using a combination of pressure, temperature, acidity, and nisin. They exposed the samples (buffer at pH 4 to 7, inoculated with spores) to 405 MPa at different temperatures (25 to 90°C) for 15 or 30 min. After pressure treatment, spores were pour plated into agar with or without nisin. They reported an increase in spore inactivation with a decrease in pH. Also, sterilization was achieved using higher temperatures and pressures; e.g., pressurization to 405 MPa at 45°C for 30 min resulted in complete inhibition of *C. sporogenes* at pH 4, while 90°C yielded complete inhibition over a pH range of 4 to 6. *B. subtilis* was completely inhibited by pressurization to 405 MPa at 70°C for 15 min at pH 4. Addition of nisin resulted in further reduction of microbial numbers. Unfortunately, this study was limited to only one pressure and the D-value was not documented. The use of other combinations of pressures and temperatures may have given more promising results.

Maggi et al. (1996) studied the combined effect of pressure and temperature on the inactivation of *C. sporogenes* PA 3679 (ATCC 7953) spores in liquid media at pH 7.0 (beef or carrot broth medium, and phosphate buffer). They reported that 1500 MPa at 20°C for 5 min resulted in no spore inactivation, whereas 1500 MPa at 60°C fully inactivated the spores; a pressure of just 800 MPa at 80 to 90°C resulted in sterilization of beef or carrot broth. Rovere et al. (1996) studied the effect of high pressure (up to 1500 MPa) and temperature (20 to 88°C) on the destruction of *C. sporogenes* strain PA 3679, ATCC 7955, and reported that total bacterial spore inactivation could be obtained using a combined pressure (1000 MPa) and temperature (50 to 60°C) treatment. Butz et al. (1996) also reported a combined pressure–temperature treatment for the inactivation of *B. nivea* DSM 1824 ascopores. Researchers (Awao and Taki, 1990; Mallidis and Drizou, 1991; Taki et al., 1991; Seyderhelm and Knorr, 1992; Clouston and Wills, 1969; Gould and Sale, 1970; Nishi et al., 1994) have also attempted to inactivate bacterial spores (*B. stearothermophilus*, *B. licheniformis*, *B. cereus*, *B. coagulans*, *C. botulinum*) using combined high pressure (>0.7 MPa) and moderate temperatures (>50°C) and found satisfactory results. However, for complete spore inactivation, pressures >100 MPa in combination with temperatures of 60 to 80°C are desired.

Sonoike et al. (1992) examined the death rates of *Lactobacillus casei* Y1T9018 and *E. coli* JCM1649 under various temperatures (0 to 60°C) and pressures (0.1 to 400 MPa). They reported that death rates of both strains decreased with rising temperatures under a high pressure and that contours of constant death rates of both strains on the pressure–temperature plane were elliptical and similar to those of the free-energy difference for pressure–temperature-reversible denaturation of proteins.

Hashizume et al. (1995) studied the inactivation of yeast using HPP at low temperatures by subjecting *S. cerevisiae* IFO 0234 to a pressure range of 120 to 300 MPa at –20 to 50°C. After performing regression analysis of 43 inactivation rates, they reported that the same degree of inactivation was achieved at higher pressures and higher temperatures compared with low pressures and low (subzero) temperatures (e.g., the inactivation effect at 190 MPa and –20°C was similar at 320 MPa and room temperature). They concluded that high-pressure treatment applied at subzero temperatures requires lower pressures than when conducted at high temperatures to achieve the same degree of microbial inactivation. Only a few studies have been performed using HPP at subzero temperatures, so further studies are needed in this area to demonstrate the robustness of HPP at low temperatures, which may lead to interesting results.

COMBINED PROCESS: PRESSURE AND OTHERS

Crawford et al. (1996) studied the combined use of pressure and irradiation to destroy *C. sporogenes* spores in chicken breast. They reported a 5-log reduction at ambient temperature (25°C) with a pressure of 689 MPa applied for 60 min; heating samples at 80°C for 20 min resulted in the lowest number of survivors. They also reported that a 3.0-kGy irradiation treatment before and after pressurization at 80°C for 1, 10, and 20 min did not show any significant differences in spore numbers between samples that were pressurized and then irradiated or vice versa. However, the irradiation D-value of *C. sporogenes* decreased from 4.1 to 2 kGy at high pressures (>600 MPa at 80°C for 20 min); their research showed that high pressure reduced the irradiation dose required to produce chicken with an extended shelf-life. Crawford et al. concluded that pretreatment with irradiation (prior to HPP) is a useful technique for inactivating *C. sporogenes* spores, reducing the radiation dose required to eliminate the spores by irradiation alone.

Fornari et al. (1995) studied the inactivation of *Bacillus* spp. using a combination of pressure, time, and temperature. They studied four *Bacillus* spp. (*B. cereus*, SSICA/DA1 from wheat flour, *B. licheniformis* SSICA/DA2 from spices, *B. coag-*

ulans SSICA 1881 from tuna in tomato sauce, and *B. stearothermophilus* SSICA/T460 from spoiled canned peas) by subjecting prepared samples to a pressure range of 200 to 900 MPa for 1 to 10 min at 20, 50, 60, or 70°C. They also examined the effect of pressure cycling on spore inactivation (pressure treatment of 200 to 500 MPa followed by 900 MPa). They found that *B. cereus* spp. were more sensitive to pressure treatment. (Inactivation of 4 × 10 + 5 endospores/mL was achieved at ambient temperature by treatment at 200 MPa for 1 min followed by 900 MPa for 1 min.) However, for other species, a combination of higher pressure and moderate temperature was needed for significant reduction (*B. licheniformis* was inactivated at 800 MPa for 5 min at 60°C; *B. coagulans* was reduced to 10 to 4 endospores/mL at 900 MPa for 5 min at 70°C; and *B. stearothermophilus* was inactivated at 70 MPa for 5 min at 70°C).

Aleman et al. (1996) studied the effects of pulsed and static HPP on fruit preservation by inactivating *S. cerevisiae* 2407–1a in unsweetened pineapple juice. They applied sinusoidal and step-pressure pulses and compared the inactivation effects with static pressure treatments. They reported that no inactivation was observed after the application of 40 to 4000 fast sinusoidal pulses (10 cycles/sec) at 4 to 400 sec over a pressure range of 235 to 270 MPa; however, static pressure treatments of 270 MPa at 40 and 400 sec gave 0.7 and 5.1 decimal reductions, respectively. Also, slower 0- to 270-MPa step pulses at 0.1 (10 pulses), 1 (100 pulses), and 2 (200 pulses) cycles/sec with total time of 100 sec resulted in 3.3, 3.5, and 3.3 decimal reductions, respectively. They also reported that the ratio of on-pressure time to off-pressure time affected inactivation (e.g., on-pressure time of 0.6 sec and off-pressure time of 0.2 sec resulted in a four-decimal reduction in 100 sec). Aleman et al. concluded that slower step pulses resulted in increased effectiveness of HPP as more reduction in microbial numbers was observed in less time in step-pressure processing. However, studies need to be done using step-pressure processing for inactivating other microorganisms and documenting their destruction kinetics. More studies are needed along the same lines for inactivation of other baroresistant microbes (e.g., Gram-positive bacteria, spores) using cycled pressure treatment. However, the high resistance of bacterial spores to HPP is still a major outstanding issue and is the subject of a variety of reports (Shimada, 1992; Crawford et al., 1996; Maggi et al., 1996; Roberts and Hoover, 1996).

Shimada (1992) reported that the combined treatment of high pressure and alternating current yields lethal damage in *E. coli* and *B. subtilis* spores. Earlier, Shimada and Shimahara (1985, 1987) found that the exposure of *E. coli* cells to an AC of 50 Hz caused the release of intracellular materials located in the nucleus region within the cells, causing a decrease in the resistance to basic dyes. This was believed to have resulted from loss and (or) denaturation of cellular components responsible for the normal function of the cell membrane, suggesting that the lethal damage to microorganisms may be enhanced when the organisms are exposed to AC before or after the pressure treatment.

Shimada (1992) subjected *E. coli* cells to 300 MPa for 10 min immediately after AC exposure and *B. subitlis* suspension to 400 MPa for 30 min before AC exposure. AC exposure was carried out at 0.6 A/cm^2 for *E. coli* cells at 35°C for 2 h and at 1

A/cm^2 for spores at 50°C for 5 h. He found that the surviving fractions of *E. coli* cells and *B. subtilis* spores treated with AC and pressure were significantly reduced. It was also found that the susceptibility of *E. coli* cells and *B. subtilis* spores to some chemicals increased after the combination treatment, suggesting that the combined use of pressure and AC also lowers the tolerance level of microorganisms to other challenges.

Response to pressure cycling (Knorr, 1994; Honma and Haga, 1991), ultrasound with pressure (Knorr, 1994), and additives plus pressure (Knorr, 1994; Papineau et al., 1991; Popper and Knorr, 1990) have been studied, and significant interactive effects on microbial inactivation have been found. In addition, sensory and functional characteristics of foods were enhanced. Knorr (1994) reported that neither ultrasonic nor high-pressure treatment alone was capable of inactivating *Rhodoturola rubra*; however, pretreatment of samples with ultrasonic waves (100 W/cm^2, 25°C, 25 min) followed by HPP (400 MPa, 25°C, 15°C) resulted in complete inactivation of *R. rubra*. Popper and Knorr (1990) demonstrated the effectiveness of combinations of enzymes like lysozyme, lactoperoxidase, and glucose oxidase on the inactivation of microbes at atmospheric pressure. It is highly likely that the combination of enzyme pretreatment with HPP may result in significant inactivation of microbes even at low pressures.

The work done in combined pressure–temperature, pressure–AC exposure, pressure–cycling, and pressure–ultrasound areas is at a preliminary level and the exact mechanisms of spore inactivation by such combined processes are not well known. No doubt, combined high pressure and moderate temperature (60 to 80°C) or high pressure and AC exposure will have a beneficial impact on the sensory characteristics of heat-sensitive products, yet no such processes can be commercialized until microbial safety can be guaranteed. In most of the studies, researchers did not evaluate the effect of pressure-induced "adiabatic" heating on sample temperatures, nor did they examine spore inactivation during pressure "come-up" time, both of which may affect inactivation kinetics significantly. Also, the operational efficiency of a high-pressure food processor is very important during combined treatments; therefore, proper equipment maintenance and effective training of personnel are prerequisite for such studies. A detailed evaluation of the database accumulated from studies that used a combination of pressure and other treatments (temperature, AC exposure, cycling) is needed prior to commercialization of HPP. Also, corresponding D-values for specific microorganisms need to be documented and validated to standardize HPP before it can be commercialized.

OTHER CHALLENGES

Tewari et al. (1999b) have surveyed applications of HPP on microbial and enzyme inactivation as well as on its potential application in food industries related to beverage, dairy, egg, meat, and starch. Potential commercial applications of HPP are being outlined (Table 9.2). Spore-forming pathogens can be inactivated using HPP, provided high pressures are employed along with high temperatures and/or AC exposure. Thus far, no concerted effort has been made to validate HPP procedures for complete elimination of pathogenic bacteria. It is important that standard oper-

TABLE 9.2
Potential Commercial Applications of High-Pressure Processing for Retention of Sensory and Nutritional Characteristics of Fruits and Vegetables

Product	Pressure (MPa)	Holding Time (min)	Temperature (°C)	Microorganisms Tested	Other Parameters Tested
Potato cubes	400	15	5 to 50		Microbial safety, softness, functionality
Chopped tomatoes	400, 600, or 800	3, 5, or 7			Color, sugar content, pH
Apricot nectar, distilled water	600 to 900	1 to 20	20	*Byssochlamys fulva*, *B. nivea*, *Neosartorya fischeri*, *Talaromyces flavus*	
Jams					Quality volatile flavor components, anthocyanins, browning index, furfural, sucrose, vitamin C, microbiological stabilization
White- and red-grape	304 to 811	1 to 5	25		Microbiological stabilization
Angelica keiskei juice	0.01	7	25		Sensory and shelf-life
Fresh apples, pears, bananas, parsley, potatoes, celery, carrot juice, apple juice, vitaminized carrot, mixed apple and vitaminized carrot juice	300, 370	6, 15	25		Preservation, aroma, flavor, microbial quality
Citrus juice	500 to 700	1 to 1.5	0 to 5		Freshness
Orange juice	350	1	30		Microbial activity and chemical composition
Vegetable juices (beets, carrots, cauliflower, spinach, tomatoes, kohlrabi, grapefruit, strawberries	400, 600	10	35		Antimutagenic activity
Guava purée	400, 600	15	25		Quality and shelf-life
Extra virgin olive and seed oils (grape seed, sunflower, soybean, peanut, and maize)	700	10	25		Lipid oxidation

Partially adapted from Tewari, G. et al., *Sciences des Aliments*, 19:619–661, 1999.

ating criteria and processing conditions be established to ensure the reliability of HPP as an alternative to thermal processing.

Thermal processing has developed standards like D- and Z- values, holding times, and minimum temperatures for different foods; likewise, standards need to be set for HPP. Very few studies on the HPP inactivation of microorganisms discuss HPP effects in terms of the D-value associated with each particular strain of organism. More research is needed along these lines so that a database of HPP D-values for different microorganisms can be prepared. Without a well-documented database of D-values, there is no means to effectively compare the results of experiments performed with different microorganisms under different processing conditions. Comparable standards might indicate the resistance of various microorganisms to pressure and temperature combinations used in HPP. Different D-values at constant values of temperature and pressure may be used as an indication of resistance of a particular bacterial species with respect to another species. Thus, an effort should be made to understand and derive D- and Z-values so that minimum processing conditions may be developed for contaminants in target foods. Because the general norm followed in the food industry is to subject food products to treatments that ensure a 12-D reduction of microorganisms, to ensure that HPP is effective in inactivating microorganisms, treatments that result in 12-D reduction must be another goal.

Perhaps it is not surprising that no database of D- and Z-values for microbial species has been developed for HPP. It must be kept in mind that these parameters have been derived for temperature and time effects together, without regard to pressures. Therefore, even if one has these values for various microorganisms, the precise parameters of pressure, temperature, and time required to eliminate the bacteria completely from the food by HPP cannot be predicted. Additionally, microbial death during come-up phase, as well as due to adiabatic heating, must be incorporated while calculating these values. To fully validate HPP, a relationship must be developed between the pressure of processing and the reduction in microbial population under treatment over a fixed time and temperature. The methods adopted to derive this parameter may be different but the end result must provide an effective means of predicting bacterial reductions due to HPP. This area requires further research.

CONCLUSIONS

Driven by consumers demanding convenience as well as nutritional and fresh foods, the food industry is looking for potential applications of nonconventional food preservation technologies. Extensive work has been done both at pilot and commercial levels by employing nonthermal food preservation technologies. Much remains to be answered before these techniques can provide shelf-life, low-acid foods; however, these techniques can be well adapted for high acids and refrigerated foods. Among nonthermal techniques, high-pressure processing has gained extensive attention and is able to deliver quite a few commercial products. However, extensive guidelines are required for nonthermal processes so that they can be adapted commercially with ease. Additionally, food packaging interactions utilizing these nonconventional techniques are missing and require significant work; otherwise, the potential of nonthermal techniques to deliver better and fresh quality products is compromised.

REFERENCES

Aleman, G.D. et al., 1996, Pulsed ultra high pressure treatments for pasteurization of pineapple juice, *J. Food Sci.*, 61:388–390.

Alpas, H., Kalchayanand, N., and Bozugluf, R.B., 1998, Interaction of pressure, time and temperature of pressurization on viability loss of *Listeria innocua*, *World J. Microbiol. Biotech.*, 14:251–253.

Awao, T. and Taki, Y., 1990, Sterilization of spores of genus *Bacillus* by hydrostatic pressure, *Kagaku kogaku*, 54:283–285.

Becker, R.L., 1983, Absence of induced radioactivity in irradiated foods, in *Recent Advances in Food Irradiation,* Elias, P.S. and Cohen, A.L., Eds., Amsterdam, The Netherlands: Elsevier Biomedical Press, 285–313.

Butz, P. et al., 1996, High pressure inactivation of *Byssochlamys nivea* ascospores and other heat resistant moulds, *Lebensm. Wiss. U Technol.*, 29:404–410.

Castro, A.J., Barbosa-Canovas, G.V., and Swanson, B.G., 1993, Microbial inactivation of foods by pulsed electric field, *J. Food Process. Preservation*, 17:47–73.

Cheftel, J.C., 1995, Review: high pressure, microbial inactivation and food preservation, *Food Sci. Technol. Int.*, 1:75–90.

Clouston, J.G. and Willis, P.A., 1969, Initiation of germination and inactivation of *Bacillus pumilus* spores by hydrostatic pressure, *J. Bacteriol.*, 97:684–690.

Crawford, Y.J. et al., 1996, Use of high hydrostatic pressure and irradiation to eliminate *Clostridium sporogenes* spores in chicken breast, *J. Food Prot.*, 59:711–715.

Dunn, J., Ott, T., and Clark, W., 1995, Pulsed-light treatment of food and packaging, *Food Technol.*, 49(9):95–98.

Farr, D., 1990, High pressure technology in the food industry, *Trends Food Sci. Technol.*, 1:14–16.

Fornari, C. et al., 1995, Inactivation of *Bacillus* endospores by high-pressure treatment, *Ind. Conserv.*, 70:259–265.

Gould, G.W. and Sale, A.J.H., 1970, Initiation of germination of bacterial spores by hydrostatic pressure, *J. Gen. Microbiol.*, 60:335–346.

Hashizume, C., Kimura, K., and Hayashi, R., 1995, Kinetic analysis of yeast inactivation by high pressure treatment at low temperatures, *Biosci. Biotech.*, 59:1455–1458.

Hauben, K.J.A. et al., 1997, *Escherichia coli* mutants resistant to inactivation by high hydrostatic pressure, *Appl. Environ. Microbiol.*, 63:945–950.

Hayakawa, I. et al., 1994, Oscillatory compared with high pressure sterilization on *Bacillus stearothermophilus* spores, *J. Food Sci.*, 59:164–167.

Heremans, K., 1992, From living systems to biomolecules, in *High Pressure and Biotechnology,* Balny, C. et al., Eds., London, U.K.: John Libbey and Co. Ltd., 37–44.

Hite, B.N., Giddings, N.J., and Weakly, C.E., 1914, The effect of certain microorganisms encountered in the preservation of fruits and vegetables, *Bull. W. Va. Agric. Exp. Stn.*, 14:63–67.

Ho, S.Y. and Mittal, G.S., 2000, High voltage pulsed electric field for liquid food pasteurization, *Food Rev. Int.*, 16(4):395–434.

Ho, S.Y. et al., 1995, Inactivation of *Pseudomonas fluorescens* by high voltage electric pulses, *J. Food Sci.*, 60:1337–1340, 1343.

Honma, K. and Haga, N., 1991, Effects of high pressure treatment on sterilization and physical effect of high pressure treatment on sterilization and physical properties of egg white, in *High Pressure Science for Food*, Hayashi, R., Ed., Kyoto, Japan: San-Ei Pub. Co., 317–324.

Hoover, D.G. et al., 1989, Biological effects of high hydrostatic pressure on food microorganisms, *Food Technol.*, 43:99–107.

Knorr, D. et al., 1994, Food applications of high electric field pulse, *Trends Food Sci. Technol.*, 5:71–75.

Knorr, D., 1994, Hydrostatic pressure treatment of food: microbiology, in *New Methods of Food Preservation*, Gould, G.W., Ed., London: Blackie Academic and Professional, 159–175.

Leadley, C.E. and Williams, A., 1997, High pressure processing of food and drink — an overview of recent developments and future potential, New Technologies, Bulletin No. 14, March, Campden and Chorleywood Food and Drink Research Association, Chipping Campden, U.K.

MacDonald, A.G., 1992, Effect of high hydrostatic pressure on natural and artificial membranes, in *High Pressure and Biotechnology,* Balny, C. et al., Eds., London: John Libbey and Co. Ltd., 67–75.

Maggi, A. et al., 1995, Use of high pressure for inactivation of butyric clostridia in tomato serum, *Ind. Conserve*, 70:289–293.

Maggi, A. et al., 1996, Effects of combined high pressure-temperature treatments on *Clostridium sporogenes* spores in liquid media, *Ind. Conserve.*, 71:8–14.

Maggi, A. et al., 1994, High pressure treatments of ascopores of heat-resistant moulds and patulin in apricot nectar and water, *Ind. Conserve.*, 69:26–29.

Mallidis, C.G. and Drizou, D., 1991, Effect of simultaneous application of heat and pressure on the survival of bacterial spores, *J. Appl. Bacteriol.*, 71:285–288.

Mertens, B. and Knorr, D., 1992, Developments of nonthermal processes for food preservation, *Food Technol.*, 46(5):124–126, 132.

Mittal, G.S., 1993, Electric pulse technology for food pasteurization, in *Food Processing and Preservation. Proceedings of Latin American & Caribbean Workshop on Food Preservation*, Sankat, C.K. and V. Maharaj, V., Eds., University of West Indies: St. Augustine, Trinidad, 213–223.

Nishi, K., Kato, R., and Tomita, R., 1994, Activation of *Bacillus* spp. spores by hydrostatic pressure, *Nippon Skokuhin Kogyo Gakkaishi*, 41:542–549.

Palou, E. et al., 1996, Combined effect of high hydrostatic pressure and water activity on *Zygoaccharomyces bailii* inhibition, in *Institute of Food Technologists Ann. Meet., Book of Abstracts*, 57.

Papineau, A.M. et al., 1991, Antimicrobial effect of water-soluble chitosans with high hydrostatic pressure, *Food Biotechnol.*, 5:45–47.

Patterson, M.F. et al., 1995, Sensitivity of vegetative pathogens to high hydrostatic pressure treatment in phosphate-buffer saline and foods, *J. Food Prot.*, 58:524–529.

Popper, L. and Knorr, D., 1990, Applications of high-pressure homogenisation for food preservation, *Food Technol.*, 44:84–89.

Pothakumary, U.R., Monsalve-Gonzalez, A., and Barbosa-Canovas, G.V., 1995a, Inactivation of *Escherichia coli* and *Staphylococcus aureus* in model foods by pulsed electric field technology, *Food Res. Int.*, 28:167–171.

Pothakumary, U.R. et al., 1995b, High voltage pulsed electric field inactivation of *Bacillus subtilis* and *Lactobacillus delbrueckii*, *Spanish J. Food Sci. Technol.*, 35:101–107.

Pothakumary, U.R., Barbosa-Canovas, G.V., and Swanson, B.G., 1995c, Magnetic-field inactivation of microorganisms and generation of biological changes, *Food Technol.*, 47(12):85–93.

Qin, B.L. et al., 1994, Inactivation of microorganisms by pulsed electric fields of different voltage waveforms, *IEEE Trans. Dielectric Electr. Insulation*, 1:1047–1057.

Qin, B.L. et al., 1995, Pulsed electric field treatment chamber design for liquid food pasteurization using a finite element method, *Trans. Am. Soc. Agricul. Eng.*, 38:557–565.

Qiu, X. et al., 1998, An integrated PEF pilot plant for continuous nonthermal pasteurization of fresh orange juice, *Trans. Am. Soc. Agricul. Eng.*, 41:1069–1074.

Roberts, C.M. and Hoover, D.G., 1996, Sensitivity of spores of *Clostridium sporogenes* PA3679 and *B. subtilis* 168 to combination of high hydrostatic pressure, heat, acidity, and nisin, in *Institute of Food Technologists Ann. Meet., Book of Abstracts*, 174–175.
Rovere, P., Tosoratti, D., and Maggi, A., 1996, Sterilizing trails to 15 000 bar to obtain microbiological and enzymatic stability, *Ind. Alimentari*., 35:1062–1065.
Sale, A.J.H., Gould, G.W., and Hamilton.W.A., 1970, Inactivation of bacterial spores by hydrostatic pressure, *J. Gen. Microbiol.*, 60:323–334.
Seyderhelm, I. and Knorr, D., 1992, Reduction of *Bacillus stearothermophilus* spores by combined high pressure and temperature treatments, ZFL (*J. Food Ind.*), 43:17–20.
Shimada, K., 1992, Effect of combination treatment with high pressure and alternating current on the lethal damage of *Escherichia coli* cells and *Bacillus subtilis* spores, in *High Pressure and Biotechnology*, Balny, C. et al., Eds., London: John Libbey and Co., Ltd., 49–51.
Shimada, K. and Shimahara, K., 1985, Leakage of cellular contents and morphological changes in resting *Escherichia coli* B cells exposed to an alternating current, *Agr. Biol. Chem.*, 49:3605–3607.
Shimada, K. and Shimahara, K., 1987, Effect of alternating current exposure on the resistivity of resting *Escherichia coli* B cells to crystal violet and other basic dyes, *J. Appl. Bacteriol.*, 62:261–268.
Sonoike, K. et al., 1992, Effect of pressure and temperature on the death rates of *Lactobacillus casei* and *Escherichia coli*, in *High Pressure and Biotechnology*, Balny, C. et al., Eds., London: John Libbey and Co. Ltd., 297–301.
Taki, Y. et al., 1991, Sterilization of *Bacillus* spp. spores by hydrostatic pressure, in *High Pressure Science for Food,* Hayashi, R., Ed., Kyoto, Japan: San-Ei Pub. Co., 217–224.
Tewari, G. and Jayas, D.S., 1999, Aseptic processing and packaging: promising packaging technique for food processing industries*, Indian Food Ind.*, 18:23–33.
Tewari, G., Jayas, D.S., and Holley, R.A., 1999b, High pressure processing of foods: an overview, *Sci. des Aliments*, 19:619–661.
Thayer, D.W., 1990, Food irradiation: benefits and concerns, *J. Food Qual.*, 13:147–169.
Thayer, D.W., 1994, Wholesomeness of irradiated foods, *Food Technol.*, 48(5):132–134, 136.
Thayer, D.W., Fox, J.B., Jr., and Lakritz, L., 1991, Effects of ionizing radiation on vitamins, in *Food Irradiation,* Thorne, S., Ed., London, UK: Elsevier Applied Science Publishers, 285–325.
Timson, W.J. and Short, A.J., 1965, Resistance of microorganisms to hydrostatic pressure, *Biotechnol. Bioeng*., 7:139–159.
Ye, H.Y. et al., 1996, Pressure tolerance of microbes and techniques for high pressure of acid food (jams, etc.) at room temperature, *Food Sci. China*, 17:30–34.
Zhang, Q., Barbosa-Canovas, G.V., and Swanson, B.G., 1992, Engineering aspects of pulsed electric field pasteurization, *J. Food Eng*., 25:261–281.

10 Modified Atmosphere Packaging for Shelf-Life Extension

James T.C. Yuan

CONTENTS

1-58716-041-2/03/$0.00+$1.50

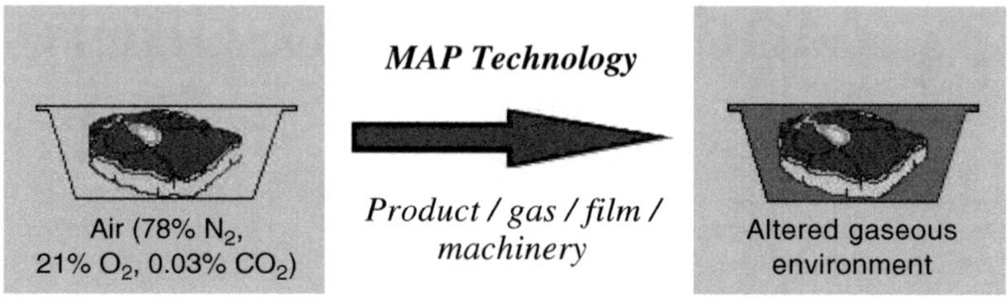

FIGURE 10.1 Modified atmosphere packaging (MAP) entails replacing the air in a package headspace with a specific gas or gas mixture (courtesy of Air Liquide America Corp).

INTRODUCTION

Modified atmosphere packaging (MAP) is not a novel topic; there are many excellent reviews of MAP and related technologies (Mathlouthi, 1994; Paine and Paine, 1992; Brody, 1989). However, due to consumer demands for fresh, close-to-fresh, or minimally processed foods (MPF), as well as the desire for more healthy, tasty, and safer foods, MAP has been elevated to a new degree of importance. This chapter will provide an overview of traditional MAP technology and its impact on shelf-life extension of products, transitions to novel packaging technology, and applications for ensuring the safety of minimally processed foods.

CLASSICAL MAP

The basic concept of MAP is depicted in Figure 10.1. Food is simply packaged in a sealed container with the surrounding gaseous environment different from air.

MAP Gases

The most commonly used gases in MAP are:

1. Nitrogen (N_2)
    - Sparingly soluble in both water and fats
    - No intrinsic effect on microorganisms
    - Primarily used to displace oxygen (O_2) as an "inert" gas resulting in reduction of oxidative reactions and inhibition of aerobic microorganisms
    - Used as a "filler" gas (physically protects food products by prevention of the package collapse, such as potato chips).
2. Oxygen (O_2)
    - Preserves color ("bloom") of red meat
    - Sustains the respiration of fresh produce
3. Carbon dioxide (CO_2)
    - Highly soluble in both water and fat
    - May cause package collapse and acidic taste at high concentrations due to the formation of carbonic acid in foods

- Known to have a bacterial inhibitory effect. While the detailed mechanisms on how CO_2 suppresses or kills microorganisms are not fully understood, there are few hypotheses on this scenario (Löwenadler, 1994):
  - When CO_2 comes in contact with membrane proteins having free amino groups on their surfaces, it changes ionic charges from positive to negative and vice versa. The new charge can interrupt the transport of specific ions needed for maintaining homeostasis in the cytoplasm such as the exchange of Na^+ and K^+ ions.
  - CO_2 plays a role on cytoplasmic enzymes by either inducing or repressing enzyme synthesis and enzymatic reaction rate. Not all enzymes are inhibited by CO_2, however.
  - In regard to the mechanism of CO_2 and metabolism, CO_2 permeates the membrane and reacts with water in the cytoplasm to form bicarbonate and hydrogen ions. The hydrogen ions acidify the inside of the cell and the organism requires cellular energy to pump the protons back out. The added requirement for this energy creates a burden on the cells, thus inhibiting their growth.

   All of the theorized CO_2 mechanisms vary among different bacterial species because species react differently to specific growth requirements in the medium, environmental parameters, substrates, etc. In most cases, the growth rate and lag phase of the spoilage bacteria correlate well with CO_2, while Gram-negative bacteria are more sensitive to CO_2 than Gram-positive bacteria are (Devlieghere and Debevere, 2000).

4. Argon (Ar) and other noble gases
   - Functions of Ar and other noble gases are quite similar to that of nitrogen. However, because their molecular weights are heavier than nitrogen, the inert effects and inactivating efficiencies are better (Spencer, 1995).

Depending on the product, packaging gases could involve any combinations of these gases at various ratios. In minimally processed foods other than produce, the packaging system relies mainly on the gas mixture to deliver the defined function (e.g., inhibit spoilage microorganisms, enhance color). Once the gas mixture is decided, the film is usually a barrier type, that minimizes gas exchange. In the produce category, the bag or top lidding film is usually permeable to oxygen, carbon dioxide, and water vapor. Permeable films have various oxygen transmission rates (OTR) to control the O_2/CO_2 exchange rate between the container and environment. There are two ways of modifying the surrounding atmosphere of MPF in the package: passive or active MAP.

Passive MAP

Passive MAP typically applies to fresh or minimally processed produce. It consists of packaging the produce product under ambient conditions (air: $N_2/O_2 = 78/21$) using a permeable film. During the course of storage, the produce product continues to respire by consuming O_2 and producing CO_2 and the atmospheric environment changes accordingly (high in CO_2 and low in O_2). Depending on the film permeability, the atmospheric content in the container will eventually reach a state of equilibrium, extending the keeping quality by a few extra days.

However, it takes time before the atmospheric content in the container reaches the state of equilibrium and it is difficult to obtain optimum results for keeping quality. Also, different produce products have different respiration rates, so it is difficult to optimize the keeping quality by using the film alone.

Active MAP

Active MAP is similar to the passive MAP process except that a special gas mixture is introduced into the container before it is sealed. The gas mixture has three main functions:

1. Control the produce product's respiration rate. Instead of relying solely on film permeability, a favorable atmospheric environment is provided to the product at time 0 to regulate the respiration rate and provide more favorable results (Diaz and Hotchkiss, 1996). Figure 10.2 illustrates the benefit of active MAP; the appropriate combination of a gas mixture (N_2/CO_2 = 80%/20%) and a permeable film (OTR = 1500/m^2) was applied to the packaged fresh-cut pineapple (approximately 1 × 0.5 in. size) on day 0. The respiration rate (measured by CO_2 concentration in the headspace) of pineapple was controlled at a constant rate so that the shelf-life was noticeably better than the control (the same film was applied, but packaged in air). The film permeability remains a crucial parameter because the atmospheric environment needs to be maintained during storage.

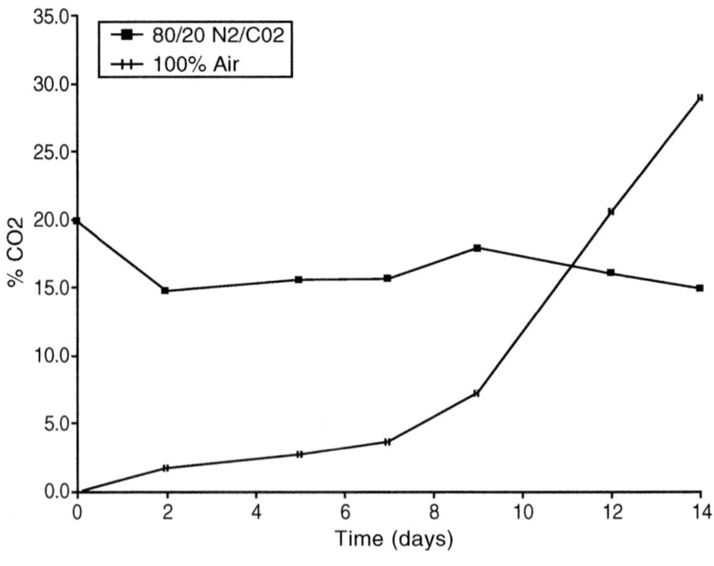

FIGURE 10.2 Carbon dioxide production rates of fresh-cut pineapples stored under MAP vs. air at refrigerated (2 ± 1°C) temperature (courtesy of Air Liquide).

FIGURE 10.3 Quality attribute (color) of fresh ground meat stored under MAP vs. air at refrigerated (4 ± 1°C) temperature after 5 days (courtesy of Air Liquide America Corp).

2. Maintain or enhance the product's attributes. For example, O_2 level is essential to make the color of the meat product bloom (Figure 10.3). On the other hand, any oxygen-sensitive products, e.g., potato chips or deep fried foods, should avoid oxygen in the package.
3. Retard the growth rate of spoilage microorganisms. The effectiveness of the gas or gas mixture depends on numerous factors. One factor could trigger the others, and it takes more skill to optimize those factors. Generally speaking, CO_2 at a higher level (more than 20%) can slow down spoilage bacteria and improve keeping quality. One example is salmon fillets packaged with 70% CO_2 and balanced with N_2 (Ohta and Sasaki, 1995). Although elevated CO_2 provides a favorable environment for the growth of beneficial lactic acid bacteria in the biopreservation technology, a high concentration of CO_2 (>50%) also has inhibitory effects on them. In contrast, lactic acid bacteria are undesirable in many packaged MFP and are considered spoilage microorganisms. As a result, high CO_2 has its applications in cheese products (Eliot et al., 1998) and other MAP packaged foods. Fedio et al. (1994) also reported that CO_2 inhibited spoilage microorganisms (yeast and fungi) in cottage cheese. Figure 10.4 demonstrates the benefit of using active MAP to retard the growth of spoilage microorganisms during storage of fresh bakery products. Based on the results, N_2/CO_2:50%/50% substantially suppressed the growth of molds on pound cakes by 99.9% following the storage at room temperature.

Studies show that CO_2 has little impact on some food-borne pathogens, such as *Clostridum botulinum*, *Yersinia enterocolitica*, *Salmonella typhimurium*, *Escherichia coli*, *Staphylococcus aureus*, *Listeria. monocytogenes*, and *Campylobacter* (Hintlian and Hotchkiss, 1986; Silliker and Wolfe, 1980). However, different results showing that 75% CO_2 (balance N_2) was effective on *S. typhimurium* and *P. fragi* were reported by Lee and Cash (1998). Also, Juneja et al. (1996) found 25 to 50% CO_2/20% O_2/30 to 55% N_2 inhibited *Clostridium perfringens* at 4°C.

For produce products, the proper gas mixture can significantly slow down the respiration rates, but it is the permeability of the film that maintains the equilibrated package atmosphere. Also, in order to attain the desired effect on a packaged product, the gas mixture headspace volume must be carefully considered. As illustrated in

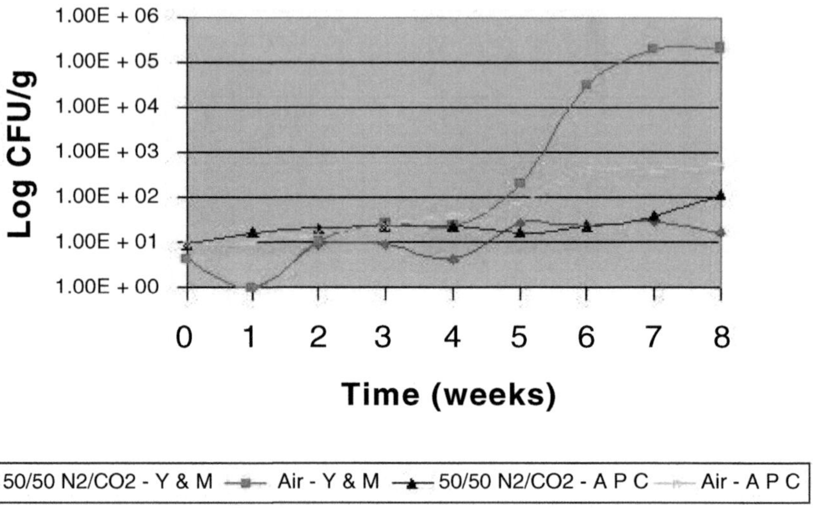

FIGURE 10.4 Microbiological results (APC and Y&M [yeast and mold]) of fresh pound cakes (without any preservatives) stored under MAP vs. air at ambient (18 ±2°C) temperature (courtesy of American Air Liquide Corp.).

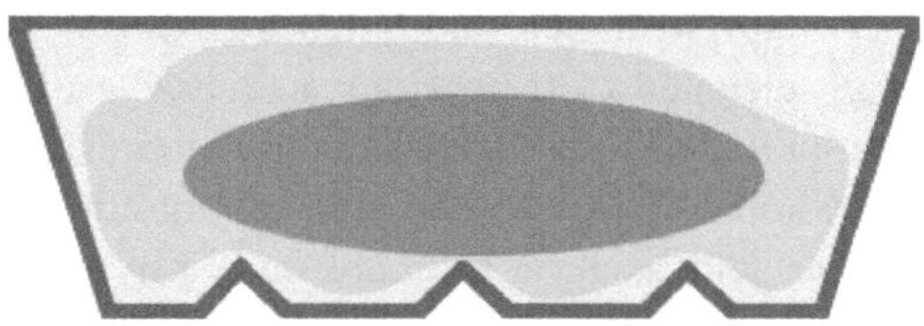

FIGURE 10.5 Illustration of the headspace-to-product ratio in the package (courtesy of Air Liquide America Corp.).

Figure 10.5, in ideal conditions the product is surrounded by the gas atmosphere in the package. If the headspace-to-product (HTP) ratio is too low, the gas mixture may not manifest its expected effect on the product. If the HTP is too high, then the package will appear "empty" to consumers and display space may become limited in retail areas. Normally, HTP = 1:1 works out best for packaged MPF products.

Because each minimally processed product has its own characteristics (e.g., respiration rate), no "one-size-fits-all" solution exists in terms of packaging. Each solution must be customized for each product and then validated.

Shelf-Life vs. Keeping Quality

The main goal for MAP of minimally processed produce is to extend keeping quality or shelf-life. "Keeping quality" is more commonly used than "shelf-life" because of consumer demand for product freshness. It is often difficult to label as "fresh" a product that has remained in refrigerated storage for 2 weeks. However, the extension of shelf-life remains a key objective for food processors due to economic reasons. Figure 10.6 illustrates the concept of extending keeping quality vs. shelf-life.

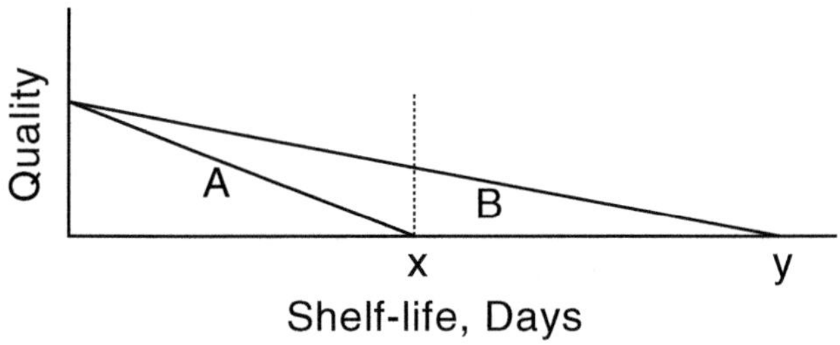

FIGURE 10.6 Shelf-life vs. quality extension.

For any given minimally processed food, line A represents *x* days of shelf-life based on the current packaging format and line B represents extended *y* days of shelf-life by using MAP technology. Properly engineered, MAP can extend the shelf-life of a product or better preserve its freshness and quality within the same shelf-life. A vertical line drawn on day *x* clearly shows the quality difference between line A and line B. MPF processors can decide their preference for extra days of shelf-life or better quality based on their business strategies and objectives.

NOVEL MAP

Advancements in packaging technology have allowed food processors to provide consumers with safer and fresher foods. Because all foods vary in their physical and chemical properties, food safety is dependent on many factors, from inherent ones such as indigenous microflora, to environmental consequences such as temperature and handling by processors, distributors, and consumers. Once the product is packaged for distribution, it is difficult to determine if product safety will be maintained when it reaches the consumer. Within this time frame, an undetected leak in the package could have permitted the entry of air or temperature abuse could have occurred due to careless actions or ignorance. Because a minimally processed food receives no terminal step, anything can add concern.

The leading pathogen of concern is *C. botulinum*. Many research publications have demonstrated that if *C. botulinum* is present in the MAP product, the contaminated product could be potentially hazardous (Lawlor et al., 2000; Phillips Daifas, D. et al., 1999). Similar findings have been published for other human pathogens in MAP products (Mano et al., 2000; Whitley et al., 2000; Lyver et al., 1998). In the new MAP era, it is necessary to break the perceptional bottleneck that MAP deals only with gas mixtures and films. The new thought is to incorporate the functional intrinsic and extrinsic elements to address the food safety issue proactively and to minimize or eliminate food safety risks. The intrinsic elements, such as intelligent or active packaging, are directly related to the packaging. The extrinsic elements cover the processing environment that will impact the packaging.

Intrinsic Elements

Intelligent or active packaging technologies (Day, 2000; Rooney, 1995), collectively called "interactive packaging," have been introduced for use in conjunction with existing packaging techniques such as MAP. Interactive packaging concepts involve various technologies incorporated into the package or the packaging film to maintain or monitor the quality or safety of the product. These may interact with the food products, alter the packaging atmosphere, or respond to environmental changes and communicate to consumers. For example, active packaging prolongs shelf-life because the removal of oxygen inhibits the growth of aerobic spoilage microorganisms. With the aid of intelligent packaging, it is possible not only to extend the shelf-life further, but also to warn the consumer if a product has spoiled, thus allowing proper decision-making. Many factors can promote the growth of spoilage bacteria. Therefore many interactive packaging applications, such as oxygen scavengers, ethylene scavengers, moisture absorbent pads, and antimicrobial films, balance out these factors.

Scavengers

The function of a scavenger is to remove an unwanted element in the package. Oxygen, ethylene, moisture, and taint are several targets of scavengers.

Oxygen Scavengers

Oxygen is a major cause of food spoilage, especially with MPF, in that it causes oxidation of fats and lipids, rancidity, microbial growth of aerobes, loss of taste and flavor, discoloration, and much more. Removal of oxygen in the package headspace is critical in preventing these problems and maintaining freshness for longer time periods. Every food is different in its rate of deterioration and its sensitivity to oxygen. However, some oxygen needs to be incorporated or allowed to permeate through the film for respiring fruits and vegetables in order to prevent an anaerobic environment, which can be as physiologically harmful to tissue metabolism.

Ethylene Scavengers

Ethylene, a plant hormone, has significant effects on fruits and vegetables, even in very low doses. Ethylene brings about plant stress, faster aging, and faster ripening. With these physiological changes, opportunistic microbes take advantage of the nutrients that they need to survive. When exposed, ethylene binds to receptors, which then undergo a conformational change leading to tissue metabolism. Ethylene accelerates the respiration rate, resulting in shorter shelf-life; therefore, storing fruits and vegetables in refrigeration temperatures is recommended in order to lower the respiration rate. Ethylene production is also enhanced with increasing amounts of oxygen and it is important to reduce oxygen, but not to levels that inhibit fruit and vegetable respiration or produce anaerobic conditions. Carbon dioxide may inhibit the production of ethylene because it binds ethylene-binding sites of specific enzymes, thus preventing ethylene-induced conformational changes of the tissue. Consequently, reduced oxygen and ethylene concentrations, combined with increased CO_2 levels, may avoid the accumulation of ethylene and in turn prevent accelerated deterioration of the plant tissues.

Ethylene scavengers include those that use potassium permanganate, which oxidizes ethylene to acetate and ethanol and further to carbon dioxide and water.

Potassium permanganate is absorbed onto celite, silica gel, or another inert material and inserted into the packaging film or included in the package as a sachet. Although potassium permanganate may work efficiently, its disadvantages are its purple color and toxicity. Ethylene can also be absorbed on a solid surface, for example, activated carbon scavengers, in which adsorption of ethylene gas by various metal catalysts can occur on activated carbon. Activated earth-type scavengers include adsorption of ethylene by finely dispersed minerals. These minerals are embedded into the polyethylene films, which are then used to package the product. However, the minerals may produce micropores in the film, thus enabling permeability of the gases. This may not be a positive effect because ethylene production is stimulated by oxygen concentration.

Moisture Control

The moisture content inside the package needs to be controlled because too little will cause the product to dry out and too much will promote microbial growth and cell collapse. When processed meats are packaged and refrigerated, condensation can build up to the point where water droplets from the film fall onto the meat, increasing the water content, lowering the pH, and creating conditions for microbial growth. Another problem of moisture buildup is fogging of the films where the food is not even visible.

To control the moisture level, a series of absorbent pads to control surplus moisture in fresh-cut produce and fresh meats has been developed. These pads contain a layer of gel that traps liquid and microbes, thus restricting growth.

Freshness Indicators

Freshness indicators indicate spoilage due to temperature abuse, package leaks, and anything else that would make the product microbiologically unsafe to eat. There are many commercial applications of freshness indicators, ranging from chemical to biological products of the organism. If a metabolite from a microorganism is detected, the metabolite-specific indicator will be activated and produce a visible signal to show that spoilage has taken place. In this respect, freshness indicators can be especially useful to the consumer.

One example of freshness indicators is the FreshTag™ indicator (Check-it ApS, Denmark), which consists of a plastic chip incorporating a reagent-containing wick. As volatile amines pass through the wick, a bright pink color is developed along the wick to indicate quality deterioration of fresh seafood products.

Toxin Guard™, another type of freshness indicator developed by Toxin Alert Inc. (Ontario, Canada), detects the presence of pathogenic bacteria with the help of labeled, immobilized antibodies.

Time–Temperature Indicators

Similar to freshness indicators, time–temperature indicators (TTIs) are another type of visible indicator devices that provide irreversible evidence of a temperature threshold being exceeded over a given time. The reaction can be a physical, chemical, or enzymatic change, which is usually observed through a color change. Some TTIs can indicate the partial history.

Antimicrobial Films for Food Packaging

Antimicrobial film technology looks promising; however, much research is still taking place in order to validate the efficacy of embedded antimicrobials interacting with polymer films, gaseous headspace, and the food itself. Many potentially active compounds can be used as antimicrobials, such as Ag-based compounds, bacteriocins such as nisin, enzymes such as lysozyme, and organic compounds such as triclosan. As antimicrobials, they work directly on the bacterial cell wall by coming into contact and breaking down components leading to cell rupture. Although antimicrobials have been established, the difficult task is to understand the compatibility of the packaging material with the antimicrobial, the heat lability, and the interactions of antimicrobials with food constituents. Generally speaking, the microorganisms usually remain on the surface of the food product. The antimicrobial activity is activated, rather than in constant action to maintain food safety. For example, if elevated temperatures or excess moisture is detected, then activation will take place. Two advantages of this technology are the decreased use of food additives and the combination of packaging with preservation.

For example, Microban®, by Microban Products Co. (Huntersville, North Carolina), has incorporated triclosan into plastic that performs effective antibacterial action following the migration to the surface of the food. The triclosan comes into contact with the cell wall and destroys the cell's ability to function, grow, and reproduce. Triclosan has a broad-spectrum antimicrobial effect, but when compared in an *in vitro* assay with an actual simulation of *Listeria* on chicken breast at 7°C, the triclosan failed to work, whereas kill rate was dramatic *in vitro*. This indicates that antimicrobials do work, but when in contact with food, they may lose their inhibitory activity.

Another example of antimicrobial films is the Microsphere™ System created by Bernard Technologies, Inc. (Chicago, Illinois). Microsphere powder is stable when stored dry and is activated when exposed to threshold levels of humidity in ambient air. When activated, it produces sodium dioxide from sodium chlorite, an oxidizer and a well-known antimicrobial agent. Sodium dioxide inactivates the functional enzymes of the cell and thus kills the cell. Microsphere powder releases controlled doses of NaO_2; because it is effective at low levels, it inhibits a wide range of microorganisms. Even though it releases sodium dioxide, chlorine is not released as a by-product which could be detrimental to the consumer. This system is dependent upon the humidity level, with increasing antimicrobial activity coincident with increasing humidity.

Currently, commercially available antimicrobial films have not been widely accepted, mainly because of the limited features combined with their cost. Nevertheless, as mentioned earlier, antimicrobial film is a promising technology, and can be a major contributor to food safety. Many research works are underway to screen and investigate effective antimicrobial compounds and their potential commercial feasibility.

Edible Coatings and Films

Edible coatings and films are used to inhibit the migration of moisture, oxygen, etc. while also serving as antimicrobials, antioxidants, and flavor components. Edible coatings are applied directly onto the food by a liquid film–forming solution or by molten compounds, by brushing, spraying, dipping, etc. Edible films are formed separately and applied to the food independently. Not only can the coatings and films be consumed, but they can also serve as conventional means for food safety by providing barriers to moisture and serving as vehicles for active components. Using edible coatings directly on food is a type of barrier allowing a reduction in the final packaging of the product. Since the coating is applied directly, the product can remain fresh through constant barrier protection even after the package is opened.

Aroma Enhancement

Not only can antimicrobials be incorporated into the packaging film, but many companies have now also developed ways to incorporate aromas to stimulate the senses. The film could contain volatile chemicals that are released and, as a result, create a pleasant odor. Some garbage bags have this technology to overcome the strong unpleasant odors of garbage decay. This application may be useful in the food industry in including pleasant food odors prior to packaging that would enhance the consumer's appetite.

Extrinsic Elements

MAP, even with the most advanced packaging technology, is an integral part of the entire manufacturing system and cannot be treated as an isolated solution to ensure food safety or to improve product quality. Research has revealed that the quality of the product before packaging and its storage temperature are two key elements in determining its eventual shelf-life. The benefits of MAP technology can only be obtained when storage temperature is well controlled and quality of raw product is good. Good sanitation practices, good manufacturing practices (GMPs), strict temperature management, and other existing hygiene systems should still be in place to ensure ultimate outcome from MAP. Therefore, the simplest ways to introduce extrinsic elements are C (clean), C (cold), and P (practice hurdles).

Clean

Keep it clean! A food manufacturing system follows the "garbage in, garbage out" principle. Minimal processing can take care of pathogens only to a certain extent. Thus, effective sanitation practices, workers' hygiene, and GMP are the foundations of MPF safety. Without those implemented practices, food safety programs will not work well.

An air curtain and positive air pressure are typically maintained in the processing line. Production personnel apply good hygiene practices so that contamination from air- and human-borne microorganisms can be reduced.

Cold

Keep it cold! A common misperception is that MAP can replace refrigeration to preserve foods. MAP is effective only when the temperature is maintained well below a predetermined appropriate level because the effectiveness of MAP has a reciprocal relationship with temperature (Harrison et al., 2000; Ozbas et al., 1997; Harris and Barakat, 1995; Mano et al., 1995; Marshall et al., 1991). Consequently, serious considerations for maintaining the cold chain intact from preparation to handling to distribution are a prerequisite for the success of MAP. The cold chain covers the chilling rate and temperature control. Chilling rate affects the rate of bacterial growth and the rate of cell deterioration, which then impact sensory properties of food products.

Chilling extends the life of food products in good condition by retarding the rate of deterioration. Chilling cannot improve the quality of a poor product or stop the process of spoilage.

Practice Hurdles

This concept is probably as old as classical MAP. Nevertheless, it plays a more viable role than ever in ensuring food safety. Hurdle and barrier technology means that multiple factors or technologies are employed to control microorganisms in foods effectively. The practice has been applied to some foods for over a century. A large number of factors can be applied to food systems as hurdles, and more minimally processed foods are likely to embody this concept to ensure food safety. Some alternative technologies being studied with MAP include gamma-irradiation (Thayer and Boyd, 2000; Grant and Patterson, 1991a,b) and high-pressure processing (Amanatidou et al., 2000). Both were found to have strong synergies with reductions in test microorganisms and quality enhancements, but further optimization needs to be studied before universal acceptance. Brody (1996) mentioned the potential integration of MAP into aseptic processing systems and good temperature control to deliver better quality, as well as more convenient and safer, food products.

Another way of applying effective hurdles is to combine process and product design factors, such as adding antimicrobial compounds (Szabo and Cahill, 1998), salt dipping (Pastoriza et al., 1998), nisin (Fang and Lin, 1994a,b), or organic acid (Zeitoun and Debevere, 1991). Also, barriers like water activity (a_w) and storage temperature (Ellis et al., 1994) can be effective in controlling and minimizing the growth of food-borne pathogens.

CONCLUSION

MAP does not need to be limited to the gas mixture and film. There is no magic solution to remedy food safety from packaging technology alone. Therefore, it is necessary to continue researching packaging technologies from a broad perspective. From the center of packaging, it is beneficial to know which technology can better be integrated into packaging materials through gases or in polymers. From the point of external processing world, it is important to know how to keep raw products clean and how other modern technologies can be synergized with packaging. Most of all, in order to make MPF safer, processors must know their products, know the potential microbial hazard, know processes, apply effective hurdles, and validate the entire system through well-designed challenge tests.

ACKNOWLEDGMENTS

The author wishes to thank Christine Boisrobert and Sejal Thakkar (Air Liquide America Corp.) for their collaborative help in writing this chapter.

REFERENCES

Amanatidou, A. et al., 2000, Effect of combined application of high pressure treatment and modified atmospheres on the shelf life of fresh Atlantic salmon, *Innovative Food Sci. Emerging Technol.*, 1(2000):87–98.

Brody, A.L., 1996, Integrating aseptic and modified atmosphere packaging to fulfill a vision of tomorrow, *Food Technol.*, 50(4):56–66.

Brody, A.L., 1989, *Controlled/Modified Atmosphere/Vacuum Packaging of Foods*, Trumbull, CT: Food & Nutrition Press.

Day, B., 2000, Conference Proceedings, International Conference on Active and Intelligent Packaging, September 7–8, Gloucestershire, U.K., Campden & Chorleywood Food and Drink Research Association, Chipping Campden, U.K.

Devlieghere, F. and Debevere, J., 2000, Influence of dissolved carbon dioxide on the growth of spoliage bacteria, *Lebensmittle-Wissenschaft und Technologie,* 33(8):531–537.

Diaz, C. and Hotchkiss, J.H., 1996, Comparative growth of *Escherichia coli* O157:H7, spoilage organisms and shelf-life of shredded iceberg lettuce stored under modified atmospheres, *J. Sci. Food Agric.*, 70(4):433–438.

Eliot, S.C., Vuillemard, J.C., and Emond, J.P., 1998, Stability of shredded mozzarella cheese under modified atmospheres, *J. Food Sci.*, 63(6):1075–1080.

Ellis, W.O. et al., 1994, Growth of and aflatoxin production by *Aspergillus flavus* in peanuts stored under modified atmosphere packaging (MAP) conditions, *Int. J. Food Microbiol.*, 22(2/3):173–187.

Fang, T.J. and Lin, L.-W., 1994a, Inactivation of *Listeria monocytogenes* on raw pork treated with modified atmosphere packaging and nisin, *J. Food Drug Anal.*, 2(3):189–200.

Fang, T.J. and Lin, L.-W., 1994b, Growth of *Listeria monocytogenes* on cooked pork in a modified atmosphere packaging/nisin combination system, *J. Food Prot.*, 57(6):479–485.

Fedio, W.M., Macleod, A., and Ozimek, L., 1994, The effect of modified atmosphere packaging on the growth of microorganisms in cottage cheese, *Milchwissenschaft*, 49(11):622–629.

Grant, I.R. and Patterson, M.F., 1991a, Effect of irradiation and modified atmosphere packaging on the microbiological safety of minced pork stored under temperature abuse conditions, *Int. J. Food Sci. Technol.*, 26(5):521–533.

Grant, I.R. and Patterson, M.F., 1991b, Effect of irradiation and modified atmosphere packaging on the microbiological and sensory quality of pork stored at refrigeration temperatures, *Int. J. Food Sci. Technol.*, 26(5):507–519.

Harris, L.J. and Barakat, R.K., 1995, Growth of *Listeria monocytogenes* and *Yersinia enterocolitica* on cooked poultry stored under modified atmosphere at 3.5, 6.5 and 10 °C, *J. Food Prot.*, 58(Suppl.):38.

Harrison, W.A., Peters, A.C., and Fielding, L.M., 2000, Growth of *Listeria monocytogenes* and *Yersinia enterocolitica* colonies under modified atmosphere at 4 and 8°C using a model food system, *J. Appl. Microbiol.*, 88(1):38–43.

Hintlian, C.B. and Hotchkiss, J.H., 1986, The safety of modified atmosphere packaging: a review, *Food Technol.*, 40(12):70–76.

Juneja, V.K., Marmer, B.S., and Call, J.E., 1996, Influence of modified atmosphere packaging on growth of *Clostridium perfringens* in cooked turkey, *J. Food Safety*, 16(2):141–150.

Lawlor, K.A. et al., 2000, Nonproteolytic *Clostridium botulinum* toxigenesis in cooked turkey stored under modified atmospheres, *J. Food Prot.*, 63(11):1511–1516.

Lee, C.H. and Cash, J.N., 1998, Comparative growth rates of bacteria on minimally processed meat-vegetable product under modified atmospheres, *Food Sci. Biotechnol.*, 7(1):6–12.

Löwenadler, J., 1994, Modified atmosphere packaging, carbon dioxide, its interactions with micro-organisms and application as a food preservative: a review, SIK Report No. 603, The Swedish Institute for Food Research, Göteborg, Sweden.

Lyver, A. et al., 1998, Challenge studies with *Listeria monocytogenes* in a value-added seafood product stored under modified atmospheres, *Food Microbiol.*, 15(4):379–389.

Mano, S.B. et al., 1995, Growth/survival of natural flora and *Listeria monocytogenes* on refrigerated uncooked pork and turkey packaged in modified atmospheres, *J. Food Safety*, 15(4):305–319.

Mano, S.B., Ordonez, J.A., and Garcia de Fernando, G.D., 2000, Growth/survival of natural flora and *Aeromonas hydrophila* on refrigerated uncooked pork and turkey packaged in modified atmospheres, *Food Microbiol.*, 17(6):657–669.

Marshall, D.L. et al., 1991, Comparative growth of *Listeria monocytogenes* and *Pseudomonas fluorescens* on precooked chicken nuggets stored under modified atmospheres, *J. Food Prot.*, 54(11):841–843.

Mathlouthi, M., 1994, *Food Packaging and Preservation*, Glasgow, U.K.: Blakie Academic & Professional.

Ohta, T. and Sasaki, S., 1995, Modified atmosphere storage of chum salmon (*Oncorhynchus keta*) fillets, *J. Jpn. Soc. Food Sci. Technol.*, 42(7):536–539.

Ozbas, Z.Y., Vural, H., and Aytac, S.A., 1997, Effects of modified atmosphere and vacuum packaging on the growth of spoilage and inoculated pathogenic bacteria on fresh poultry, *Fleischwirtschaft*, 77(12):1111–1116.

Paine, F.A. and Paine, H.Y., 1992, *A Handbook of Food Packaging*, 2nd ed., Glasgow, U.K.: Blakie Academic & Professional.

Pastoriza, L. et al., 1998, Influence of sodium chloride and modified atmosphere packaging on microbiological, chemical and sensorial properties in ice storage of slices of hake (*Merluccius merluccius*), *Food Chem.*, 61(1/2):23–28.

Phillips Daifas, D. et al., 1999, Effect of pH and CO_2 on growth and toxin production by *Clostridium botulinum* in English-style crumpets packaged under modified atmospheres, *J. Food Prot.*, 62(10):1157–1161.

Rooney, M.L., 1995, *Active Food Packaging*, Glasgow, U.K.: Blackie Academic & Professional.

Silliker, J.H. and Wolfe, S.K., 1980, Microbiological safety considerations in controlled-atmosphere storage of meats, *Food Technol.*, 34(3): 59–63.

Spencer, K., 1995, Conference Proceedings of Modified Atmosphere Packaging (MAP) and Related Technologies, September 6–7, Chipping Campden, UK, C-11.

Szabo, E.A. and Cahill, M.E., 1998, The combined effects of modified atmosphere, temperature, nisin and ALTA™ M2341 on the growth of *Listeria monocytogenes*, *Int. J. Food Microbiol.*, 43(1/2):21–31.

Thayer, D.W. and Boyd, G., 2000, Reduction of normal flora by irradiation and its effect on the ability of *Listeria monocytogenes* to multiply on ground turkey stored at 7°C when packaged under a modified atmosphere, *J. Food Prot.*, 63(12):1702–1706.

Whitley, E., Muir, D., and Waites, W.M., 2000, The growth of *Listeria monocytogenes* in cheese packed under a modified atmosphere, *J. Appl. Microbiol.*, 88(1):52–57.

Zeitoun, A.A.M. and Debevere, J.M., 1991, Inhibition, survival and growth of *Listeria monocytogenes* on poultry as influenced by buffered lactic acid treatment and modified atmosphere packaging, *Int. J. Food Microbiol.*, 14(2):161–169.

11 Washing and Sanitizing Raw Materials for Minimally Processed Fruit and Vegetable Products*

Gerald M. Sapers

CONTENTS

* Mention of brand or firm name does not constitute an endorsement by the U.S. Department of Agriculture above others of a similar nature not mentioned.

1-58716-041-2/03/$0.00+$1.50

INTRODUCTION

Guidelines for packing fresh or minimally processed fruits and vegetables generally specify a washing or sanitizing step to remove dirt, pesticide residues, and microorganisms responsible for quality loss and decay. Additionally, this step is used to precool cut produce and remove cell exudates that adhere to product cut surfaces and may support microbial growth or result in discoloration. The U.S. Food and Drug Administration's *Guide* calls for removing "as much dirt and mud as practicable from the produce before it leaves the field." The produce should be cleaned (washed and rinsed with processing water of such quality that it does not contaminate the produce) to be visually free of dust, dirt, and other debris and sanitized (treated "by a process that is effective in destroying or substantially reducing the numbers of microorganisms of public health concern, as well as other undesirable microorganisms, without adversely affecting the quality of the product or its safety for the consumer") (FDA, 1998).

In recent years, increasing attention has been focused on the microbiological safety of fruits and vegetables and, in particular, on interventions to kill or remove human pathogens from fresh produce. The number of outbreaks of food-borne illness associated with fresh produce appears to be increasing (Tauxe et al., 1997; NACMCF, 1999; DeWaal et al., 2000). In the U.S., the majority of reported cases involved a limited number of commodities: alfalfa sprouts, lettuce or other salad greens, melons, unpasteurized juices including fresh apple and orange juices, tomatoes, berries, cabbage, and unspecified fruits (NACMCF, 1999; DeWaal et al., 2000). Some of these have been minimally processed products (Francis et al., 1999). The causative organisms, where identified, included *Escherichia coli* O157:H7, *Salmonella* species, *Shigella* species, *Cyclospora cayetanensis*, Hepatitis A virus, Norwalk-like virus, and, rarely, *Listeria monocytogenes* and *Clostridium botulinum* (Beuchat, 1996; NACMCF, 1999). A key goal of washing and sanitizing fresh or minimally processed fruits and vegetables, therefore, is the removal or inactivation of pathogenic microorganisms.

Until recently, relatively little information was published concerning the efficacy of washing and sanitizing treatments in reducing microbial populations on produce. The limited data published suggested that conventional washing and sanitizing methods, even using some of the newer sanitizing agents such as chlorine dioxide, ozone, and peroxyacetic acid, were not capable of reducing microbial populations by more than 90 or 99% (Beuchat, 1998; Brackett, 1999). Although such reductions represent a large decrease in the numbers of microorganisms present on the commodity and may result in significant improvements in product quality and shelf-life, they are not equivalent to surface pasteurization and may be inadequate to ensure product safety. Thus, there is a need to examine the factors that limit the efficacy of washing in reducing microbial populations on produce and to devise means of overcoming such limitations. New washing and sanitizing treatments must not only be effective, but they also must be compatible with commercial packing and processing practices and technical capabilities. New treatments must be affordable and safe to carry out, have no adverse effect on quality, and be approved by applicable regulatory agencies. Acceptance by industry also may depend on regulatory constraints regarding use of the term "fresh," conformance to organic food product labeling requirements, and perceived consumer attitudes regarding irradiation and "chemicals."

The problem of decontaminating produce by washing or application of chemical sanitizing treatments cannot be understood in isolation; we must take into account not only the size of the microbial load on fruits and vegetables but also the way in which microorganisms attach and survive on produce surfaces. The condition of the attached microflora will depend, in many cases, on how and when the produce became contaminated. Generally, it is more difficult to decontaminate produce than it is to avoid contamination. Therefore, pre- and postharvest interventions that reduce the risk of contamination will make the job of sanitizing produce easier.

In this chapter, the focus will be on bacterial contamination of produce, especially contamination with human pathogens. Major factors that limit the efficacy of conventional washing methods and their relationship to the circumstances of contamination will be examined. The ability of conventional washing and sanitizing agents and produce washing equipment to reduce microbial loads will be reviewed. Finally, new developments in washing and sanitizing technology and the prospects for significant improvements in decontamination efficacy will be discussed.

Nonthermal physical treatments such as exposure to ionizing radiation, high intensity pulsed light, and high pressure have the capability to sterilize or pasteurize some commodities and may have some application to minimally processed fruit and vegetable products. However, since these processes are substantially different from washing and sanitizing treatments, they will not be discussed in this chapter. Similarly, a number of natural products have antimicrobial activity. However, in most cases, these agents do not exert their effects within the time frame of a wash but act as preservatives, inactivating microorganisms or suppressing their growth during storage. Such applications also lie outside the scope of this chapter.

FACTORS LIMITING THE EFFICACY OF WASHING

Sources of Microbial Contamination

Preharvest Contamination

Ultimately, human pathogens associated with produce can be traced to human or animal fecal contamination. Produce is grown on farms, not in clean rooms, and it is not possible to exclude animals completely from the fields and orchards where produce originates. Human pathogens associated with food-borne illness have been found in cattle (Faith et al., 1996), sheep (Kudva et al., 1996), deer (Rice et al., 1995), wild birds (Wallace et al., 1997), and amphibians (Parish, 1997). They might be carried by insects (Janisiewicz et al., 1999a; Iwasa et al., 1999; Kobayashi et al., 1999). Outbreaks of *E. coli* O157:H7 associated with apple cider have been attributed to cattle grazing in apple orchards (Besser et al., 1993). Contamination of produce might result from fertilizing with improperly composted manure in which human pathogens survive. Low-growing fruits and vegetables might be splashed with contaminated muddy water during a heavy rainstorm. The following examples, taken from recent field studies carried out by scientists at the USDA's Eastern Regional Research Center, illustrate the contamination problem.

Alfalfa sprouts have a history of involvement in outbreaks of illness caused by *Salmonella* and occasionally by *E. coli* O157:H7 (Taormina et al., 1999). In an investigation of the microbiological safety of alfalfa sprouts, an example was encountered of probable fecal contamination of alfalfa seeds, probably due to the practice of allowing cattle to graze in alfalfa fields after the last harvest of the growing season. Seed and debris fractions (weed seeds, plant fragments, insect parts, and soil) obtained from commercial seed cleaning equipment were found to be positive for generic *E. coli* and confirmed *Salmonella* (serotypes Bredeney and Worthington). Contamination apparently occurred when these organisms survived the winter in cattle feces deposited in the field and became attached to the new crop during the following spring (Fett and Sapers, 1997).

Water used for irrigation or to make up pesticide or other sprays might represent another potential source of fecal contamination if it is obtained from a contaminated pond, stream, or irrigation canal where animals drink or where run-off from adjacent fields or pastures occurs. Generic *E. coli* was detected in each of these water sources at orchard locations associated with *E. coli* contamination of apples intended for cider production (Riordan et al., 2001).

Produce might be contaminated by windblown dust from a nearby pasture or feedlot that contains particles of desiccated cattle feces and human pathogens. Evidence of this scenario was seen in apples obtained from an orchard located downwind of a feedlot (Fett and Sapers, 1997). Soil from the orchard tested positive for coliforms and generic *E. coli*. The apples were visibly dirty in the stem cavity area, and many apples tested positive for coliforms and generic *E. coli*.

Contamination from microorganisms in dust or irrigation water could complicate the problem of decontamination by favoring microbial attachment in relatively inaccessible sites such as upward facing calyx or stem areas of apples where dust and water can be trapped. Microbial penetration and growth within punctures resulting from bird pecks or hail damage are possible consequences of such contamination. The moist environment and nutrient availability prevailing in such attachment sites also might favor biofilm formation, which would increase microbial resistance to washing and sanitizing agents.

Growers should make every effort to exclude animals or their feces from their fields and orchards by use of suitable fencing or bird-repellent devices, by avoiding exposure to contaminated dust and irrigation sources, and by following other pertinent good agricultural practices and guidelines (See Interventions section below).

Postharvest Contamination

Produce might become contaminated with human pathogens as a result of contact with contaminated soil or water during harvesting, postharvest handling, or processing. Outbreaks of food-borne illness have been associated with production of unpasteurized cider from apples that had fallen on the ground (Besser et al., 1993; CDC, 1997). Bins or other containers used to hold harvested produce might be contaminated with soil, decayed fruit or vegetable fragments, or other debris. Deer, which are known sources of *E. coli* O157:H7 (Rice et al., 1995), are frequently found in orchards near wooded areas and might defecate on the ground beneath the trees where bins were stored. Stacking such bins might allow potentially contaminated dirt adhering to the bottom of one bin to fall into the bin beneath. Storing loaded bins under shade trees in the field would increase the chance of produce contamination by bird droppings. Trucks used to ship produce from the grower to the packer or processor might be contaminated with human pathogens, especially if the trucks were used previously to transport animals such as pigs or poultry and were not adequately cleaned and sanitized. Water used in drenchers, hydrocoolers, dump tanks, or flumes might be contaminated if obtained from a contaminated source. If such water were recycled without further treatment, opportunities for cross-contamination would exist. Cross-contamination also can take place in a packing or processing plant if equipment in contact with potentially contaminated produce is not regularly cleaned and sanitized.

Interventions to Avoid Postharvest Contamination

A number of interventions can be implemented to avoid postharvest contamination of produce with human pathogens. Various government agencies and industry associations have published comprehensive guidelines for the produce industry. A description of good manufacturing practices appears in the U.S. Code of Federal Regulations (21CFR 110, 1994a). The Food and Drug Administration has issued a *Guide to Minimize Microbial Food Safety Hazards for Fresh Fruits and Vegetables* (FDA, 1998). The International Fresh-cut Produce Association (IFPA) has published *Food Safety Guidelines for the Fresh-cut Produce Industry* (Zagory and Hurst, 1996). This document calls for chlorination of water used for field washing and in packing houses and provides information about the characteristics of various sanitizing agents added to fresh-cut processing water or used to sanitize plant equipment. The Codex Committee on Food Hygiene has developed a *Code of Hygienic Practice* that is now in draft form (Couture, 1999). The Canadian government has published a *Code of*

Practice for production of unpasteurized fruit juices (Canadian Food Inspection Agency, 1998).

Implementation of a hazard analysis critical control point (HACCP) plan by processors and HACCP-like plans (e.g., good agricultural practices [GAPs] and good manufacturing practices [GMPs]) by packers can result in exclusion of contaminated raw materials and implementation of effective decontamination treatments. Such plans are described in the IFPA guidelines (Zagory and Hurst, 1996), a report of the National Advisory Committee on Microbiological Criteria for Foods (NACMCF, 1997), and regulations promulgated by the FDA (FDA, 2001).

Cleaning and sanitizing represent key interventions in any program to avoid human pathogen contamination of fresh or minimally processed produce. Antimicrobial agents such as chlorine (or hypochlorous acid obtained from sodium or calcium hypochlorite or compressed chlorine gas), chlorine dioxide, ozone, or peroxyacetic acid can be added to process water used in dump tanks, flumes, and washers to reduce the bacterial load and levels of any human pathogens that might be present. This is an effective means of minimizing cross-contamination, i.e., the attachment of human pathogens suspended in the process water to uncontaminated produce passing through the water. However, as discussed above, the ability of these antimicrobial agents to kill bacteria already attached to the surface of fruits and vegetables is limited to population reductions of 90 to 99%.

Customarily, population reductions resulting from antimicrobial treatments are expressed as log reduction values rather than as percentages, to avoid the need to work with very large numbers. Log reductions are calculated by subtracting the logarithm of the microbial population surviving a wash or sanitizer treatment ($\log_{10}$CFU/g, where CFU is the number of colony forming units, i.e., the number of viable microorganisms recovered and enumerated on specific microbiological growth media) from the logarithm of the initial microbial population before treatment. Log reductions of 1, 2, 3, 4, and 5 are equivalent to percentage reductions of 90, 99, 99.9, 99.99, and 99.999, respectively. Henceforth in this chapter, log reductions will be used in place of percent reductions.

The FDA has established a 5-log reduction in the human pathogen population in fresh apple cider as a goal for pasteurization treatments (flash pasteurization, UV pasteurization) or other interventions that could be applied to the juice (FDA, 2001). A 5-log reduction in human pathogens on seeds for sprout production has been proposed by the National Advisory Committee on Microbiological Criteria for Foods but is not required by the FDA (Anonymous, 1999b; FDA, 1999). Although the 5-log reduction target might not be directly applicable to or required of minimally processed (i.e., fresh-cut) fruit and vegetable products, it does indicate the kind of population reduction goal that food safety authorities visualize for high-risk products.

Equipment used in conveying, sorting, cutting, peeling, or performing other operations on fresh produce might be sources of human pathogens if fragments of contaminated fruits and vegetables become lodged within the equipment so that microbial growth can occur or if biofilms containing human pathogens become established on food contact surfaces. When incoming produce passes through the packing or processing line, cross-contamination can occur as fragments detach and contact fruit or vegetable surfaces. To avoid this situation, equipment must be cleaned

TABLE 11.1
Annual Buyers' Guides as Sources for Washing and Sanitizing Agents and Equipment

Category	Food Quality Buyer's Guide[a]	Food Processing Guide and Directory[b]	IFPA Membership Directory and Buyer's Guide[c]	Prepared Foods Food Industry Sourcebook[d]
Surfactants and detergents	X	X		X
Sanitizers	X	X	X	X
Chlorine	X	X		
Ozone		X		
Chlorine dioxide		X	X	
Produce washing equipment		X	X	
Sanitizing equipment	X	X	X	X
Cleaning equipment	X	X	X	X
Sanitation, GMP, and HACCP consultants	X	X	X	X
Contract laboratory services	X	X	X	X
Microbiological supplies and test kits	X	X	X	X

[a] Supplement to *Food Quality Magazine*, Carpe Diem Communications, Inc., Yardley, PA. 1999/2000. Online guide at www.foodquality.com.
[b] Supplement to *Food Processing*, Putnam Publishing Co., Itasca, IL, October, 2000. Online guide at www.foodprocessing.com.
[c] Published by International Fresh-cut Produce Association, Alexandria, VA. 2000/2001. See www.fresh-cuts.org.
[d] Supplement to *Prepared Foods*, Des Plaines, IL, 1996. Online guide at www.preparedfoods.com.

frequently with suitable detergents to remove produce fragments and biofilms and then sanitized to kill microbial contaminants. Approved sanitizers suitable for equipment and food contact surfaces include chlorine-based sanitizers, iodine-based sanitizers, quaternary ammonium compounds, and acid-anionic sanitizers (Zagory and Hurst, 1996; 21CFR178.1010, 2000). Sources of these products can be located in directories of suppliers to the food industry (Table 11.1).

Microbial Attachment to Produce

The condition and location of microorganisms on produce surfaces affect their resistance to detachment by washing agents and to inactivation by antimicrobial agents. Microbial resistance to sanitizing washes will depend in part on whether the

TABLE 11.2
Effect of Interval between Inoculation of Apples with *E. Coli* (ATCC 25922) and Water Washing on Bacterial Population Reduction[a]

Time after Inoculation (hr)	$Log_{10}CFU/g^b$ 4°C Inoculated Control	After Wash	20°C Inoculated Control	After Wash	$Log_{10}CFU/g$ Reduction from Wash[f] 4°C	20°C
0.5	4.40^c	3.46^c	$4.35^{c,d}$	3.38^e	0.94*	0.97**
24	$3.89^{c,d}$	3.22^c	4.80^c	$4.33^{c,d}$	NS	0.47**
48	$3.88^{c,d}$	3.97^c	4.06^d	4.65^c	NS	–0.59**
72	3.66^d	3.64^c	$4.18^{c,d}$	$3.88^{d,e}$	NS	NS

[a] From Sapers, G.M. et al., *J. Food Sci.*, 65:529–532, 2000a.
[b] Mean of duplicate trials.
[c-e] Within the same column, means with no letter in common are significantly different ($p<0.05$) by Bonferroni LSD.
[f] Significance of $log_{10}CFU/g$ reduction tested by ANOVA: * ($p<0.05$), ** ($p<0.01$), NS = not significant.

microbial populations have become firmly attached, are concentrated in inaccessible sites, have penetrated into the interior of the commodity, or have become incorporated into biofilms. Each of these factors will affect the success of decontamination.

Rapidity of Attachment

One of the characteristics of bacterial attachment to fruits and vegetables is the rapidity of attachment to the commodity surface. The effectiveness of washing will depend on the time interval between contamination and washing. Data obtained with apples that were artificially inoculated with *E. coli* and then held for various times before washing with water indicate that an interval of 30 min between inoculation and washing resulted in a 1-log population reduction (Table 11.2). However, after 24 h, essentially all of the bacteria were firmly attached and could not be removed by washing (Sapers et al., 2000a). Similar results were obtained with cantaloupe artificially inoculated with a nonpathogenic *E. coli* or *Salmonella stanley* and then washed with water, 1000 ppm chlorine (as sodium hypochlorite), or 5% hydrogen peroxide at various intervals after inoculation (Ukuku et al., 2001; Ukuku and Sapers, 2001). Immediately after inoculation, the attached *E. coli* and *S. stanley* populations were detached and/or killed by washing with the chlorine or hydrogen peroxide solutions (but not by washing with water), resulting in reductions exceeding 3.5 to 4.5 log_{10} CFU/cm^2. However, after 72 h, washing with these antimicrobial agents was much less effective in reducing the bacterial populations, resulting in reductions less than 3 log_{10} CFU/cm^2 for *E. coli* and less than 2 log_{10} CFU/cm^2 for *S. stanley.* Studies with pepper disks have shown that initial attachment of *Salmonella* to cut surfaces is very rapid and is directly related to the inoculum population but not to exposure time (0.5 to 16 min) (Liao and Cooke, 2001).

TABLE 11.3
Distribution of *E. Coli* (ATCC 25922) on Surfaces of Inoculated Apples before and after Washing with 5% H_2O_2 at 50°C[a]

	Log_{10} (CFU/cm^2)[b]			
	24 h after Inoculation		72 h after Inoculation	
Location	Inoculated	Washed[c]	Inoculated	Washed[c]
Skin on wedges	4.77[e]	2.05[e]	4.37[e]	1.63[e]
Skin at calyx end of core	7.26[e]	5.20[d]	6.79[d]	4.46[d]
Skin on stem end of core	6.63[d]	5.06[d]	5.61[d]	4.89[d]

[a] From Sapers, G.M. et al., *J. Food Sci.*, 65:529–532, 2000a.
[b] Based on calculated surface area of skin.
[c] Washed 1 min in 5% H_2O_2 at 50°C.
[d-e] Within the same column, means with no letter in common are significantly different ($p<0.05$) by Bonferroni LSD.

Attachment in Inaccessible Sites

When bacteria attach to the surfaces of fruits and vegetables, they tend to concentrate where there are more binding sites. Attachment also might be in stomata (Seo and Frank, 1999), indentations, or other natural irregularities on the intact surface where bacteria could lodge. Bacteria also might attach at cut surfaces (Takeuchi et al., 2000; Liao and Cooke, 2001) or in punctures or cracks in the external surface (Burnett et al., 2000).

Data obtained with apples artificially inoculated with *E. coli* suggest greater attachment to skin in the calyx and stem areas than elsewhere on the apple (Table 11.3). When the inoculated apples were held 24 h and then washed with 5% hydrogen peroxide, many more survivors per square cm of skin surface were in the calyx and stem areas than elsewhere on the apple surface (Sapers et al., 2000a). Bacteria in the stem and calyx areas may adhere better to the irregular skin surface and to the flower parts of the calyx than to the smooth skin surface. Bacteria in these locations are inaccessible or at least less accessible to the washing agent, which requires physical contact to be effective. High levels of bacteria are found in the calyx and stem areas of naturally contaminated apples (Riordan et al., 2001).

Salmonella chester attach preferentially to the cut surfaces of apple and green pepper disks, where they survive washing to a much greater extent than bacteria on the intact apple and pepper surfaces (Liao and Sapers, 2000; Liao and Cooke, 2001). Since fresh-cut fruits and vegetables provide extensive cut surfaces for bacterial attachment, it is especially important with these products that exposure to human pathogens be avoided.

Attachment and Growth in Punctures

Some commodities such as certain apple and pear varieties, zucchini squash, potatoes, and carrots often have punctures, cuts, or splits that might be sites for bacterial

TABLE 11.4
Efficacy of H_2O_2-Based Washes for Decontamination of Punctured Golden Delicious Apples Inoculated with *E. Coli* (ATCC 25922)[a]

Treatment[b]	Log_{10} CFU/g Reduction[c]	
	No Puncture	Punctured[d]
5% H_2O_2	2.34	0.58
1% APL-Kleen® 245; 5% H_2O_2[e]	2.83	1.62

[a] From Sapers, G.M. et al., *J. Food Sci.*, 65:529–532, 2000a.
[b] 1-min wash at 50°C.
[c] Means of duplicate trials; based on control populations of 4.88 log_{10}CFU/g.
[d] 1-cm deep puncture made with 3.7-mm diameter sterile nail on top of apple 2 to 3 cm from stem.
[e] Two-stage treatment.

attachment. *E. coli* can grow within punctures in artificially inoculated apples in spite of the fact that this organism normally will not grow in highly acidic apple juice (Sapers et al., 2000a). Presumably, the bacteria are able to create a more hospitable microenvironment within the confines of the puncture. Other investigators have reported growth of *E. coli* in wounds on apples (Janisiewicz et al., 1999a, b). Once the bacteria have become established within a puncture, they are very difficult to kill (Table 11.4). A 5% hydrogen peroxide wash reduced the *E. coli* population on inoculated apples with punctures by only 0.6 log, compared to a 2.3-log reduction on inoculated apples without punctures (Sapers et al., 2000a).

Biofilms

Attached bacteria might grow and form biofilms, bacterial communities adherent to a surface and each other by a self-produced polysaccharide matrix (Zottola, 1994; Costerton, 1995; Carmichael et al., 1999). Alternatively, introduced bacteria may become part of existing biofilms produced by the native microflora. In this state, bacteria are very resistant to detachment or inactivation by antimicrobial washes. Human pathogens such as *E. coli* O157:H7, *Salmonella* spp., and *L. monocytogenes*, as well as other bacteria such as *Pseudomonas* and *Erwinia* spp., can form biofilms on inert surfaces (Somers et al., 1994). Biofilms on plant surfaces can comprise mixed populations of Gram-positive and Gram-negative bacteria, yeasts, and filamentous fungi (Morris et al., 1997).

Examples of typical biofilms found in the calyx of an apple and on the surface of an alfalfa sprout are shown in Figures 11.1a and 11.1b, respectively. The presence of biofilms on surfaces of fruits and vegetables greatly limits the ability to decontaminate them successfully (Carmichael et al., 1999; Fett, 2000). Similarly, biofilm formation on processing equipment complicates effective cleaning and sanitation (Zottola, 1994). Trisodium phosphate, applied as a 2 to 8% solution, has been shown

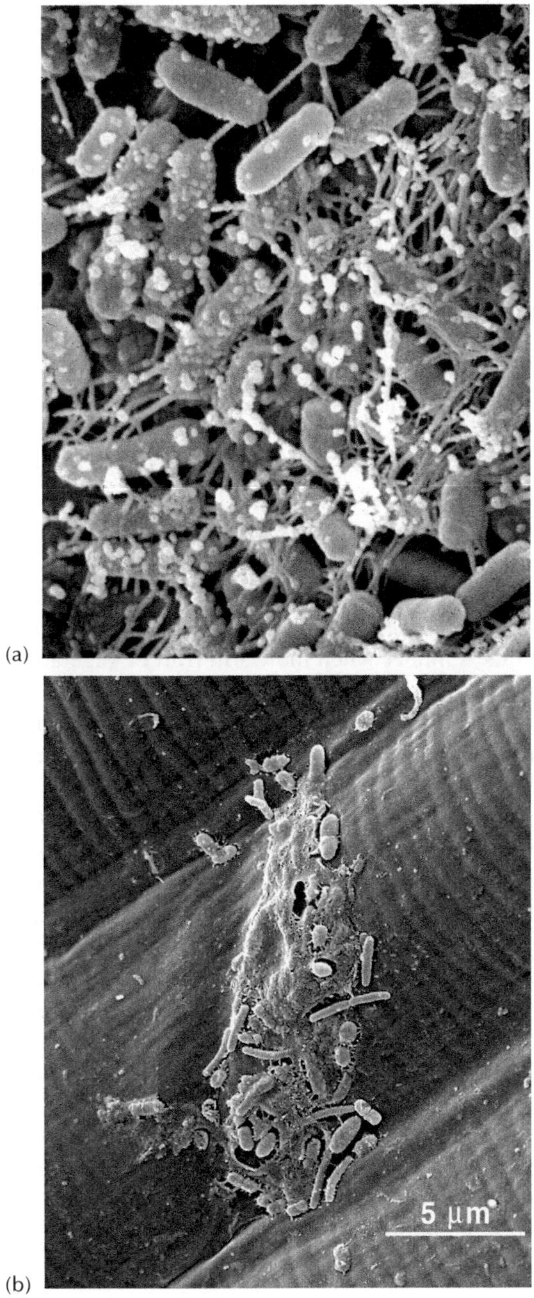

FIGURE 11.1 Micrographs of bacterial biofilms on produce: (a) apple calyx; (b) alfalfa sprouts.

to be effective against biofilm cells of *E. coli* O157:H7 and *Campylobacter jejuni* on stainless steel. Biofilm cells of *Salmonella typhimurium* and *L. monocytogenes* on stainless steel and Buna-N rubber were more resistant to trisodium phosphate treatments (Somers et al., 1994).

Internalization and Infiltration of Bacteria within Produce

If human pathogens can penetrate into the interior of a fruit or vegetable, they will survive surface decontamination treatments such as washing and surface pasteurization with hot water or steam. Internalization of bacteria can result from infiltration due to handling conditions during packing or processing (Bartz and Showalter, 1981; Bartz, 1982; Buchanan et al., 1999). Internalization also can occur naturally due to contamination of the flowering plant or during fruit development (Samish et al., 1963). It is not unusual to find a peach with a moldy pit or some other fruit or vegetable with a latent internal infection. Internalization of *E. coli* O157:H7 has been reported in lettuce (Seo and Frank, 1999; Takeuchi and Frank, 2000) and radish sprouts (Itoh et al., 1998), while other bacterial species have been detected within cucumbers and tomatoes (Samish et al., 1963; Meneley and Stanghellini, 1974; Daeschel and Fleming, 1981; Breidt et al., 2001).

Infiltration of bacteria into produce can occur in warm commodities with internal air spaces when they are placed in colder water, perhaps in a dump tank or flume. This can happen in the summer when the warm fruit or vegetable is brought in from the field and is washed with cool water obtained from a well. As the fruit or vegetable cools, the internal gas contracts, thereby creating a partial vacuum that will draw in water through pores, channels, or punctures. If the water contains human pathogens or microorganisms capable of causing spoilage, they also will be drawn into the commodity (Bartz and Showalter, 1981; Buchanan et al., 1999). Commercial experience has shown that tomato dump tank water temperature should be about 5°C higher than the fruit temperature to minimize infiltration (Hurst, 2001). Infiltration can occur when the commodity is at the bottom of a deep tank, but in this case, the driving force is hydrostatic pressure (Bartz, 1982; Sugar and Spotts, 1993). One can speculate that infiltration also might occur in the absence of an external driving force if the bacteria on the commodity surface are motile, i.e., they can move of their own accord, and they follow a channel into the interior of a fruit or vegetable when adequate surface water is available. In some apple cultivars, an open channel from the calyx leads directly to the core (Miller, 1959).

Infiltration of *Erwinia carotovora* subsp. *carotovora*, a spoilage organism, and *Salmonella montevideo*, a human pathogen, has been demonstrated in tomatoes (Bartz and Showalter, 1981; Zhuang et al., 1995). Infiltration of *E. coli* O157:H7 into Golden Delicious apples was demonstrated in the author's laboratory (Buchanan et al., 1999).

Microbial Interactions Favoring Pathogen Growth

When human pathogens contaminate a fruit or vegetable, they interact not only with the commodity surface but with the native microbial populations on the commodity, as well. In some cases, the interaction might be antagonistic, resulting in a decline in the pathogen population. In other cases, a synergistic or commensal relationship might favor pathogen survival or growth.

In studies carried out in the author's laboratory, Wells and Butterfield (1997) reported a strong statistical association between the occurrence of bacterial soft rot and the presence of presumptive *Salmonella*. In further testing, they found that about 30% of isolates from these samples were confirmed as *Salmonella*. It is not clear whether the association resulted from some positive interaction between *Salmonella* and the decay organisms or from the greater availability of nutrients in the macerated tissue that resulted from soft rot. Liao and Sapers (1999) reported an antagonistic relationship between soft-rotting bacteria on potato slices and *L. monocytogenes* strain Scott A. *Pseudomonas fluorescens* suppressed growth of *Listeria*, but *E. carotovora*, another soft-rotting species, permitted growth of *Listeria*. Thus, one cannot generalize about interactions between human pathogens and the endogenous microflora.

In a study of *E. coli* O157:H7 survival and growth in puncture wounds on apples, Riordan et al. (2000) reported that the human pathogen would die off quickly if the wound were infected with the common apple pathogen, *Penicillium expansum*. This was due in part to pH reduction caused by the fungus. However, if the wound were infected by *Glomerella cingulata*, another fungus associated with apple decay, the pH increased greatly, and *E. coli* O157:H7 grew and survived over a long period of time. *E. coli* O157:H7 grew equally well if the wound was not infected with a fungal pathogen; however, such growth was difficult to demonstrate because of unintended infection of inoculated wounds with native *Penicillium*. These results suggest a potentially hazardous situation in which apples punctured in a fall from a tree might be simultaneously infected with *E. coli* O157:H7 and *G. cingulata*. The decayed apple might eventually contain a population of *E. coli* O157:H7 sufficient to contaminate a large volume of apple cider.

These interactions have several important consequences. In the case of an antagonistic relationship, if a commodity is sanitized effectively so that the native antagonistic bacterial population is greatly reduced, a surviving human pathogen or newly introduced contaminant might get a foothold and grow without competition. The product might develop a dangerous level of the pathogen without ever showing visible spoilage. This was observed with vacuum-packed sliced potatoes artificially inoculated with *L. monocytogenes*, during storage at 15°C, an abusive condition (Juneja et al., 1998). That is why there is a concern about the development of washing treatments that extend product shelf-life by greatly reducing the population of spoilage organisms. One might be making the commodity "too clean." A second consequence is the possibility that a decayed fruit or vegetable might contain localized but not necessarily conspicuous areas where the population of a human pathogen is very high as a result of a synergistic or commensal interaction with the decay organism. This fruit or vegetable might be incorporated into a product such as unpasteurized cider if inspecting and sorting procedures are inadequate.

Detection and Removal of Produce with Defects

Fruits and vegetables with punctures, cuts, or decayed areas are at risk of being contaminated with human pathogens; thus, it would be desirable to cull defective produce from packing or processing lines to reduce the risk of contamination. Inspection and sorting of incoming raw material are standard operations in packing and processing plants, but these operations are subject to human error, especially at high production rates, when inspectors are fatigued or when defects are inconspicuous. Small companies may not be set up properly for inspection and sorting or may lack the personnel to carry out these operations. Apples with decay spots and punctures have been observed in use for cider production because inspection was inadequate.

Technology for automated defect detection and sorting of produce is under development, but a number of technical problems must be solved before the equipment becomes available. This technology also may be too costly for the small packer or processor. Thus, produce packers and processors must ensure that they have adequate resources to inspect and sort raw material properly to exclude fruits or vegetables with potentially contaminated defects.

CONVENTIONAL WASHING TECHNOLOGY

Equipment and Modes of Operation

A number of types of commercial washers are used to wash fruits and vegetables (Figure 11.2). For commodities such as apples and potatoes, a brush washer should be used. Two types are available: flat-bed washers in which rotating brushes are arranged in a horizontal plane perpendicular to the flow of product, and U-bed washers in which the rotating brushes are arranged in a U-shaped configuration parallel to the flow of product. Brushes are tailored to the characteristics of specific commodities and are designed to scrub off adhering soil without damaging the product skin. Various other types of washers are available for leafy vegetables, broccoli, root vegetables, corn, etc. These include reel washers, pressure washers, hydro air agitation wash tanks, and immersion pipeline washers. (See Table 11.1 for sources.) While these units are designed to remove soil and pesticide residues, not much is known about their efficacy in removing bacterial contaminants.

To obtain such information, Sapers and colleagues conducted a series of washing trials with inoculated apples, using conventional and experimental wash solutions applied with commercial brush washers. Preliminary washing trials carried out with a U-bed brush washer indicated that population reductions were less than 1 log when apples inoculated with a nonpathogenic *E. coli* were washed with water, 200 ppm Cl_2 (as sodium hypochlorite), a commercial acidic detergent formulation, 8% trisodium phosphate, or 5% hydrogen peroxide applied at 20 or 50°C (Sapers and Jantschke, 1998). Under laboratory conditions in which the apples were submerged in the washing agents, some of these treatments had produced population reductions between 2 and 3 logs.

Further washing trials carried out with a commercial flat-bed brush washer confirmed these results (Annous et al., 2001). Population reductions for apples inoculated with a nonpathogenic *E. coli* were less than 1 log with the same washing agents used in the earlier trials (Table 11.5). The poor performance of both brush washers was attributed to deficiencies in equipment design resulting in insufficient exposure of inoculated apple surfaces to the washing agents, especially in the

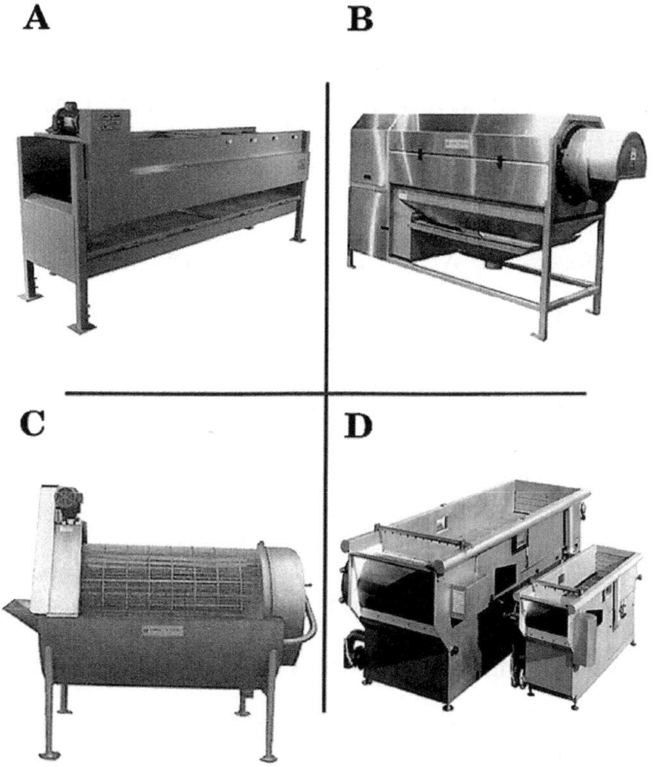

FIGURE 11.2 Commercial washing equipment for fruits and vegetables: (a) flat-bed brush washer; (b) U-bed brush washer; (c) rotary washer; (d) pressure washer.

TABLE 11.5
Decontamination of Apples Inoculated with *E. Coli* (Strain K12) with Sanitizing Washes Applied in a Flat-Bed Brush Washer [a]

Wash Treatment	Temperature (°C)	*E. coli* ($\log_{10}$ CFU/g)[b] Before Dump Tank	After Dump Tank	After Brush Washer
Water	20	5.49 ± 0.09	4.92 ± 0.37	4.81 ± 0.26
	50	5.49 ± 0.09	5.03 ± 0.15	4.59 ± 0.08
200 ppm Cl_2	20	5.87 ± 0.07	5.45 ± 0.05	5.64 ± 0.23
8% Na_3PO_4	20	5.49 ± 0.09	5.02 ± 0.43	4.98 ± 0.02
	50	5.49 ± 0.09	5.02 ± 0.08	4.75 ± 0.45
1% acidic detergent	50	5.87 ± 0.07	5.49 ± 0.03	5.42 ± 0.50
5% H_2O_2	20	5.87 ± 0.07	5.46 ± 0.40	5.27 ± 0.09
	50	5.87 ± 0.07	5.54 ± 0.31	5.49 ± 0.10

[a] From Annous, B.A. et al., *J. Food Prot.*, 64:159–163, 2000.

[b] Mean of 4 determinations ± standard deviation.

inaccessible calyx and stem areas, where contact with brushes was minimal. Bacterial adherence to apple surfaces, biofilm formation during the 18- to 24-h interval between inoculation and washing, and internalization might have contributed to the poor results, as discussed previously. However, results suggest that although brush washing may be needed for cleaning some commodities, application of sanitizing solutions to produce by full immersion in a dip tank would be more effective than brush washing in reducing microbial populations.

Washing and Sanitizing Agents for Fruits and Vegetables

Incoming fruits and vegetables are often immersed in water as they enter a packing or processing location. This may be in a cooler, drencher, dump tank, flume, or washer. Depending on its use, the water might contain soil, leaves, or other debris from harvesting and microorganisms associated with these materials. Sanitizing agents are usually added to process water to reduce the microbial population and prevent cross-contamination of the product. The most widely used sanitizing agents are highly effective in killing microorganisms suspended in water, in contrast to their limited efficacy against microorganisms attached to produce surfaces. The following sections look at the advantages and limitations of a number of agents used to sanitize fruits and vegetables (Table 11.6).

Chlorine as a Sanitizing Agent for Fruits and Vegetables

Chlorine is the most widely used among the washing and sanitizing agents available for fresh produce; it is generally assumed that chlorine is highly effective in reducing bacterial populations on commodity surfaces (Beuchat, 1998; Brackett, 1999). However, published data indicate that the most that can be expected at permitted concentrations is a 1- to 2-log population reduction (Brackett, 1987; Zhuang et al., 1995; Wei et al., 1995; Zhang and Farber, 1996; Beuchat et al., 1998; Sapers et al., 1999b; Pirovani et al., 2000). This is due in part to the inaccessibility of attached microorganisms and resistance of bacteria within biofilms but also to the rapid breakdown of chlorine in the presence of organic matter in soil and on product surfaces. Typically, chlorine is applied at a concentration no greater than 200 ppm, and solutions are adjusted to a pH of 6.5 to 7.5 to provide a high concentration of hypochlorous acid, the active form. Some improvement in the efficacy of chlorine can be obtained by addition of a wetting agent (Spotts and Cervantes, 1987). Washing formulations containing sodium hypochlorite, buffers, and surfactants are available commercially (Park et al., 1991; Wartanessian, 1997; Tenzer, 1997).

Another means of improving the efficacy of chlorine treatments is to monitor the oxidation-reduction potential (ORP) of the process water and to use this value as a means of controlling the treatment by addition of more hypochlorite or pH adjustment. An ORP value of 650 mv is recommended (Suslow et al., 2000). Commercial systems for controlling chlorine treatments, based on ORP measurements, are available (Vogel, 1999) (Table 11.1).

The use of electrolyzed water as a sanitizing agent for produce represents a special case of chlorination (Izumi, 1999). When water containing a small amount

TABLE 11.6
Advantages and Disadvantages of Commercially Available Sanitizing Agents for Washing Fresh Fruits and Vegetables

Sanitizing Agent	Use Level (ppm)	Advantages	Disadvantages
Chlorine	50 to 200	Easy to apply Inexpensive Effective against all microbial forms Not affected by hard water Easy to monitor FDA approved	Decomposed by organic matter Reaction products may be hazardous Corrosive to metals Irritating to skin Activity pH-dependent Population reductions limited to <1 to 2 logs
Ozone	0.1 to 2.5	More potent antimicrobial than chlorine No chlorinated reaction products formed Economical to operate Self-affirmed GRAS, but FDA review possible Activity not pH-dependent	Requires on-site generation Requires good ventilation Phytotoxic at high concentrations Corrosive to metals Difficult to monitor Higher capital cost than chlorine No residual effect Population reductions limited to <1 to 2 logs
Chlorine dioxide	1 to 5	More potent than chlorine Activity not pH-dependent Fewer chlorinated reaction products formed than with Cl_2 Effective against biofilms FDA approved Residual antimicrobial action Less corrosive than Cl_2 or O_3	Must be generated on-site Explosive at high concentrations Not permitted for cut fruits and vegetables Population reductions limited to <1 to 2 logs Generating systems expensive
Peroxyacetic acid	≤80	Broad spectrum antimicrobial action No pH control required Low reactivity with soil Effective against biofilms FDA approved No hazardous breakdown products No on-site generation required Monitoring not difficult Available at safe concentration	Population reduction limited to <1 to 2 logs Strong oxidant; concentrated solutions may be hazardous

of sodium chloride is subjected to electrolysis, hypochlorous acid is generated at a concentration of 10 to 100 ppm available chlorine. The solution can be highly acidic (pH <3.0) or alkaline (pH ≥ 11.0), depending on the configuration of the system. Electrolyzed water is sometimes referred to as functional water or high oxidation potential (HOP) water. An oxidation-reduction potential of 1000 to 1150 mv is desired (Kawamoto, 1999).

The results of electrolyzed water treatments have been mixed. In a comparison of acidic electrolyzed water and acidified chlorine treatments applied to inoculated lettuce leaves, Park and co-workers (2001) reported reductions of 2.49 and 2.22 $\log_{10}$CFU/lettuce leaf for *E. coli* O157:H7 and *L. monocytogenes*, respectively. However, the difference between the two treatments was not significant. One study claimed a 3.7- to 4.6-log reduction of *E. coli* O157:H7 on apples treated with pH 2.6 electrolyzed water (Horton et al., 1999), but another study could demonstrate only a 1-log reduction in the microbial population on fresh-cut vegetables (Izumi, 1999). Such inconsistencies may be attributed to experimental differences in the interval between contamination and treatment, leading to differences in degree and strength of bacterial attachment and the opportunity for biofilm formation.

The reaction of chlorine with organic residues can result in the formation of potentially mutagenic or carcinogenic reaction products (Chang et al., 1988; Hidaka et al., 1992). This is a cause for concern because some restrictions in the use of chlorine might eventually be implemented by regulatory agencies.

Detergent Formulations and Other Commercial Produce Washes

Numerous commercial washing formulations have been developed for fruits and vegetables. (See Table 11.1 for sources.) They include various surfactants, combinations of surfactants with organic or mineral acids, and alkaline formulations based on sodium hydroxide. Relatively little information on their efficacy in reducing microbial populations on fruits and vegetables has been published. Wright et al. (2000) reported similar population reductions with a commercial phosphoric acid fruit wash and with a 200 ppm hypochlorite wash, each applied to apples inoculated with *E. coli* O157:H7. Sapers et al. (1999b) tested the efficacy of some commercial washing formulations in decontaminating apples artificially inoculated with nonpathogenic *E. coli* and found that these formulations were generally similar to chlorine, achieving a 1- to 2-log population reduction. When these products were applied at 50°C instead of at ambient temperature, a log reduction of about 2.5 could be obtained.

Ozone

Ozone is one of several new sanitizing agents for produce introduced in recent years (Graham, 1997; Xu, 1999). The efficacy of ozone in killing human pathogens and other microorganisms in water is well established (Wickramanayake, 1991; Restaino et al., 1995; Kim and Yousef, 2000), and it is widely used in bottled water purification systems as an alternative to chlorine (Graham, 1997). Ozone is effective in reducing bacterial populations in flume and wash water and may have some applications as a chlorine replacement in reducing microbial populations on produce (Achen and Yousef, 1999; Kim et al., 1999; Smilanick et al., 2000). Use levels of 0.5 to 4.0 μg/ml are recommended for wash water and 0.1 μg/ml for flume water (Zagory and Hurst, 1996; Strasser, 1998). However, ozone treatment (5.5 μg/ml water for 5 min) was ineffective in reducing postharvest fungal decay of pears (Spotts and Cervantes, 1992). Treatment of lettuce, inoculated with *P. fluorescens*, with ozone (10 μg/ml for 1 min) achieved less than a 1-log population reduction (Kim et al., 1999).

One of the major advantages claimed for ozone is the absence of potentially toxic reaction products. However, ozone must be adequately vented to avoid worker exposure and it must be generated on-site by passing air or oxygen through a corona discharge or UV light (Xu, 1999). A number of commercial systems for generating ozonated water for produce washing are available. (See Table 11.1 for sources of ozone generators.) An independent expert committee, sponsored by the Electric Power Research Institute, recommended that use of ozone as a disinfectant or sanitizer for foods be classified as "generally recognized as safe" (GRAS) (Graham, 1997). However, FDA may require further review to determine whether ozone treatments affect nutrient levels (Anonymous, 1999a).

Chlorine Dioxide

Chlorine dioxide also is used as an antimicrobial agent for produce washing. It is efficacious against many classes of microorganisms at lower concentrations than would be required with chlorine (Dychdala, 1991). It can reduce microbial populations in dump tank and wash water, but tests with cucumbers resulted in less than a 1-log population reduction on product surfaces (Costilow et al., 1984). In tests conducted in an apple packinghouse, addition of chlorine dioxide (3 to 5 μg/ml) to dump tank water reduced the population of filamentous fungi (Roberts and Reymond, 1994). Treatment of pears inoculated with *Botrytis cinerea*, *Mucor piriformis*, or *P. expansum* with 10 μg/ml chlorine dioxide for 10 min suppressed decay, but addition of 0.5 μg/ml of chlorine dioxide to flume water did not reduce decay of inoculated fruit (Spotts and Peters, 1980). A chlorine dioxide product, Oxine® (Bio-Cide International, Inc., Norman, OK), reduced the population of *E. coli* O157:H7 on inoculated apples by only 2.5 logs at 80 μg/ml, 16 times the recommended concentration (Wisniewsky et al., 2000).

Chlorine dioxide is approved for use on uncut produce (21CFR173.325, 2000). It produces fewer potentially carcinogenic chlorinated reaction products than chlorine (Tsai et al., 1995; Rittman, 1997); however, it is explosive and must be generated on site. A number of companies produce systems for in-plant generation of chlorine dioxide from stable precursors comprising sodium hypochlorite; hydrochloric, citric, or phosphoric acid; and sodium chlorite. (See Table 11.1 for sources of chlorine dioxide generators.)

Peroxyacetic Acid

Peroxyacetic acid, or peracetic acid as it is sometimes called, is actually an equilibrium mixture of the peroxy compound, hydrogen peroxide, and acetic acid (Ecolab,

1997, 2000). The superior antimicrobial properties of peroxyacetic acid are well known (Block, 1991). This agent is recommended for use in treating process water, but one of the major suppliers is also claiming substantial reductions in microbial populations on produce surfaces. However, the best that could be claimed with cut corn was a 1-log reduction (Ecolab, 1997). Population reductions for aerobic bacteria, coliforms, and yeasts and molds on fresh-cut celery, cabbage, and potatoes treated with 80 ppm peroxyacetic acid were less than 1.5 logs (Hilgren and Salverda, 2000). Several published studies have looked at the efficacy of peroxyacetic acid against *E. coli* O157:H7 on inoculated apples.

In one study in which apples were washed about 30 min after inoculation, the commercial peroxyacetic acid formulation (80 ppm peroxyacetic acid) reduced the *E. coli* population by about 2 logs, compared to a water wash (Wright et al., 2000). In another study, where the inoculated apples were held for 24 h before washing, the peroxyacetic acid treatment reduced the *E. coli* population by less than 1 log at the recommended concentration (80 ppm) and by 3 logs at 16 times the recommended concentration (Wisniewsky et al., 2000). Similar results with apples inoculated with a nonpathogenic *E. coli* have been reported (Sapers et al., 1999b). Like ozone and chlorine dioxide, peroxyacetic acid is effective in killing pathogenic bacteria in suspension at lower concentrations than would be required with chlorine (Block, 1991). Addition of octanoic acid to peroxyacetic acid solutions increased efficacy in killing yeasts and molds in fresh-cut vegetable process waters but had little effect on population reductions on fresh-cut vegetables (Hilgren and Salverda, 2000). Peroxyacetic acid is approved for addition to wash water (21CFR173.315, 2000). It decomposes into acetic acid, water, and oxygen, all harmless residuals. It is a strong oxidizing agent and can be hazardous to handle at high concentrations but not at strengths marketed to the produce industry.

NEW TECHNOLOGY FOR DECONTAMINATION OF PRODUCE

NOVEL WASHING AND SANITIZING AGENTS

Hydrogen Peroxide

The author has had extensive experience with hydrogen peroxide, which is also a strong oxidizing agent effective against a wide range of bacteria but less active against fungi (Block, 1991). Hydrogen peroxide vapor treatments have been used to inhibit postharvest decay in grapes (Forney et al., 1991), melons (Aharoni et al., 1994), and other commodities and to disinfect prunes (Simmons et al., 1997). However, vapor treatments tend to be slow and can cause injury to some commodities such as mushrooms, raspberries, and strawberries (Sapers and Simmons, 1998). Dilute hydrogen peroxide solutions are effective in washing mushrooms (McConnell, 1991; Sapers et al., 1994, 1999a, 2001a), controlling postharvest decay of vegetables (Fallik et al., 1994), extending the shelf-life of fresh-cut vegetables and melons (Sapers and Simmons, 1998), and decontaminating apples containing nonpathogenic *E. coli* (Sapers et al., 1999b; 2000a).

Studies in this laboratory have shown that 5% hydrogen peroxide solutions, alone or combined with commercial surfactants, can achieve substantially higher log reductions for inoculated apples than 200 ppm chlorine. When applied at a temperature of 50 to 60°C, reductions as great as 3 to 4 $\log_{10}$ CFU/g have been obtained (Sapers et al., 1999b). With cantaloupe melon, a 5% hydrogen peroxide wash applied at 50 or 60°C to the whole melon prior to rind removal was superior to chlorine in extending the shelf-life of fresh-cut melon cubes. Visual observations of spoilage were consistent with the microbiological data showing suppression of bacterial growth following the peroxide treatment, perhaps indicative of injury to spoilage-causing bacteria (Sapers et al., 2001b). The hydrogen peroxide wash was highly effective in inactivating nonpathogenic *E. coli* and *S. Stanley* on inoculated cantaloupe within 24 h of inoculation but only partially effective if the inoculated melons were stored for several days prior to washing (Ukuku and Sapers, 2000; Ukuku et al., 2001). These efficacy data and the apple results probably indicate attachment of *E. coli* and *S. Stanley* to inaccessible sites and biofilm formation.

Hydrogen peroxide is GRAS for some food applications (21CFR184.1366, 1994b) but has not yet been approved as an antimicrobial wash for produce. It produces no residue because it is broken down to water and oxygen by catalase, an enzyme found throughout the plant kingdom. However, hydrogen peroxide is phytotoxic to some commodities, causing browning in lettuce and bleaching of anthocyanins in mechanically damaged berries (Sapers and Simmons, 1998). It is hazardous in high concentrations and must be handled with care. Numerous suppliers can provide technical information about hydrogen peroxide applications (see Table 11.1 for sources), but prior FDA approval would be required for it to be used in washing fresh produce.

Trisodium Phosphate

Trisodium phosphate solutions are effective in decontamination of animal carcasses (Lillard, 1994; Dickson et al., 1994). This treatment is effective against *E. coli* and *Salmonella* and has been used experimentally as a wash to decontaminate tomatoes (Zhuang and Beuchat, 1996) and apples (Sapers et al., 1999b). It has been shown to be effective against bacteria in biofilms on stainless steel or buna-N rubber chips (Somers et al., 1994); however, 2% trisodium phosphate was ineffective in killing *L. monocytogene* on lettuce (Zhang and Farber, 1996). Trisodium phosphate is classified as GRAS (21CFR182.1778, 2000) when used in accordance with good manufacturing practice and is marketed by Rhodia Food (Rhodia, 1998) under the brand names AvGard™ and Assur-Rinse® for treatment of poultry and beef. Experimental applications for disinfection of fruits and vegetables have employed concentrations as high as 12% (Rhodia, 1998).

Organic Acids

Organic acids such as lactic and acetic acids are effective agents against bacteria (Foegeding and Busta, 1991). Lactic acid dips and sprays are used commercially to decontaminate carcasses containing *E. coli* O157:H7, *L. monocytogenes*, and *Sal-*

monella (Purac, 1997; Castillo et al., 2001). Microbial population reductions on treated beef retail cuts were less than 1 log, but outgrowth during storage was suppressed by a residual effect (Kotula and Thelappurate, 1994). Lactic acid rinses are being recommended for decontamination of fruits and vegetables. Total plate count reductions of 1 to 1.5 logs were reported on endive rinsed with 1.9% lactic acid for 1.5 min (Purac, 1997). Acetic acid has been tested as an antimicrobial agent for apples. In one study, a 5% acetic acid wash was reported to reduce the population of *E. coli* O157:H7 on inoculated apples by about 3 logs. However, these apples were inoculated only 30 min prior to treatment, probably providing insufficient time for strong bacterial attachment and possible biofilm formation (Wright et al., 2000). In another study, apples that had been inoculated with *E. coli* O157:H7 and air dried for 30 min were treated with 5% acetic acid at 55°C for as long as 25 min. Although the *E. coli* population was greatly reduced in the apple skin and stem areas, as many as 3 to 4 logs survived in the calyx tissue (Delaquis et al., 2000). It is not clear whether organic acid treatments would produce off-flavors in treated produce.

Other Experimental Washes

Peroxidase-generated iodine, which has been used to kill *Salmonella* on chicken breast skin (Bianchi et al., 1994), may be applicable to produce, but efficacy data are lacking. Dipping for 10 min in a saturated solution of calcinated calcium, an alkaline product obtained by ohmic heating of oyster shells, was reported to reduce bacterial populations in cucumbers and radish sprouts (Isshiki and Azuma, 1995). Copper and silver ions are known to exert antimicrobial activity (Hurst, 1991) and have been used to disinfect swimming pool water (Tew, 2000). Addition of 0.50 ppm copper and 0.035 to 0.05 ppm silver ions to water used in produce packing lines and dump tanks has been recommended (Tew, 2000), but the regulatory status of such treatments is not known. Grapefruit seed extract has been reported to be effective in inhibiting postharvest decay of fruits and vegetables (Cho et al., 1994), and grapefruit oil is used as a constituent of a produce wash intended for home use (Procter & Gamble, 2001). Allylisothiocyanate (AITC), a food preservative derived from cruciferous vegetables (i.e., *Wasabi* horseradish, mustard) and marketed in Japan under the name Wasaouro®, may have application as an antimicrobial agent for minimally processed fruits and vegetables (Tokuoka and Isshiki, 1994; Delaquis and Mazza, 1995). AITC may be applied as a dip or in the vapor phase. This product is marketed in the U.S. by Midori Pharmerica Corp. (150 East 52 St., New York, NY).

Novel Means of Applying Sanitizing Agents

Vacuum Infiltration

One means of improving contact between a sanitizing agent and bacteria attached in inaccessible sites on produce surfaces is by application of the treatment solution under vacuum. This procedure might be expected to remove gas barriers that block penetration of the sanitizing agent into pores, punctures, or sites like the apple calyx channel. By applying a 5% hydrogen peroxide solution to inoculated apples under vacuum, the author was able to obtain greater population reductions in the calyx

TABLE 11.7
Surviving Bacterial Populations in Excised Calyx and Stem Areas of Golden Delicious Apples Inoculated with *E. Coli* (ATCC 25922) and Decontaminated by Washing or Vacuum Infiltration with 5% H_2O_2, with or without Stem Removal[a]

Experiment	Treatment	Stem Removal	Population Reduction[b] Log_{10} CFU/g Stem Tissue[c]	Calyx Tissue[c]
A	5% H_2O_2 at 60°C	No	2.96	2.40
		Yes	2.97	3.54
	5% H_2O_2 vac. infilt. at 45°C	No	1.69	5.25
		Yes	3.07	5.16
B	5% H_2O_2 at 60°C	No	3.30	2.76
		Yes	2.70	2.82
	5% H_2O_2 vac. infilt. at 45°C	No	1.23	3.79
		Yes	3.34	4.81

[a] Apples washed with 5% H_2O_2 at 60°C for 2 min or vacuum infiltrated with 5% H_2O_2 at 45°C and 100 mm Hg for 3 min; apples rinsed with H_2O after treatment.

[b] Based on inoculated control (without stem removal) *E. coli* population of 5.29 and 6.38 $\log_{10}$CFU/g in stem and calyx areas, respectively, for Expt. A and 5.09 and 6.53 $\log_{10}$CFU/g in stem and calyx areas, respectively, for Expt. B.

[c] Stem and calyx tissues excised aseptically from individual apples, weighed, and homogenized for enumeration of surviving *E. coli*.

area than were possible without the vacuum (Table 11.7). Treatment temperature and vacuum level were selected to minimize boiling of the hydrogen peroxide solution under vacuum, which might interfere with the treatment. Removal of stems prior to vacuum infiltration of hydrogen peroxide solution substantially improved treatment efficacy in the stem area. This may be due to the greater exposure of bacteria attached at the base of the stem to the sanitizing agent.

One of the limitations of this approach is the presence of sanitizer residue in the pores, channels, and punctures following infiltration treatment. This would not be an issue with hydrogen peroxide, which breaks down to oxygen and water soon after treatment. To carry out the vacuum infiltration treatment, processors would require a vacuum chamber, perhaps similar to that used for commercial vacuum infiltration of sugar, citric acid, and ascorbic acid into apple slices (Hall, 1989). This would be a batch process. Our observations indicate that vacuum infiltration of hydrogen peroxide is noninjurious to apples and might be applicable to apples intended for fresh market as well as for processing (Sapers et al., 2002).

Vapor-Phase Treatments

A vapor-phase antimicrobial treatment might be more effective than a wash in killing microbial contaminants attached in inaccessible sites. In studies carried out previously, the author and other investigators found that vapor-phase applications of hydrogen peroxide were effective in suppressing bacterial spoilage of mushrooms and other commodities but required exposure times as long as 60 min (Sapers and Simmons, 1998). With a more efficient means of vapor production to reduce the required treatment time, this use of hydrogen peroxide might have some application in disinfection of raw material for minimally processed fruits and vegetables.

Acetic acid vapor treatment of cabbage, mung bean seeds, and grapes has been reported to reduce microbial populations and prevent decay (Sholberg and Gaunce, 1995, 1996; Sholberg et al., 1996, 1998; Delaquis et al., 1997, 1999). A 5-log reduction has been obtained in the bacterial population of apples artificially contaminated with nonpathogenic *E. coli* and exposed to acetic acid vapor at 50°C (Sapers and Sites, 2001). The treated apples showed slight browning, which was indicative of injury and might limit use of this technology.

Vapor-phase disinfection of produce with chlorine dioxide (Han et al., 2000) has been investigated. Acetaldehyde has been reported to have antimicrobial properties and might have potential in inhibiting postharvest spoilage (Aharoni et al., 1973). It is not clear whether this treatment might be effective against human pathogens or would impart off-flavors to the treated commodities.

Surface Pasteurization

The surface of fresh fruits and vegetables might be pasteurized with hot water or steam, provided that required exposure times and temperatures were not capable of causing heat injury to the product, resulting in shortened shelf-life or altered flavor, color, or texture. Exposure of apples to water at temperatures above 65°C will cause the skin to brown, and exposure at 80°C for more than 30 sec will soften the outer 1 or 2 mm of flesh (Sapers et al., 2002). This would rule out treatments at temperatures greater than 65°C for fruit intended for fresh market. Such treatments would not be applicable to apples intended for cider production under newly promulgated FDA regulations (FDA, 2001). Fresh cantaloupes were found to tolerate exposure to water or 5% hydrogen peroxide at 80°C for 3 min with no adverse effects on appearance, flavor, or texture, initially or after storage at 4°C for 30 days (Sapers et al., 2000b).

A hot water washing system has been developed and commercialized in Israel for treatment of various fruits and vegetables (Fallik et al., 1996; Porat et al., 2000). This system is claimed to reduce the population of decay organisms and extend produce shelf-life, but data on inactivation of bacteria, especially human pathogens, are not available. A hot water decontamination treatment has been developed in the U.S. for cider apples (Dean, 1998). An experimental hot water pasteurization system developed by FDA scientists gave a 2-log reduction in apples inoculated with a nonpathogenic *E. coli* and then treated at 88 to 100°C (Keller, 1999). In other studies with apples inoculated with *E. coli* O157:H7 and surface pasteurized in water at 95°C, FDA scientists demonstrated a 5-log reduction when the apples were inoculated by applying droplets of inoculum to the fruit surface. However, only a 1-log reduction could be obtained when the apples were inoculated by immersion in the inoculum. Presumably, the surviving bacteria were attached in areas protected from exposure, perhaps by internalization during inoculation (Fleischman et al., 2001).

TABLE 11.8
Surface Pasteurization at 80°C of Golden Delicious Apples and Cantaloupe Inoculated with *E. Coli* (ATCC 25922)

Commodity	Treatment	Log Reduction[a]	Appearance of Skin	Softening of Surface
Apples	1 min in H_2O	1.59 ± 0.60	Severe darkening	Slight
	1 min in 5% H_2O_2	2.96 ± 0.07	Severe darkening	Slight
Cantaloupe	30 sec in 5% H_2O_2	2.44 ± 0.04	Same as control	None
	1 min in 5% H_2O	3.05 ± 0.03	Same as control	None
	3 min in 5% H_2O_2	4.37[b]	Same as control	None
	5 min in 5% H_2O_2	4.37[b]	Same as control	None

[a] Based on counts on *E. coli* Petrifilm. Mean of duplicate trials ± standard deviation; log reductions for apples and cantaloupe expressed as $\log_{10}$CFU/g and $\log_{10}$CFU/cm^2, respectively.
[b] No survivors detected; log reduction = control population.

Several methods of surface pasteurization of citrus fruit prior to juicing have been investigated (Beasley, 1999). Spraying with water at 93 to 99°C for 60 sec was claimed to achieve a 5-log reduction in fruit inoculated with *E. coli*. Exposure to steam for 30 sec in a steam tunnel reduced the population by 3.7 logs.

Whether such treatments can kill bacteria attached within inaccessible sites such as the calyx or stem areas of apples or within punctures is not clear. In preliminary studies with inoculated apples, immersion for 3 min in 5% hydrogen peroxide at 80°C gave a 3-log reduction in the *E. coli* population (Sapers et al., 2002) (Table 11.8). The large number of surviving *E. coli* (about 2 $\log_{10}$ CFU/g) indicates that such treatments cannot be considered effective surface pasteurization. On the other hand, immersion of cantaloupe inoculated with *E. coli* (ATCC 25922) in 5% hydrogen peroxide at 80°C for 3 min resulted in a 4.4-log reduction (no survivors detected), with no indication of treatment-induced quality defects (Table 11.8). Such a treatment should be capable of decontaminating the external surface of the melon so that the flesh does not become cross-contaminated when the rind is removed during fresh-cut processing (Sapers et al., 200b).

Synergistic Treatment Combinations

Although the efficacy of individual antimicrobial treatments might be limited to 1- to 2-log reductions in microbial populations, several treatments applied sequentially might result in substantial improvements in efficacy due to synergistic interactions between treatments. An acidified surfactant treatment, applied in a brush washer to maximize soil removal, might be followed by a hydrogen peroxide treatment, applied by immersing the commodity in a dip tank. Other examples of treatment combinations with the potential for synergism include an acidified surfactant wash treatment combined with surface pasteurization, vacuum infiltration of hydrogen peroxide, or ozone solution, or with vapor-phase application of a sanitizer vapor. Research on the efficacy of such combinations would establish whether synergism could be obtained.

CONCLUSIONS

Washing and sanitizing raw material for minimally processed fruit and vegetable products are key steps in assuring microbiological quality and safety. Conventional washing and sanitizing agents and equipment for their application generally cannot achieve reductions in microbial populations greater than 1 to 2 logs (90 to 99%). Such reductions may result in improvements in product quality and shelf-life extensions, but they cannot exclude the possibility of human pathogen survival and the associated risk of food-borne illness.

The efficacy of conventional washing and sanitizing agents is limited by bacterial adherence to produce surfaces, attachment in inaccessible sites, formation of resistant biofilms, and penetration within commodities. Furthermore, conventional washing equipment may not provide sufficient exposure of contaminated produce surfaces to washing and sanitizing agents. Incremental improvements can be made in sanitizer formulation and equipment design, but these are unlikely to increase treatment efficacy greatly.

New washing technology is needed to overcome these deficiencies. New treatments must be superior in efficacy, approved by regulatory agencies, safe to apply, compatible with existing industry practices, and affordable. A number of new approaches show promise. These include washing with more powerful antimicrobial agents such as hydrogen peroxide solutions, vacuum infiltration of sanitizers, vapor-phase application of sanitizers, surface pasteurization, and use of synergistic treatment combinations. Such innovations might not be capable of achieving the greater than 5-log reductions in pathogen populations possible with true pasteurization treatments, but they might bring about large improvements in the microbiological quality and safety of minimally processed fruits and vegetables.

ACKNOWLEDGMENT

The author thanks Dr. William C. Hurst at the University of Georgia, Athens, and Drs. Pina M. Fratamico and William F. Fett at the Eastern Regional Research Center (ERRC), ARS, USDA, for their thorough and constructive reviews of this chapter, and Drs. Bassam A. Annous and William F. Fett at ERRC for providing micrographs of biofilms on apple calyx and alfalfa sprout surfaces, respectively.

REFERENCES

Achen, M. and Yousef, A.E., 2001, Efficacy of ozone against *Excherichia coli* O157:H7 on apples, *J. Food Sci.,* 66:1380–1384.

Aharoni, Y., Copel, A., and Fallik, E., 1994, The use of hydrogen peroxide to control postharvest decay on 'Galia' melons, *Ann. Appl. Biol.*, 125:189–193.

Aharoni, Y. and Stadelbacher, G.J., 1973, The toxicity of acetaldehyde vapors to postharvest pathogens of fruits and vegetables, *Phytopathology,* 63:544–545.

Annous, B.A. et al., 2001, Efficacy of washing with a commercial flat-bed brush washer, using conventional and experimental washing agents, in reducing populations of *Escherichia coli* on artificially inoculated apples, *J. Food Prot.*, 64:159–163.

Anonymous, 1999a, UV light, ozone, hydrogen peroxide use on fresh produce and juices may require expanded review by FDA, *Food Chem. News*, May 17, 18–19.

Anonymous, 1999b, Sprout industry has two months to implement new FDA guidelines, *Food Chem. News*, November 1, 15–16.

Bartz, J.A., 1982, Infiltration of tomatoes immersed at different temperatures to different depths in suspensions of *Erwinia carotovora* subsp. *carotovora*, *Plant Dis.*, 66:302–306.

Bartz, J.A. and Showalter, R.K., 1981, Infiltration of tomatoes by aqueous bacterial suspensions, *Phytopathology,* 71:515–518.

Beasley, L.R.H., 1999, Treatment options for fresh-squeezed citrus juices, presented at symposium on minimally processed fruit and vegetable juices: scientific issues to achieve a safe product, sponsored by National Center for Food Safety and Technology at Illinois Institute of Technology and Institute of Food Technologists, Chicago (O'Hare), May 12–13.

Besser, R.E. et al., 1993, An outbreak of diarrhea and hemolytic uremic syndrome from *Escherichia coli* O157:H7 in fresh pressed apple cider, *JAMA,* 269:2217–2220.

Beuchat, L., 1996, Pathogenic microorganisms associated with fresh produce*, J. Food Prot.*, 59:204–216.

Beuchat, L.R., 1998, Surface decontamination of fruits and vegetables eaten raw: a review, Food Safety Unit, World Health Organization, WHO/FSF/FOS/98.2.

Beuchat, L.R. et al., 1998, Efficacy of spray application of chlorinated water in killing pathogenic bacteria on raw apples, tomatoes, and lettuce, *J. Food Prot.*, 61:1305–1311.

Bianchi, A. et al., 1994, A peroxidase catalyzed chemical dip for the reduction of *Salmonella* on chicken breast skin, *J. Food Prot.*, 57:301–304, 326.

Block, S.S., 1991, Peroxygen compounds, in *Disinfection, Sterilization, and Preservation*, 4th ed., Block, S.S., Ed., Philadelphia: Lea & Febiger, 182–190.

Brackett. R.E., 1987, Antimicrobial effect of chlorine on *Listeria monocytogenes*, *J. Food Prot.*, 50:999–1003.

Brackett, R.E., 1999, Incidence, contributing factors, and control of bacterial pathogens in produce, *Postharvest Biol. Technol.*, 15:305–311.

Breidt, F., Hayes, J.S., and Fleming, H.P., 2001, Reduction of microflora of whole pickling cucumbers by blanching, *J. Food Sci.*, 65:1354–1358.

Buchanan, R.L. et al., 1999, Contamination of intact apples after immersion in an aqueous environment containing *Escherichia coli* O157:H7, *J. Food Prot.*, 62:444–450.

Burnett, S.L., Chen, J., and Beuchat, L.R., 2000, Attachment of *Escherichia coli* O157:H7 to the surfaces and internal structures of apples as detected by confocal scanning laser microscopy, *Appl. Environ. Microbiol.*, 66:4679–4687.

Canadian Food Inspection Agency, 1998, Code of practice for the production and distribution of unpasteurized apple and other fruit juice/cider in Canada, Plant Products Directorate, Agriculture and Agri-Food Canada.

Carmichael, I. et al., 1999, Bacterial colonization and biofilm development on minimally processed vegetables, *J. Appl. Microbiol.*, symp. suppl., 85:45S-51S.

Castillo, A. et al., 2001, Lactic acid sprays reduce bacterial pathogens on cold beef carcass surfaces and in subsequently produced ground beef, *J. Food Prot.*, 64:58–62.

CDC, 1997, Outbreaks of *Escherichia coli* O157:H7 infection and Cryptosporidiosis associated with drinking unpasteurized apple cider — Connecticut and New York, October 1996, *Morbidity Mortality Wkly. Rep.*, 46(1):4–8.

21CFR,110. 1994a, Current good manufacturing practice in manufacturing, packing, or holding human food, Code of Federal Regulations 21, Part 110, 217–227.

21CFR173.315, 2000, Chemicals used in washing or to assist in the peeling of fruits and vegetables, Code of Federal Regulations 21, Part 173, Section 173.315.

21CFR173.325, 2000, Acidified sodium chlorite solutions, Code of Federal Regulations 21, Part 173, Section 173.325.

21CFR178.1010, 2000, Sanitizing solutions, Code of Federal Regulations 21, Part 178, Section 178.1010.

21CFR182.1778, 2000, Sodium phosphate, Code of Federal Regulations 21, Part 182, Section 182.1778.

21CFR184.1366, 1994, Hydrogen peroxide, Code of Federal Regulations 21, Parts 170–199, Section 184.1366, 463.

Chang, T.-L., Streicher, R., and Zimmer, H., 1988, The interaction of aqueous solutions of chlorine with malic acid, tartaric acid, and various fruit juices. A source of mutagens, *Anal. Lett.*, 21:2049–2067.

Cho, S.-H., Chung, J.-H., and Ryu, C.-H., 1994, Inhibitory effects of natural antimicrobial agent on postharvest decay in fruits and vegetables under natural low temperatures, *J. Korean Soc. Food Nutr.*, 23:315–321.

Costerton, J.W., 1995, Overview of microbial biofilms, *J. Ind. Microbiol.*, 15:137–140.

Costilow, R., Uebersax, M.A., and Ward, P.J., 1984, Use of chlorine dioxide for controlling microorganisms during the handling and storage of fresh cucumbers, *J. Food Sci.*, 49:396–401.

Couture, H., 1999, Proposed draft code of hygienic practice for the primary production, harvesting and packaging of fresh fruits and vegetables, Codex Committee on Food Hygiene.

Daeschel, M.A. and Fleming, H.P., 1981, Entrance and growth of lactic acid bacteria in gas-exchanged, brined cucumbers, *Appl. Environ. Microbiol.*, 42:1111–1118.

Dean, L., 1998, Maryland cider maker plans to turn up the heat on his apples, *Great Lakes Fruit Growers News*, February:8.

Delaquis, P., Graham, H.S., and Hocking, R., 1997, Shelf-life of coleslaw made from shredded cabbage fumigated with gaseous acetic acid, *J. Food Process. Preserv.*, 21:129–140.

Delaquis, P.J. and Mazza, G., 1995, Antimicrobial properties of isothiocyanates in food preservation, *Food Technol.*, 49(11):73–84.

Delaquis, P., Sholberg, P.L., and Stanich, K., 1999, Disinfection of mung bean seed with vaporized acetic acid, *J. Food Prot.*, 62:953–957.

Delaquis, P.J., Ward, S.M., and Stanich, K., 2000, Evaluation of pre-pressing sanitary treatments for the destruction of *Escherichia coli* O157:H7 on apples destined for production of unpasteurized apple juice, Technical report no. 9901, Agriculture and Agri-Food Canada, Pacific Agri-Food Research Centre, Summerland, BC.

DeWaal, C.S., Alderton, L., and Jacobson, M.F., 2000, *Outbreak Alert! Closing the Gaps in Our Federal Food-Safety Net*, Center for Science in the Public Interest, Washington, D.C.

Dickson, J.S., Nettles-Cutter, C.G., and Siragusa, G.R., 1994, Antimicrobial effects of trisodium phosphate against bacteria attached to beef tissue, *J. Food Prot.*, 57:952–955.

Dychdala, G.R., 1991, Chlorine and chlorine compounds, in *Disinfection, Sterilization, and Preservation*, 4th ed., Block, S.S., Ed., Philadelphia: Lea & Febiger, 131–151.

Ecolab, Inc., 1997, Catching the wave, *Food Qual.*, 4(1):51–52.

Ecolab, Inc., 2000, Personal communications, Ecolab Food & Beverage Division, St. Paul, MN.

Faith, N.G. et al., 1996, Prevalence and clonal nature of *Escherichia coli* O157:H7 on dairy farms in Wisconsin, *Appl. Environ. Microbiol.*, 62(5):1519–1525.

Fallik, E. et al., 1994, Postharvest hydrogen peroxide treatment inhibits decay in eggplant and sweet red pepper, *Crop Prot.*, 13:451–454.

Fallik, E. et al., 1996, A method for simultaneously cleaning and disinfecting agricultural produce, Israel Patent Application No. 116965 (30/1/96).

FDA, 1998, *Guidance for Industry. Guide to Minimize Microbial Food Safety Hazards for Fresh Fruits and Vegetables*, Food Safety Initiative Staff, HFS-32, U.S. Food and Drug Administration, Center for Food Safety and Applied Nutrition, Washington, D.C., http://www.foodsafety.gov/~dms/prodguid.html.

FDA, 1999, Guidance for industry: reducing microbial food safety hazards for sprouted seeds and Guidance for industry: sampling and microbial testing of spent irrigation water during sprout production, *Fed. Regist.*, 64:57893–57902.

FDA, 2001, Hazard analysis and critical control point (HACCP); procedures for the safe and sanitary processing and importing of juice; final rule, *Fed. Regist.*, 66(13):6137–6202.

Fett, W.F., 2000, Naturally occurring biofilms on alfalfa and other types of sprouts, *J. Food Prot.*, 63(6):625–632.

Fett, W.F. and Sapers, G.M., 1997, Unpublished data, Eastern Regional Research Center, ARS, U.S. Department of Agriculture, Wyndmoor, PA.

Fleischman, G.J. et al., 2001, Hot water immersion to eliminate *Escherichia coli* O157:H7 on the surface of whole apples: thermal effects and efficacy, *J. Food Prot.*, 64:451–455.

Foegeding, P.M. and Busta, F.F., 1991, Chemical food preservatives, in *Disinfection, Sterilization, and Preservation*, 4th ed., Block, S.S., Ed., Philadelphia: Lea & Febiger, 802–832.

Forney, C.F. et al., 1991, Vapor phase hydrogen peroxide inhibits postharvest decay of table grapes, *Hort. Sci.,* 26:1512–1514.

Francis, G.A., Thomas, C., and O'Beirne, D., 1999, The microbiological safety of minimally processed vegetables, *Int. J. Food Sci. Technol.*, 34:1–22.

Graham, D.M., 1997, Use of ozone for food processing, *Food Technol.*, 51(6):72–75.

Hall, G.C., 1989, Refrigerated, frozen, and dehydrofrozen apples, in *Processed Apple Products*, Downing, D.L., Ed., New York: Van Nostrand Reinhold, 247.

Han, Y. et al., 2000, The effects of washing and chlorine dioxide gas on survival and attachment of *Escherichia coli* O157:H7 to green pepper surfaces, *Food Microbiol.*, 17:521–533.

Hidaka, T. et al., 1992, Disappearance of residual chlorine and formation of chloroform in vegetables treated with sodium hypochlorite, *Shokuhin Eiseigaku Zasshi*, 33:267–73.

Hilgren, J.D. and Salverda, J.A., 2000, Antimicrobial efficacy of a peroxyacetic/octanoic acid mixture in fresh-cut-vegetable process waters, *J. Food Sci.*, 65:1376–1379.

Horton, A.R. et al., 1999, Disinfection effects of electrolyzed water on *E. coli* O157:H7 and apple quality, Abstract 22D-6, Abstracts of Papers Presented, 1999 IFT Annual Meeting, Chicago, IL, July 24–28.

Hurst, C.J., 1991, Disinfection of drinking water, swimming pool water, and treated sewage effluents, in *Disinfection, Sterilization, and Preservation*, 4th ed., Block, S.S., Ed., Philadelphia: Lea & Febiger, 713–729.

Hurst, W.C., 2001, Private communication, The University of Georgia, Athens.

Isshiki, K. and Azuma, K., 1995, Microbial growth suppression in food using calcinated calcium, *JARQ*, 29:269–274.

Itoh, Y. et al., 1998, Enterohemorrhagic *Escherichia coli* O157:H7 present in radish sprouts, *Appl. Environ. Microbiol.*, 64:1532–1535.

Iwasa, M. et al., 1999, Detection of *Escherichia coli* O157:H7 from *Musca domestica* (*Diptera: Muscidae*) at a cattle farm in Japan, *J. Med. Entomol.*, 36(1):108–112.

Izumi, H., 1999, Electrolyzed water as a disinfectant for fresh-cut vegetables, *J. Food Sci.*, 64:536–539.

Janisiewicz, W.J. et al., 1999a, Fate of *Escherichia coli* O157:H7 on fresh-cut apple tissue and its potential for transmission by fruit flies, *Appl. Environ. Microbiol.*, 65:1–5.

Janisiewicz, W.J., Conway, W.S., and Leverentz, B., 1999b, Biological control of postharvest decays of apple can prevent growth of *Escherichia coli* O157:H7 in apple wounds, *J. Food Prot.*, 62(12):1372–1375.

Juneja, V.K., Martin, S.T., and Sapers, G.M., 1998, Control of *Listeria monocytogenes* in vacuum-packaged pre-peeled potatoes, *J. Food Sci.*, 63:911–914.

Kawamoto, K., 1999, Personal communications, Pinto Pacific (representing ARV Company, Ltd.), Burlingame, CA.

Keller, S.E., 1999, Surface heat treatment of apples used for cider production, presented at Apple Cider Food Safety Control Workshop, Food and Drug Administration, Center for Food Safety and Applied Nutrition, Washington, D.C., July 15–16.

Kim, J.-G. and Yousef, A.E., 2000, Inactivation kinetics of foodborne spoilage and pathogenic bacteria by ozone, *J. Food Sci.*, 65:521–528.

Kim, J.-G., Yousef, A.E., and Chism, G.W., 1999, Use of ozone to inactivate microorganisms on lettuce, *J. Food Safety*, 19:17–34.

Kobayashi, M. et al., 1999, Houseflies: not simple mechanical vectors of enterohemorrhagic *Escherichia coli* O157:H7, *Am. J. Trop. Med. Hyg.*, October, 61(4):625–629.

Kotula, K.L. and Thelappurate, R., 1994, Microbiological and sensory attributes of retail cuts of beef treated with acetic and lactic acid solutions, *J. Food Prot.*, 57:665–670.

Kudva, I.T., Harfield, P.G., and Hovde, C.J., 1996, *Escherichia coli* O157:H7 in microbial flora of sheep, *Appl. Environ. Microbiol.*, 34:431–433.

Liao, C.-H. and Cooke, P.H., 2001, Response to trisodium phosphate treatment of *Salmonella* Chester attached to fresh-cut green pepper slices, *Can. J. Microbiol.*, 47(1):25–32.

Liao, C.-H. and Sapers, G.M., 1999, Influence of soft rot bacteria on the growth of *Listeria monocytogenes* on potato tuber slices, *J. Food Prot.*, 62:343–348.

Liao, C.-H. and Sapers, G.M., 2000, Attachment and growth of *Salmonella* Chester on apple fruits and *in vivo* response of attached bacteria to sanitizer treatments, *J. Food Prot.*, 63(7):876–883.

Lillard, H.S., 1994, Effect of trisodium phosphate on salmonellae attached to chicken skin, *J. Food Prot.*, 57:465–469.

McConnell, A.L., 1991, Evaluation of wash treatments for the improvement of quality and shelf life of fresh mushrooms (*Agaricus bisporus*), M.S. Thesis, Department of Food Science, Pennsylvania State University.

Meneley, J.C. and Stanghellini, M.E., 1974, Detection of enteric bacteria within locular tissue of healthy cucumbers, *J. Food Sci.*, 39:1267–1268.

Miller, P.M., 1959, Open calyx tubes as a factor contributing to carpel discoloration and decay of apples, *Phytopathology,* 49:520–523.

Morris, C.E., Monier, J.-M., and Jacques, M.-A., 1997, Methods for observing microbial biofilms directly on leaf surfaces and recovering them for isolation of cultural microorganisms, *Appl. Environ. Microbiol.*, 63:1570–1576.

NACMCF, 1997, Hazard analysis and critical control point principles and application guidelines, National Advisory Committee on Microbiological Criteria for Foods, August 14.

NACMCF, 1999, Microbiological safety evaluations and recommendations on fresh produce, National Advisory Committee on Microbiological Criteria for Foods, *Food Control*, 10:117–143.

Parish, M.E., 1997, Public health and nonpasteurized fruit juices, *Crit. Rev. Microbiol.*, 23:109–119.

Park, D.L., Rua, S.M., Jr., and Acker, R.F., 1991, Direct application of a new hypochlorite sanitizer for reducing bacterial contamination on foods, *J. Food Prot.*, 54:960–965.

Park, C.-M. et al., 2001, Pathogen reduction and quality of lettuce treated with electrolyzed oxidizing and acidified chlorinated water, *J. Food Sci.*, 66:1368–1372.

Pirovani, M.E. et al., 2000, Survival of *Salmonella hadar* after washing disinfection of minimally processed spinach, *Lett. Appl. Microbiol.*, 31:143–148.

Porat, R. et al., 2000, Reduction of postharvest decay in organic citrus fruit by a short hot water brushing treatment, *Postharvest Biol. Technol.*, 18:151–157.

Procter & Gamble Co., 2001, "Fit" information and specifications, http://www.tryfit.com.

Purac America, Inc., 1997, Personal communications, Lincolnshire, IL.

Restaino, L. et al., 1995, Efficacy of ozonated water against various food-related microorganisms, *Appl. Environ. Microbiol.*, 61:3471–3475.

Rhodia Food, 1998, Personal communications, Cranbury, NJ.

Rice, D.H., Hancock, D.D., and Besser, T.E., 1995, Verotoxigenic *E. coli* O157 colonization of wild deer and range cattle, *Vet. Rec.*, 137:524.

Riordan, D.C.R., Sapers, G.M., and Annous, B.A., 2000, The survival of *E. coli* O157:H7 in the presence of *Penicillium expansum* and *Glomerella cingulata* in wounds on apple surfaces, *J. Food Prot.*, 63:1637–1642.

Riordan, D.C. et al., 2001, A study of U.S. orchards to identify potential sources of *Escherichia coli* O157:H7, *J. Food Prot.*, 64:1320–1327.

Rittman, D.D., 1997, "Can you have your cake and eat it too" with chlorine dioxide, *Water/Eng. Manage. Mag.*, April.

Roberts, R.G. and Reymond, S.T., 1994, Chlorine dioxide for reduction of postharvest pathogen inoculum during handling of tree fruits, *Appl. Environ. Microbiol.*, 60:2864–2868.

Samish, Z., Etinger-Tulczynsky, R, and Bick, M., 1963, The microflora within the tissue of fruits and vegetables, *J. Food Sci.*, 28:259–266.

Sapers, G.M. and Jantschke, M., 1998, Unpublished data, National Food Processors Association, Dublin, CA.

Sapers, G.M. et al., 1999a, Structure and composition of mushrooms as affected by hydrogen peroxide wash, *J. Food Sci.*, 64:889–892.

Sapers, G.M. et al., 2000a, Factors limiting the efficacy of hydrogen peroxide washes for decontamination of apples containing *Escherichia coli, J. Food Sci.*, 65:529–532.

Sapers, G.M., Miller, R.L., and Mattrazzo, A.M., 1999b, Effectiveness of sanitizing agents in inactivating *Escherichia coli* in Golden Delicious apples, *J. Food Sci.*, 64:734–737.

Sapers, G.M. et al., 1994, Enzymatic browning control in minimally processed mushrooms, *J. Food Sci.*, 59:1042–1047.

Sapers, G.M., Miller, R.M., and Pilizota, V., 2000b, Unpublished data, Eastern Regional Research Center, Agricultural Research Service, U.S. Department of Agriculture, Wyndmoor, PA.

Sapers, G.M. et al., 2001a, Shelf-life extension of fresh mushrooms (*Agaricus bisporus*) by application of hydrogen peroxide and browning inhibitors, *J. Food Sci.*, 66:362–366.

Sapers, G.M. et al., 2001b, Anti-microbial treatments for minimally processed cantaloupe melon, *J. Food Sci.*, 66:345–349.

Sapers, G.M. and Simmons, G.F., 1998, Hydrogen peroxide disinfection of minimally processed fruits and vegetables, *Food Technol.*, 52(2):48–52.

Sapers, G.M. and Sites, J., 2001, Unpublished data, Eastern Regional Research Center, Agricultural Research Service, U.S. Department of Agriculture, Wyndmoor, PA.

Sapers, G.M. et al., 2002, Improved antimicrobial wash treatments for decontamination of apples, *J. Food Sci.,* 67:1886–1891.

Seo, K.H. and Frank, J.F., 1999, Attachment of *Escherichia coli* O157:H7 to lettuce leaf surface and bacterial viability in response to chlorine treatment as demonstrated by using confocal scanning laser microscopy, *J. Food Prot.*, 62:3–9.

Sholberg, P.L., Delaquis, P., and Moyls, A.L., 1998, Use of acetic acid fumigation to reduce the potential for decay in harvested crops, in Recent Research Developments in Plant Pathology, *Res. Signpost*, 2:31–41.

Sholberg, P.L. and Gaunce, A.P., 1995, Fumigation of fruit with acetic acid to prevent postharvest decay, *Hortic. Sci.*, 30:1271–1275.

Sholberg, P.L. and Gaunce, A.P., 1996, Fumigation of stonefruit with acetic acid to control postharvest decay, *Crop. Prot.*, 15:681–686.

Sholberg, P.L., Reynolds, A.G., and Gaunce, A.P., 1996, Fumigation of table grapes with acetic acid to prevent postharvest decay, *Plant Dis.*, 80:1425–1428.

Simmons, G.F. et al., 1997, Reduction of microbial populations on prunes by vapor-phase hydrogen peroxide, *J. Food Prot.*, 60:188–191.

Smilanick, J. et al., 2000, Control of spores of postharvest fungal pathogens of produce with ozonated water, Abstract 47–6, Abstracts of Papers Presented, 2000 IFT Annual Meeting, Dallas, TX, June 11–14.

Somers, E.B., Schoeni, J.L., and Wong, A.C.L., 1994, Effect of trisodium phosphate on biofilm and planktonic cells of *Campylobacter jejuni*, *Escherichia coli* O157:H7, *Listeria monocytogenes* and *Salmonella typhimurium*, *Int. J. Food Microbiol.*, 22:269–276.

Spotts, R.A. and Cervantes, L.A., 1987, Effects of the nonionic surfactant Ag-98 on three decay fungi of Anjou pear, *Plant Dis.*, 71:240–242.

Spotts, R.A. and Cervantes, L.A., 1992, Effect of ozonated water on postharvest pathogens of pear in laboratory and packinghouse tests, *Plant Dis.*, 76:256–259.

Spotts, R.A. and Peters, B.B., 1980, Chlorine and chlorine dioxide for control of d'Anjou pear decay, *Plant Dis.*, 64:1095–1097.

Strasser, J., 1998, Ozone applications in apple processing, Tech Application, Electric Power Research Institute, Inc., Palo Alto, CA.

Sugar, D. and Spotts, R.A., 1993, The importance of wounds in infection of pear fruit by *Phialophora malorum* and the role of hydrostatic pressure in spore penetration of wounds, *Phytopathology*, 83:1083–1086.

Suslow, T.V. et al., 2000, Oxidation-reduction potential (ORP) monitoring of heated chlorinated water for improving disinfection of minimally processed green onions, Abstract 65A-8, Abstracts of Papers Presented, 2000 IFT Annual Meeting, Dallas, TX, June 11–14.

Takeuchi, K. and Frank, J.F., 2000, Penetration of *Escherichia coli* O157:H7 into lettuce tissues as affected by inoculum size and temperature and the effect of chlorine treatment on cell viability, *J. Food Prot.*, 63(4):434–440.

Takeuchi, K. et al., 2000, Comparison of the attachment of *Escherichia coli* O157:H7, *Listeria monocytogenes, Salmonella typhimurium*, and *Pseudomonas fluorescens* to lettuce leaves, *J. Food Prot.*, 63:1433–1437.

Taormina, P.J., Beuchat, L.R., and Slutsker, L., 1999, Infections associated with eating seed sprouts: an international concern, *Emerging Infect. Dis.*, 5(5):626–634.

Tauxe, R. et al., 1997, Microbiological hazards and emerging issues associated with produce. A preliminary report to the National Advisory Committee on Microbiologic Criteria for Foods, *J. Food Prot.*, 60(11):1400–1408.

Tenzer, A.I., 1997, Personal communications, Bonagra Technologies and Services, Bakersfield, CA.

Tew, H. 2000, Personal communications, Tew Manufacturing Corp., Penfield, NY.

Tokuoka, K. and Isshiki, K., 1994, Possibility of application of allylisothiocyanate vapor for food preservation, *Nippon Shokuhin Kogyo Gakkaishi*, 41:595–599.

Tsai, L.-S., Higby, R., and Schade, J., 1995, Disinfection of poultry chiller water with chlorine dioxide: consumption and byproduct formation, *J. Agric. Food Chem.*, 43:2768–2773.

Ukuku, D.O., Pilizota, V., and Sapers, G.M., 2001, Influence of washing treatment on native microflora and *Escherichia coli* 25922 populations of inoculated cantaloupes, *J. Food Safety*, 21:31–47.

Ukuku, D.O. and Sapers, G.M., 2001, Effect of sanitizer treatments on *Salmonella stanley* attached to the surface of cantaloupe and cell transfer to fresh-cut tissues during cutting practices, *J. Food Prot.*, 64:1286–1291.

Vogel, K., 1999, Personal communications, Bio-Safe System, Pulse Instruments, Woodland Hills, CA.

Wallace, J.S., Cheasty, T., and Jones, K., 1997, Isolation of Vero cytotoxin-producing *Escherichia coli* O157 from wild birds, *J. Appl. Microbiol.*, 82:399–404.

Wartanessian, S., 1997, Personal communications, DECCO Dept., Elf Atochem North America, Inc., Monrovia, CA.

Wei, C.I. et al., 1995, Growth and survival of *Salmonella montevideo* on tomatoes and disinfection with chlorinated water, *J. Food Prot.*, 58:829–836.

Wells, J.M. and Butterfield, J.E., 1997, *Salmonella* contamination associated with bacterial soft rot of fresh fruits and vegetables in the marketplace, *Plant Dis.*, 81:867–872.

Wickramanayake, G.B., 1991, Disinfection and sterilization by ozone, in *Disinfection, Sterilization, and Preservation*, 4th ed., Block, S.S., Ed., Philadelphia: Lea & Febiger, 182–190.

Wisniewsky, M.A. et al., 2000, Reduction of *Escherichia coli* O157:H7 counts on whole fresh apples by treatment with sanitizers, *J. Food Prot.*, 63:703–708.

Wright, J.R. et al., 2000, Reduction of *Escherichia coli* O157:H7 on apples using wash and chemical sanitizer treatments, *Dairy, Food Environ. Sanit.*, 20:120–126.

Xu, L., 1999, Use of ozone to improve the safety of fresh fruits and vegetables, *Food Technol.*, 53(10):58–61, 63.

Zagory, D. and Hurst, W.C., Eds., 1996, *Food Safety Guidelines for the Fresh-cut Produce Industry*, Alexandria, VA: International Fresh-cut Produce Association.

Zhang, S. and Farber, J.M., 1996, The effects of various disinfectants against *Listeria monocytogenes* on fresh-cut vegetables, *Food Microbiol.*, 13:311–321.

Zhuang, R.-Y. and Beuchat, L.R., 1996, Effectiveness of trisodium phosphate for killing *Salmonella montevideo* on tomatoes, *Lett. Appl. Microbiol.*, 232:97–100.

Zhuang, R.-Y., Beuchat, L.R., and Angulo, F.J., 1995, Fate of *Salmonella montevideo* on and in raw tomatoes as affected by temperature and treatment with chlorine, *Appl. Environ. Microbiol.*, 61:2127–2131.

Zottola, E.A., 1994, Microbial attachment and biofilm formation: a new problem for the food industry?, *Food Technol.*, 48(7):107–114.

12 Microbial Safety, Quality, and Sensory Aspects of Fresh-Cut Fruits and Vegetables

Hong Zhuang, M. Margaret Barth, and Thomas R. Hankinson

CONTENTS

INTRODUCTION

Fresh-cut produce has been thought to be one of the fastest growing convenience foods in history. Fresh-cut fruits and vegetables offer a number of advantages over bulk produce including cost control, waste reduction, variety and selection, consistent quality and safety, and less in-store labor (Konczal et al., 1992). However, the fresh-cut industry and the evolving processes used to sustain freshness are in their infancy and face considerable challenges — from raw material processing to dining tables and from food safety to food quality.

MICROBIOLOGICAL SAFETY OF FRESH-CUT PRODUCE

Food biological safety is still the top priority of the fresh-cut industry.

1-58716-041-2/03/$0.00+$1.50

Although fruits and vegetables are among the safest of foods, it has been demonstrated that, under certain circumstances, many fresh-cut products serve as vehicles for food-borne pathogenic microorganisms and result in disease. In 2000, more than 500 people were sickened and the life of a 3-year-old girl was claimed by *Escherichia coli* O157:H7 following ingestion of beef-contaminated salad bar watermelon at two Sizzler restaurants in Wisconsin. Outbreaks of *Salmonella* infections were also traced to precut watermelons in 1955, 1979, and 1991 (Blostein, 1993; CDC, 1979; Gayler et al., 1955). In 1990, sliced cantaloupe from multiple sources in Mexico and Central America was linked to 295 cases of *Salmonella chester* infections in 30 states in the U.S. (Ries et al., 1990). In 1991, more than 400 cases of *Salmonella poona* infections were linked to pre-sliced cantaloupe that originated in Texas or Mexico (CDC, 1991). In 1990 (Wood et al., 1991) and 1993 (CDC, 1993), an outbreak of *Salmonella* spp. infections was epidemiologically linked to consumption of sliced tomatoes. In 1983 and 1989, an outbreak of *Shigella sonnei* was associated with shredded lettuce (Hurst, 1995). In 1981, a *Listeria monocytogenes* outbreak was linked to coleslaw (Hurst, 1995). In 1987, a *Clostridium botulinum* outbreak was suspected to result from consumption of packaged cabbage (Solomon et al., 1990). Recently, Fresh Products Northwest recalled its Crunch Pak Fresh Sliced Apple packages because the product might have been contaminated with *L. monocytogenes* (SafetyAlerts, 2001a). Boskovich Farms Fresh Cut Division recalled halved red bell peppers, since the product was contaminated with *L. monocytogenes* (SafetyAlerts, 2001b). These cases strongly suggest that consumption of fresh-cut fruits and vegetables can result in food-borne disease outbreaks.

Food-borne disease can cause catastrophic economic problems for the companies and individuals involved. It has been estimated that food-borne disease costs an average $679 for each case in the U.S. (Todd, 1989) and around $1000 for each case in Canada (Mayers and Couture, 1999). These figures do not include the cost to food companies resulting from litigation, recall procedures, and lost sales due to adverse publicity. Costs can easily run into millions of dollars per company and force companies to close. This was clearly demonstrated by the mad cow disease incident in the U.K.; its impact on exports of British meat has been estimated to cost more than $10 billion in losses. The *Salmonella* outbreaks in Victoria, Australia, in 1997 cost the Australian small goods industry more than $400 million. In the U.S., Sara Lee has agreed to pay a $200,000 fine and to spend $3 million on food safety research as well as $1.2 million to settle a civil law suit of *Listeria*-contaminated hot dogs and deli meats by Bil Mar Foods in 1998, even though tests of Bil Mar food products conducted by the U.S. Department of Agriculture during 1998 were negative for *L. monocytogenes*. Following the *E. coli* O111 food poisoning incident in 1995, Garibaldi mettwurst sales throughout Australia fell by 40%. There were reports of 400 to 500 small goods producers going out of business as a result (ANZFA, 2001). Since 1995, a series of food-borne illness outbreaks has been linked with fresh sprouts in the U.S., causing users of sprouts to discontinue placing them in salad bars, sandwiches, and other foods. The sprout industry lost significant business and these losses are continuing (Anonymous, 2000).

The working definitions for fresh-cut produce can vary by fresh-cut processors and researchers depending on their specific goals. However, no matter what definition is used, fresh-cut produce must be cleaned, washed, and ready to eat. Consumers eat fresh-cut produce directly out of the bag. Foods eaten raw are potentially more hazardous than foods cooked immediately prior to consumption, which deactivates many of the toxins produced by microbes, including C. *botulinum*, and kills infectious pathogens such as *E. coli* O157:H7. Eating fresh-cuts without washing and cooking indicates that most consumers purchase fresh-cuts with full trust that they will be safe. This gives the fresh-cut industry an ethical responsibility to ensure that full risk reduction is made.

Fresh-cuts are not subject to the thermal or other preservative treatments designed to reduce or eliminate microbial load that some other processed foods receive. A good hazard analysis critical control point (HACCP) plan should effectively prevent the presence of bacterial pathogens, but it cannot provide 100% assurance that the fresh-cut product is pathogen-free. For example, chlorine, the most widely used sanitizer in the fresh-cut industry, is highly effective in killing microorganisms in produce wash solutions. Chlorinated water has been considered as an HACCP (Hazard Analysis Critical Control Point) for managing human pathogens in finished fresh-cut products. Clean, chlorinated water can reduce microbial populations on produce surfaces by not more than 10- to 100-fold when properly applied. This is due in part to the survival of bacteria within biofilms and also to the rapid breakdown of chlorine in the presence of organic matter in processing flumes or wash tanks. Researchers have shown that 200 ppm of chlorine, the recommended maximal concentration used in a fresh-cut flume system, resulted in 1- to 2-log reduction of *E. coli* O157:H7 on lettuce (Rodriguez et al., 2001), 2-log reduction of *L. monocytogenes* on brussel sprouts (Brackett, 1987), 1.3- to 1.7-log reduction of *L. monocytogenes* on lettuce (Zhang and Farber, 1996), and 1.2-log reduction of *S. montevideo* on tomatoes (Zhuang et al., 1995). However, Barnard and Jackson (1984) found that chlorination was not very effective in destroying *Giardia* cysts. Golden et al. (1993) showed that chlorinated water reduced but did not eliminate *Salmonella* once it was on the rind of melon. They concluded that chlorine was only a risk reduction factor, and other preventive measures were needed to further reduce the risk of *Salmonella* on melon rind.

Package integrity and refrigeration are also important control points for fresh-cut shelf-life. However, they are not a good defense against pathogens such as *Salmonella spp., L. monocytogenes, Aeromonas hydrophila,* and *Yersinia enterocolitica* (Sumner and Peters, 1997). Berrang et al. (1989, 1990) showed that modified atmosphere packaging preserved the sensory quality and limited microbial growth of vegetables at refrigeration temperatures. However, human pathogens, such as *L. monocytogenes* and *A. hydrophila*, could actually thrive under these conditions. Piagentini et al. (1997) inoculated shredded cabbage packaged in monooriented polypropylene bags with *S. hadar* and stored the cabbage at 4, 12, and 20°C, respectively, for 10 days. *S. hadar* was found to survive and proliferate on minimally processed cabbage and thereby posed a potential hazard to consumers.

Most pathogens are not native to fresh-cut produce; however, more and more evidence shows that fresh-cut produce can harbor potential pathogens. Food-borne pathogens can survive and even grow on fresh-cut fruits and vegetables packed under MA (Modified Atmosphere) during refrigerated storage. Odumeru et al. (1997) studied the microbiological food safety of fresh-cut vegetables, including chopped lettuce, salad mix, carrot sticks, cauliflower florets, sliced celery, coleslaw mix, broccoli florets, and sliced green peppers before and after processing. They found that, overall, *L. monocytogenes* was detected in 13 of 120 (10.8%) fresh-cut vegetables stored for up to 11 days at 10°C and 5 of 176 (2.8%) samples stored at 4°C. *E. coli* was detected in 2 of the 120 (1.7%) processed vegetables after day 7 of storage at 10°C and 1 of the 65 (1.5%) unprocessed vegetables from the same batches of vegetables used for processing. Steinbruegge et al. (1988) and Beuchat and Brackett (1990) demonstrated that *L. moncytogenes* can grow on fresh-cut lettuce.

In chopped tomatoes, *L. monocytogenes* was able to maintain original population density numbers for up to 2 weeks of storage at 10 or 21°C (Beuchat and Brackett, 1991). Lin and Wei (1997) showed that *Salmonella* spp. on the surface of tomatoes, inoculated into the tomato meat during slicing, could subsequently grow. Zhuang et al. (1995) found that *Salmonella* spp. survived on the surface of mature intact tomatoes and grew rapidly (reaching 10^8/g within 24 h) in chopped ripe tomatoes at ambient temperature, thus reflecting the importance of cold temperature chain management. The growth of *C. botulinum* was also demonstrated in fresh tomatoes (Draughon et al., 1988). In experiments on shredded iceberg lettuce artificially contaminated with *S. lexington, E. coli*, or *L. monocytogenes*, Klepzig et al. (1999) found that a transfer to 25°C resulted in an increase in counts of all three pathogens. Transfer to 15°C resulted in growth of *E. coli* and *S. lexington* after 12 h. Hao et al. (1999) detected botulinal toxin in broccoli florets packaged in bags with an OTR of 3 and stored at 13°C for 21 days and in bags with an OTR (Oxygen Transmission Rate) of 4 and 3 and stored at 21°C for 10 days. Sugiyama and Yang (1975) reported that *C. botulinum* not only grew but also produced toxin in modified atmosphere packed mushroom.

Some strains of *C. botulinum* can grow at temperatures as low as 3.5°C (Hauschild, 1989). *A. hydrophila* can survive or grow well in vegetables stored at 5°C or less (Berrang et al., 1989). Badawy et al. (1985) found rotavirus could survive on vegetables, especially lettuce, for up to a month during refrigerated storage. On eggs of *A. lumbricoides,* protozoan parasites are very resistant to harsh conditions; they are resistant to drying and many chemical disinfectants and have been known to survive refrigerated storage (Brackett, 1994). Although *Shigella* does not grow at refrigeration temperatures, it can survive for extended times under these conditions (Morris, 1984).

Many different food systems are grouped into the enormous category called fresh-cut produce. Each and every commodity can have its unique food safety challenges and critical control points for addressing them. Farber et al. (1998) inoculated wholesale and retail packaged vegetables and salads, including whole rutabagas, butternut squash, onions, packaged Caesar salad, cut carrots, coleslaw mix, and stir-fry vegetables, with a mixture of strains of *L. monocytogenes* and incubated at 4 or 10°C. They found that *L. monocytogenes* population levels remained constant on all fresh-cut vegetables stored at 4°C for 9 days except for carrots and butternut squash; populations declined on carrots and increased on butternut squash. Fresh-cut vegetables stored at 10°C supported growth of *L. monocytogenes* except for chopped carrots, where the population decreased approximately 2 log units over a 9-day storage period; butternut squash supported the highest rate of cell growth. Jacxsens et al. (1999) found that the inoculated pathogens were more influenced by the type of vegetable than by the type of atmosphere. A decline in numbers of *L. monocytogenes* was detected on Brussels sprouts and on carrots.

Aeromonas spp. had a higher growth rate than *L. monocytogenes* on shredded chicory endives and shredded iceberg lettuce at 7°C.

From biological and food-safety perspectives, fresh-cut fruit is different from fresh-cut vegetables. Low-acid fruit will have food-safety challenges different from those of high-acid fruit, and leafy vegetables will have challenges different from those of fleshy vegetables. In addition to distinct variations in the physiology, biochemistry, and microbiology of the many food systems in fresh-cut produce, different processing steps are associated with some categories that could have a major impact on food safety.

Finally, no good indicator or rapid assays for the detection of food-borne pathogens have been developed for fresh-cut produce that might be contaminated by pathogens. Human pathogen testing of food is usually conducted at an outside contract laboratory because of the potential for enriching and growing human pathogens in food-processing environments. Cultural and biochemical methods for the isolation, identification, and enumeration of *Listeria, E. coli* O157:H7, *Salmonella*, and *Campylobacter* spp. from foods are laborious and can take up to 4 days or longer (USDA-FSIS, 1998). Unlike shelf-stable food or frozen foods, because fresh-cut produce is perishable, the products cannot be held until test results are reported. Rapid assays have been constructed using antibody or DNA probes for each of the pathogens. Highly sensitive PCR (Polymerase Chain Reaction) assays are currently available for *Listeria, E. coli* O157:H7, and *Salmonella* (Wang et al., 1997). Highly sensitive ELISA (Enzyme-Linked Immunosorbent Assays) are currently available for *Listeria* (Feldsine et al., 1997), *E. coli* O157:H7 (Fratamico and Strobaugh, 1998), *Salmonella* (Gangar et al., 1998), and *Campylobacter* (Brooks et al., 1998). However, none of these methods is able to enumerate these target pathogens accurately and quantitatively.

A certain minimum amount of testing for pathogens on finished product should be conducted to detect potentially catastrophic pathogen contamination. However, product safety can never be 100% assured by testing finished product unless 100% of the finished product is tested. The major drawback to testing finished product for human pathogens is that one often cannot take a statistically significant sample from a production lot, as it may be nonhomogenous or may produce a sample that is too large to deal with practically. For example, in the 1998 Sara Lee *Listeria* incident, although tests of Bil Mar food products conducted by the United States Department of Agriculture during 1998 were all negative for *L. monocytogenes*, the investigation still indicated that hot dogs and deli meats made by Bil Mar Foods of Sara Lee were the source of an outbreak of listeriosis. Therefore, sampling finished product for the presence of human pathogens often relies on finding a needle in a haystack. This means negative test results for the presence of human pathogens will not assure safety. It may simply mean that one did not take a large enough sample to detect a low-frequency contamination event.

MICROBIOLOGICAL QUALITY OF FRESH-CUT PRODUCE

Microbial spoilage appears to be one of the major causes of quality loss of fresh-cut produce by formation of off-flavor, fermented aroma, and tissue decay. The shelf-

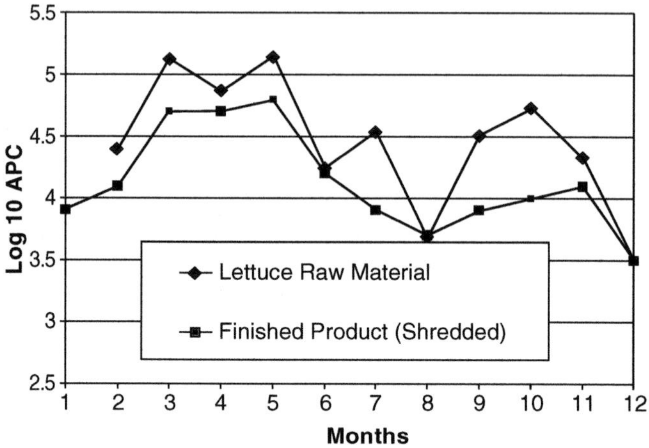

FIGURE 12.1 Changes in average aerobic bacterial counts on raw material and finished products (shredded) of lettuce throughout a year.

life of many food products may be accurately predicted by quantifying the population of microbes present on the food product. Bolin et al. (1977) demonstrated that the initial microbial level dramatically influenced the rate of spoilage of shredded lettuce. When shredded lettuce contained high initial microbial counts (5.1×10^6 CFU/g), the shelf-life was only 2 to 3 days. Under the same package and storage conditions, the low microbial count (5.8×10^3 CFU/g) resulted in 18 days of shelf-life. The size of the initial microbial load, expressed as the microbial population density in colony forming units (CFUs) per gram of fresh weight, in turn, depends on initial microbial population density on raw materials (Figure 12.1) and effects of processing on microbial population density (Figures 12.2 and 12.3).

Figure 12.1 shows a relationship between the initial population density on lettuce raw material and the initial microbial load on the finished product. The changes in the average microbial population on finished product (shredded lettuce) closely (correlation coefficient = 0.87) followed the initial average population density on lettuce raw material through a 12-month period. The effect of processing on microbial population is demonstrated in Figures 12.2 and 12.3. Figure 12.2 illustrates the impact of chlorination and subsequent processing on the total aerobic population on product during 2 hours of processing. Overall, the process results in only a net fourfold reduction in the aerobic population as broccoli is prepared from raw heads through to finished florets. Figure 12.3 illustrates the impact of processing when chlorine is replaced with acidified sodium chlorite (SANOVA®, Alcide Corporation). In this case, the net reduction potential across the floret process is 57-fold. It appears from these results that SANOVA yields superior results by providing a greater reduction in population density at the point of treatment and also maintaining a lower population density at each subsequent process step. However, low initial microbial population density does not guarantee low microbial population during storage. The subsequent microbiological population growth can be signifi-

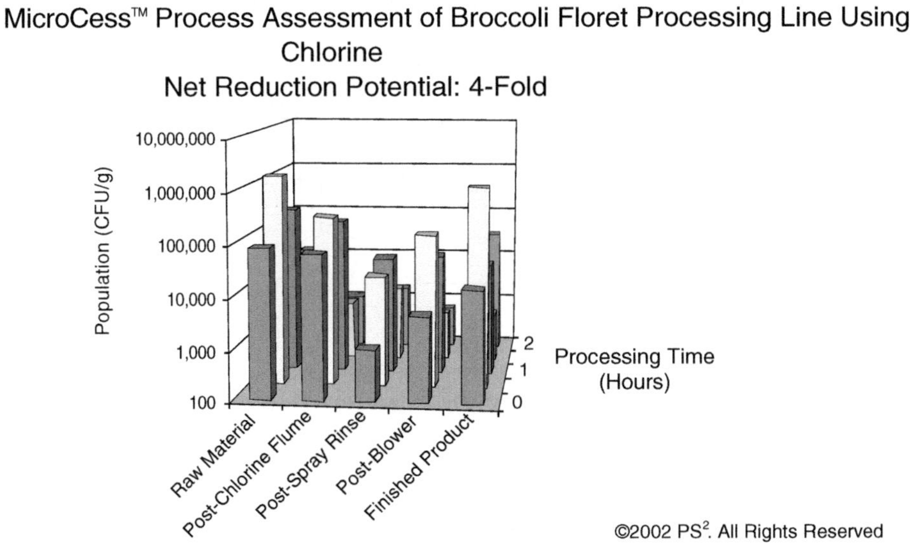

FIGURE 12.2 Assessment of broccoli floret processing line using chlorine. Net reduction potential: fourfold.

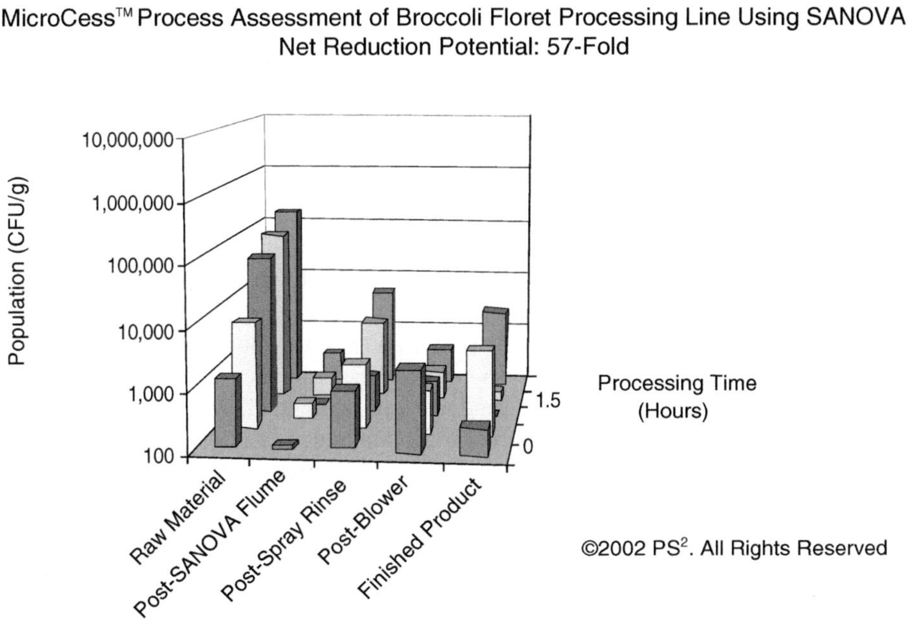

FIGURE 12.3 Assessment of broccoli floret processing line using SANOVA. Net reduction potential: 57-fold.

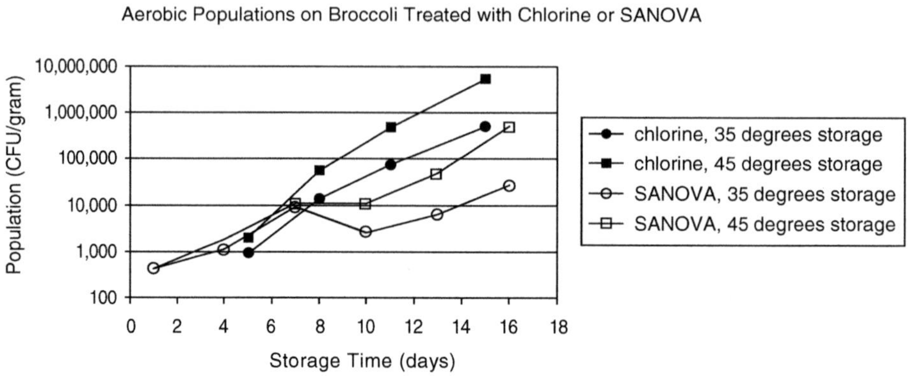

FIGURE 12.4 Aerobic populations on broccoli treated with chlorine or SANOVA.

cantly affected by storage temperature and a combination of processing method and storage temperature.

Figure 12.4 shows the effect of storage temperature on microbial population on broccoli florets. The samples were packaged and stored for 16 days at 35 and 45°F. During this shelf-life study, total aerobic bacteria and total yeasts and molds were monitored by opening and assaying a fresh package every 2 to 3 days. The response of the aerobic bacteria population is shown in Figure 12.4. Population growth was more rapid at 45°F and resulted in a population density approximately tenfold higher than found on florets treated with the same sanitizing agent and stored at 35°F. Sanitizer treatment had a similar impact on population density toward the end of the study: aerobic bacteria on SANOVA-treated florets were about tenfold lower than populations on chlorine-treated florets at a given storage temperature. Taken together, one can conclude that selecting the better sanitizer combined with proper finished product storage can result in at least a 100-fold improvement in bacterial load during the critical latter stages of shelf-life.

French legislation specifies a maximum of 5×10^5 CFU/g (5.7 $\log_{10}$) at production and 5×10^7 (7.7 $\log_{10}$) at the use-by date for fresh-cut products (Nguyen-The and Carlin, 1994). In the U.S., the meat industry uses 8.0 $\log_{10}$ to indicate spoilage. These recommendations have informally been adopted by some food service companies as their specification for fresh-cut produce; however, fresh produce often becomes heavily contaminated with sand and soil in the field. While some wash and disinfection technologies reduce microbial contamination, it is not unusual to harbor a bacterial population ranging from hundreds of thousands to millions per gram before (Figure 12.5, Table 12.1) and after produce is processed.

Microbial population density was easily over 1,000,000 CFU/g (6 $\log_{10}$) on the outer leaves of leaf vegetables before processing (Figure 12.5). The average population of raw materials from bulb vegetables, such as yellow onion, was 5.9 $\log_{10}$ (Table 12.1). Babic and Watada (1996) showed that the initial number of microorganisms ranged 10^7 to 10^8 CFU/g for mesophile, psychrophile, and *Pseudomonadaceae* on fresh-cut, washed spinach leaves. After 7 days of storage at 5°C, numbers reached 10^{10}, 4×10^8, and 5×10^8 CFU/g in air, respectively, and 10^8, 5×10^7, and

TABLE 12.1
Aerobic Plate Counts on Raw Materials of Fresh-Cut Vegetables in 2001

	Iceberg Lettuce	Tomatoes	Onions
APC ($\log_{10}$ CFU/g)	4.57 ± 1.34	4.38 ± 1.11	5.95 ± 0.59

6×10^7 CFU/g under control atmosphere (CA), respectively. Prakash et al. (2000) found that the initial load of the cut romaine was between 10^5 and 10^6. At the "best-if-used-by" day, the aerobic plate count (APC) exceeded 10^8. Hagenmaier and Baker (1998) obtained microbial count values ranging from 2×10^5 to 7.6×10^8 CFU/g on ready-to-eat salads on the day of purchase. On the expiration date, the values ranged from 1.4×10^7 to 6.5×10^8 CFU/g. Portela and Cantwell (2001) showed that

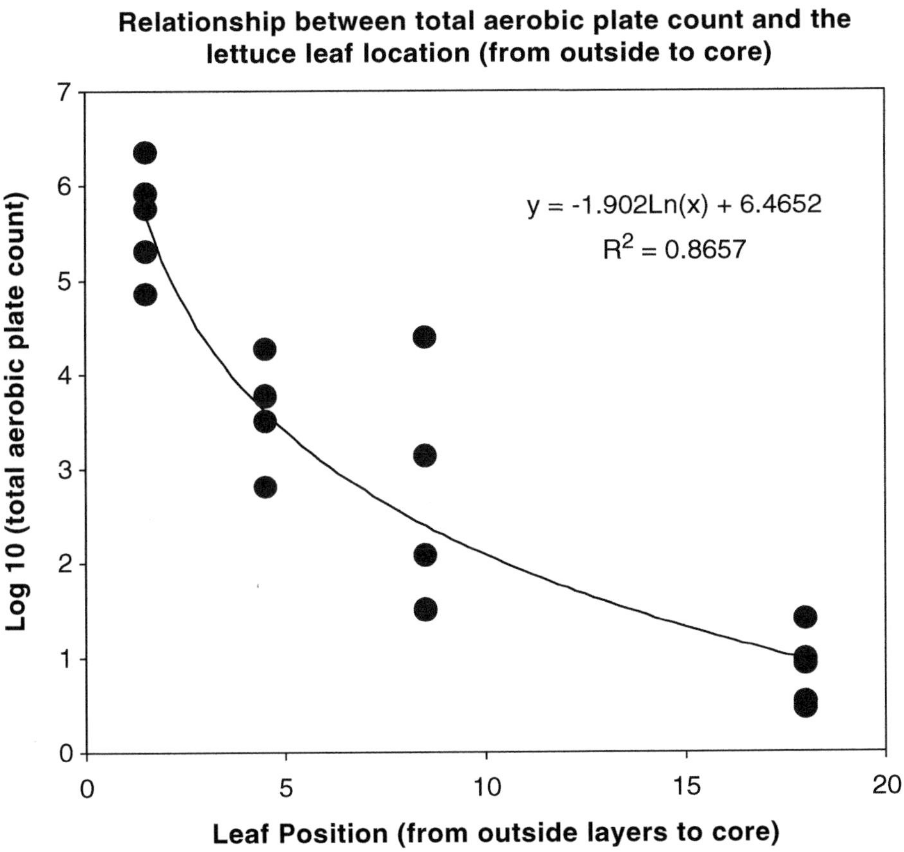

FIGURE 12.5 Relationship between total aerobic plate count and the lettuce leaf location (from outside to core).

TABLE 12.2
Aerobic Plate Counts on Different Finished Fresh-Cut Vegetables in 1999

	Log$_{10}$ APC/g Fresh Weight			
	Average ± SD	Data Points	% Population Distribution > 5.7 log	Recommended Upper Limit (95%)
Broccoli florets	4.80 ± 0.84	216	14.2	6.2
Cabbage (diced)	4.33 ± 0.65	49	1.7	5.4
Cantaloupe (cubes)	3.43 ± 0.85	93	0.4	4.8
Carrots (diced)	4.12 ± 0.57	105	0.3	5.1
Carrots (shredded)	4.08 ± 0.58	315	0.3	5.0
Celery (cross-cut)	5.10 ± 0.44	134	8.7	5.8
Green pepper (diced)	4.85 ± 0.66	2928	9.9	5.9
Green pepper (sliced)	4.83 ± 0.61	144	7.6	5.8
Honeydew (cubes)	3.04 ± 0.93	93	0.2	4.6
Lettuce (shredded)	4.25 ± 0.64	339	1.2	5.3
Tomatoes (diced)	4.78 ± 0.61	101	6.6	5.8
Yellow onions (diced)	4.15 ± 0.68	556	1.1	5.3
Yellow onions (sliced)	4.19 ± 0.68	507	1.3	5.3

the initial APC of 3.7×10^2 on cut cantaloupe counts increased 1,000,000-fold to 3.5×10^8 at the end of storage. Blanchard et al. (1996) found that psychrotroph flora on diced yellow onions increased from an initial load of 10^4 CFU/g fresh tissue to 10^8 CFU/g after 14 days at 4°C. Mesophile flora responded in a similar way. Nguyen-The (1991) observed that total mesophile flora on ready-to-use vegetables generally stabilized at about 10^8 bacteria/g, irrespective of initial population density; despite the size of this microbial population, the products usually retained a satisfactory sensory quality.

Setting one microbial upper limit for all fresh-cut products is not practical for food safety or quality. Initial microbial population density varied with commodities of finished fresh-cuts (Table 12.2) and the seasons throughout the year (Table 12.3). If the upper limit of 5×10^7 APC/g (or 5.7 log$_{10}$) and 0.95 cumulative probabilities are used for specification of microbial load of finished fresh-cut products, 38% (5 out of 13) of the listed fresh-cuts products in Table 12.2 are out of spec. These products include broccoli florets, cross-cut celery, diced green pepper, mushroom, and diced tomatoes. If 0.99 cumulative probabilities are used with the same upper limit (5.7 log$_{10}$), 8/13 products on the list could be out of specification.

For the same fresh-cut product, the initial microbial population also depended on the month of year, and the season differed with commodities (Table 12.3). The microbial population of diced tomatoes likely passed the upper limit of 5.7 log$_{10}$ in November, December, and January. However, for diced yellow onion, the microbial population likely passed 5.7 log$_{10}$ in August and September. For diced green pepper, a high microbial load was noted from May to September. For shredded lettuce, a

TABLE 12.3
Aerobic Plate Counts on Different Fresh-Cut Vegetables after Processing in 1999

	Log$_{10}$ APC/g Fresh Weight (Ave. ± SD)					
	Diced Tomatoes	Diced Yellow Onion	Diced Green Pepper	Shredded Lettuce	Whole Peeled Onions	Sliced Fuji Apples
January	5.2 ± 0.4	4.1 ± 0.6	4.7 ±0.6	3.9 ± 0.7	2.72 ± 0.2	3.35 ± 0.5
February	4.6 ± 0.5	4.2 ± 0.6	4.6 ± 0.6	4.1 ± 0.9	2.48 ± 0.5	3.98 ± 0.6
March	4.2 ± 0.5	4.3 ± 0.6	4.4 ± 0.6	4.7 ± 0.6	2.92 ± 0.5	—
April	—	4.6 ± 0.6	4.7 ± 0.6	4.7 ± 0.5	2.81 ± 0.9	4.04 ± 0.5
May	4.5 ± 0.3	3.9 ± 1.1	5.2 ± 0.4	4.8 ± 0.5	2.43 ± 0.9	4.58 ± 0.4
June	4.9 ± 0.4	4.1 ± 0.5	5.1 ± 0.5	4.2 ± 0.6	—	3.51 ± 0.8
July	4.7 ± 0.8	4.3 ± 0.5	5.2 ± 0.7	3.9 ± 0.6	2.51 ± 0.5	3.29 ± 0.4
August	4.7 ± 0.7	5.0 ± 0.6	5.4 ± 0.5	3.7 ± 0.6	2.58 ± 0.5	3.14 ± 0.9
September	4.5 ± 0.4	4.9 ± 0.7	5.2 ± 0.5	3.9 ± 0.7	2.76 ± 0.3	2.09 ± 0.7
October	4.6 ± 0.8	4.2 ± 0.5	5.0 ± 0.5	4.0 ± 0.5	2.80 ± 0.7	2.42 ± 0.5
November	4.9 ± 0.5	4.3 ± 0.6	4.7 ± 0.7	4.1 ± 0.4	2.89 ± 0.6	3.19 ± 0.7
December	5.2 ± 0.7	4.2 ± 0.6	4.9 ± 0.5	3.5 ± 0.5	—	—

high microbial reading was found from March to May. These results suggest using the different APC upper limits for different fresh-cut products (Table 12.2) and also that the upper limit should vary with seasons. For example, 5.9 log$_{10}$ CFU/g of APC should be used as a specification for diced green peppers at 0.95 cumulative probabilities overall (Table 12.2); however, it should be more than 6 log during the summer (Table 12.3). This is true only if vulnerability to pathogen contamination or spoilage correlates with changes in initial microbial load. If product shelf-life and pathogen incidence remain more or less constant, seasonal variability is not relevant.

In addition, the initial microbiological quality can be affected by cut size and sharpness of machine blade used for the processing. Barry-Ryan and O'Beirne (1998) compared the effects of slicing carrots with a razor blade, a sharp machine blade, and a blunt machine blade and found that microbial loads were reduced in the treatment with sharp slicers.

Total aerobic population density (TPC or APC) is generally used as one indicator for quality of fresh-cuts. However, fresh-cut produce can be of good quality despite very high microbial populations. Undoubtedly, the types of microorganisms, their relative proportion, the condition of the product, and the care used during processing are the most important variables in quality and food safety of fresh-cut products. Table 12.4 shows the changes in firmness, acceptability, APC, and yeast and molds of cut honeydew in two different trials during storage. The loss in sensory quality of cut honeydew was not associated with the population of APC or yeast and molds. For example, in trial 1, the total number of yeast and molds was undetectable and APC was 3.8 log$_{10}$ CFU/g fresh weight on day 13 when the quality of cut honeydew was unacceptable. However, in trial 2, the total number of yeast and mold was 2.6

TABLE 12.4
Relationship between Microbial Load and Quality of Honeydew Chunks over Storage

Storage Time	Firmness (N)	Acceptability	Log_{10} APC	Log_{10} Yeast and Molds
Trial 1				
Day 0	10.4 ± 1.3	4.25 ± 0.8	1	—
Day 7	8.2 ± 0.9	4.25 ± 0.8	2	0
Day 13	7.0 ± 1.5	2.75 ± 0.8	3.8	0
Trial 2				
Day 0	4.8 ± 0.7	3.0 ± 0.7	0	0
Day 7	5.1 ± 1.1	2.5 ± 0.5	1	2.6

Notes: 1 = highly unacceptable; 2 = unacceptable; 3 = marginally acceptable; 4 = acceptable; 5 = highly acceptable.

log_{10} (400 CFU/g) and APC was 1 log_{10} when the product was unacceptable on day 7. Instead, the firmness (or condition) of the product seemed to play a key role in the shelf-life of cut honeydew during storage.

The situation is further complicated when examining the results from a study of fresh-cuts commercially processed at different times of the year and stored at two temperatures. The experiments were conducted using shredded lettuce (Figure 12.6)

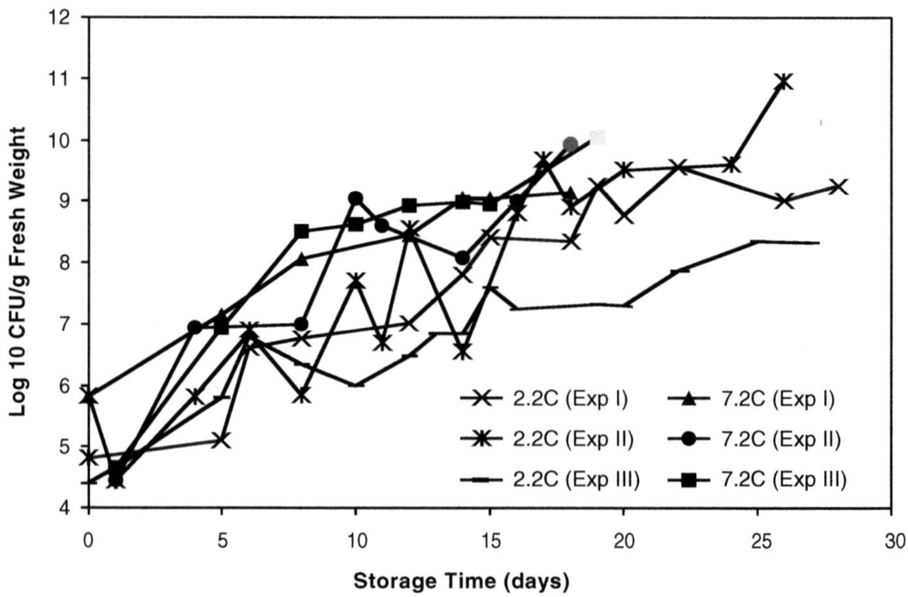

FIGURE 12.6 Changes in aerobic bacterial counts on shredded lettuce during storage.

and diced yellow onions (data not shown) from October to December of 1999. Shredded lettuce was stored at 2.2 and 7.2°C, respectively, and diced yellow onions were stored at 2.2 and 5.6°C, respectively. Microbial assays were discontinued when the visual quality of products became unacceptable. Figure 12.6 shows that shelf-life of shredded lettuce was 18 to 19 days when samples were stored at 7.2°C, compared with more than 25 days at 2.2°C. Obviously, there was no correlation between aerobic population density and product quality. For example, in experiment I, there was no difference in the APC number between the samples stored at 2.2°C and those stored at 7.2°C at the end of shelf-life (about 9 $\log_{10}$ cycles). However, in experiment II, the APC number of the sample stored at 2.2°C was 1 log cycle higher (11 $\log_{10}$ cycles) than that (10 $\log_{10}$ cycles) at 7.2°C at the end of shelf-life. In experiment III, the APC number was 2 log cycles higher at 7.2°C than that at 2.2°C (10 vs. 8 log). Similar results were found with sliced yellow onion. In trial I, aerobic populations were similar (9 $\log_{10}$) at the end of shelf-life, regardless of storage temperatures; however, trial II showed that aerobes on the samples stored at 2.2°C were at least tenfold higher (more than 10 $\log_{10}$) than the samples stored at 5.6°C. These results demonstrate that the total number of microorganisms can greatly vary when the quality of fresh-cuts fails; they further suggest that the types of microorganisms and conditions of fresh-cut products likely have more impact on shelf-life than microbial population density alone.

Often only a small population of total microbes found on fresh-cut products, at least initially, represents spoilage organisms. The majority of bacteria responsible for spoilage of fresh-cut fruits and vegetables are Gram-negative. Of these, *Erwinia* spp. is among the most aggressive. *E. carotovara* is the species most commonly associated with decay of vegetables (Lund, 1983); fluorescent *Pseudomonas* is another group of common and important spoilage organisms of refrigerated vegetables. Both groups are ubiquitous in nature and abundant on the surfaces of most produce. *Pseudomonas*, in particular, are responsible for soft rot decay of many types of vegetables including celery, potato, chicory, lettuce, Chinese chard, and cabbage (Brocklehurst and Lund, 1981). Few Gram-positive bacteria, most notably clostridia and bacilli (such as lactic acid bacteria and *C. puniceum*) can cause spoilage of vegetables under the right circumstances. These bacteria grow slowly, if at all, at refrigeration temperatures (Lund, 1982).

Although many different types of fungi can be associated with the spoilage of fruits and vegetables, relatively few cause spoilage problems on fresh-cuts. The situation is very different on many types of whole produce that are susceptive to pre- and postharvest decay fungi. So far, the relationship between the population density of these specific groups of microbes and susceptibility to spoilage of a specific fresh-cut product is still unknown. Furthermore, no simple, rapid, and practical assay has been developed to monitor these specific microbes at the commercial level.

The fresh-cut industry depends on chemically sanitized water to manage spoilage microbes and pathogens in order to produce a high-quality and safe product. However, different sanitizers provide different results, depending on the type and form of produce treated. Table 12.5 shows the effect of spray wash with three different chemical sanitizers, chlorine, ozone, and Tsunami™, on APC on different fresh-cut

TABLE 12.5
Spray Wash Efficacy of Different Antimicrobial Agents, Hypochlorite, Tsunami™, and Ozone

	Log_{10} APC/g Fresh Weight			
	Celery	**Lettuce**	**Green Pepper**	**Tomatoes**
Hypochlorite (100 ppm total)				
Pre-Wash	6.0	3.9	4.0	3.2
Post-Wash	4.8	3.2	3.5	3.5
Tsunami (20 ppm)				
Pre-Wash	6.0	4.1	3.8	3.6
Post-Wash	5.7	2.8	2.8	3.4
Ozone (2 to 5 ppm)				
Pre-Wash	5.9	4.7	3.2	4.0
Post-Wash	5.4	3.4	3.8	3.1

products. Chlorinated water resulted in more than 1 $\log_{10}$ reduction in APC population on celery sticks; however, only 0.7 $\log_{10}$ reduction was found with lettuce and 0.5 $\log_{10}$ reduction on green pepper. No reduction was noted on tomatoes. With Tsunami solution, the greatest reduction in APC after washing was found on lettuce (1.2 $\log_{10}$), followed by green pepper (1 $\log_{10}$), celery (0.3 $\log_{10}$), and tomatoes (0.2 $\log_{10}$ reduction). Produce-dependent variability also was observed with ozone, with the greatest reduction in APC on lettuce (1.3 $\log_{10}$), followed by tomatoes (0.9 $\log_{10}$) and celery (0.5 $\log_{10}$).

Table 12.6 shows the comparison of effects of chlorinated flume and Tsunami flume on the microbial population on fresh-cut products. Chlorinated water was more effective than Tsunami when diced celery passed through the flume. However, Tsunami showed better killing capability than chlorine when diced green pepper, diced cabbage, shredded carrots, and cauliflower florets were flumed. The effect of chlorine and Tsunami on the shelf-life of different fresh-cut products is shown in Table 12.7. The efficacy of these two chemical sanitizers on shelf-life differed with produce commodity. In addition, using these chemical sanitizers results in other concerns, such as potential carcinogenicity of reaction compounds when chlorine is used, pungent odor of ozone, cost-saving, different operation systems required for the different chemical sanitizers, and chemical hazard generated when two different sanitizers are mixed (such as Tsunami with chlorine).

Recent research studies revealed that the microorganisms naturally occurring on produce can play a positive role in the competitive inhibition of pathogens in fresh-cut packages during storage. Bennik et al. (1996) studied the effect of chlorine wash on quality and safety of cut endive or chicory endive during storage. *L. monocytogenes* was inoculated on cut endive leaves. The samples were then sealed in polypropylene bags in air without further treatment or rinsed either in sterile distilled water or with 10% hydrogen peroxide for 2 min. It was found that the

TABLE 12.6
Effect of Tsunami and Hypochlorite Flume Wash on Aerobic Plate Counts of Different Fresh-Cuts

	Log_{10} APC/g Fresh Weight		
	Pre-Flume	Hypochlorite	Tsunami
Broccoli floret	6.1 ± 0.1	5.9 ± 0.0	6.1 ± 0.2
Diced cabbage	5.9 ± 0.1	5.6 ± 0.2	5.2 ± 0.2
Shredded carrots	4.5 ± 0.0	4.4 ± 0.1	3.6 ± 0.0
Cauliflower floret	5.8 ± 0.0	5.4 ± 0.1	4.9 ± 0.2
Diced celery	6.0 ± 0.0	4.9 ± 0.1	5.8 ± 0.1
Diced green pepper	6.2 ± 0.3	5.9 ± 0.5	5.2 ± 0.3
Shredded red cabbage	4.0 ± 0.0	3.5 ± 0.0	3.6 ± 0.2
Sliced red onion	4.8 ± 0.1	4.5 ± 0.2	4.8 ± 0.1
Sliced zucchini	5.6 ± 0.1	5.1 ± 0.0	5.1 ± 0.0

psychrotrophic pathogen *L. monocytogenes* grew better on disinfected produce than on nondisinfected or water-rinsed produce. Modified-atmosphere conditions, which were favorable for product quality and retarded growth of spoilage microorganisms during low-temperature storage, did not inhibit the growth of *L. monocytogenes*. This finding suggests that the challenge is not for chemical sanitizers to eliminate microorganisms. Rather, the role of sanitation is to manage microorganisms so that they play a positive role in the competitive inhibition of pathogens and a reduction of spoilage types in order to enhance product quality. This field of competitive inhibition on fresh-cuts should be in focus for research and development funding now.

RAW MATERIALS — KEY FACTORS FOR QUALITY AND SHELF-LIFE OF FRESH-CUT PRODUCE

In order to provide consistent, high-quality fresh-cut products throughout the year, definitely superior raw materials are required, including specifically developed cultivars and consistent raw products. In some cases, quality criteria for raw materials are well established. However, many products require much closer investigation to determine appropriate measures for quality. All fresh-cut processors today struggle with the right raw materials for processing throughout the year.

Cultivar selection is probably the most important consideration in fresh-cut fruit and vegetable processing because cultivars can vary greatly in characteristics such as flesh texture, skin color, browning potential, chilling sensitivities, and flavors (Amiot et al., 1995; Lyons and Breidenbach, 1987). For example:

- Gorny et al. (2000) investigated the genotypic difference in shelf-life of pear slices derived from 4 cultivars: Bartlett, Bosc, Anjou, and Red Anjou. They found that fresh-cut fruit slices prepared from partially ripe

TABLE 12.7
Sensory and Microbiological Quality of Fresh-Cut Vegetables over Storage (with different water sanitizer treatments)

Shredded Red Cabbages

	Chlorine Wash (100 ppm total)			
	Day 0	**Day 4**	**Day 8**	**Day 12**
Acceptability	5	5	5	4
APC (log_{10})	3.5 ± 0.1	4.1 ± 0.0	5.3 ± 0.2	5.8 ± 0.1
	Tsunami Wash (30 ppm total)			
	Day 0	**Day 4**	**Day 8**	**Day 12**
Acceptability	5	5	5	4
APC (log_{10})	3.6 ± 0.2	4.3 ± 0.0	5.1 ± 0.3	6.0 ± 0.2

Broccoli Florets

	Chlorine Wash (100 ppm total)			
	Day 0	**Day 3**	**Day 8**	**Day 13**
Acceptability	4.5	3.75	4	3.5
APC (log_{10})	5.9 ± 0.1	5.7 ± 0.1	8.3 ± 0.3	8.3 ± 0.1
	Tsunami Wash (30 ppm total)			
	Day 0	**Day 3**	**Day 8**	**Day 13**
Acceptability	4.5	4	4	3.25
APC (log_{10})	6.1 ± 0.1	6.4 ± 0.2	7.9 ± 0.1	8.3 ± 0.3

Cauliflower Florets

	Chlorine Wash (100 ppm total)			
	Day 0	**Day 5**	**Day 8**	**Day 12**
Acceptability	4.25	4	4	3
APC (log_{10})	5.4 ± 0.1	7.7 ± 0.2	7.6 ± 0.1	10.0 ± 0.1
	Tsunami Wash (30 ppm total)			
	Day 0	**Day 5**	**Day 8**	**Day 12**
Acceptability	4.25	4	4	3.5
APC (log_{10})	4.9 ± 0.2	7.1 ± 0.1	7.9 ± 0.3	9.8 ± 0.2

Sliced Zucchini

	Chlorine Wash (100 ppm total)			
	Day 0	**Day 5**	**Day 8**	**Day 12**
Acceptability	5	4	3	2
APC (log_{10})	5.1 ± 0.0	6.5 ± 0.0	7.4 ± 0.0	7.5 ± 0.2
	Tsunami Wash (30 ppm total)			
	Day 0	**Day 5**	**Day 8**	**Day 12**
Acceptability	5	4	3	2
APC (log_{10})	5.1 ± 0.0	6.4 ± 0.0	6.6 ± 0.3	7.4 ± 0.0

Notes: 1 = highly unacceptable; 2 = unacceptable; 3 = marginally acceptable; 4 = acceptable; 5 = highly acceptable.

Bartlett pears had longer shelf-life than those from Bosc, Anjou, and Red Anjou pears.

- Banjongsinsiri et al. (2000) studied the influence of potato cultivars on the quality of pre-peeled refrigerated product. They found that the potato cultivar Atlantic was less susceptible to browning, the potato cultivar Mainstay showed the greatest browning, and the Kennebec caused a high degree of browning.
- Gorny et al. (1999) evaluated the shelf-life of slices from 13 cultivars of peaches and 8 cultivars of nectarines. Among 13 peach cultivars, Cal Red, Red Cal, and Elegant Lady slices had the longest shelf-lives, while Summer Lady and Ryan Sun slices had the shortest shelf-lives. Among the nectarine cultivars, Sparkling Red, Arctic Queen, and Zee Grand slices had the longest shelf-lives.
- Varoquaux et al. (1996) studied the influence of cultivars on the storage life of fresh-cut butterhead lettuce. They found that the cultivar Ritmo had more potassium leakage and brown stain than the cultivar Musettte and the cultivar Nancy during storage.
- In a study of sliced tomatoes, Hong and Gross (2000) noted that the cultivar Mountain Pride showed at least twice as much chilling injury (water-soaked area or translucency) than the cultivar Sunbeam following 17 days of storage at 5°C.
- In lettuce, the initial slope of phenylalanine ammonia lyase (PAL) activity was most highly correlated with the visual quality loss of processed lettuce (Lopez-Galvez et al., 1996). PAL activity was most rapidly induced and reached highest levels in butterhead lettuce, showing threefold differences in rates of induction and twofold differences in maximum PAL activity compared with levels in iceberg. Romaine, green leaf, and red leaf lettuces generally showed intermediate levels. The cumulative PAL activity over 7 days was 1.6, 2.4, 2.7, 3.2, and 4.3 for iceberg, green leaf, red leaf, romaine, and butterhead, respectively.

These results suggest that cultivars of commodities can significantly influence shelf-life and quality of finished fresh-cuts. However, for most produce commodities, little effort is expended on developing cultivars specifically to meet fresh-cut requirements, since fresh-cuts are still a relatively new product in the marketplace.

The fruit and vegetable ripeness stage at cutting also has a significant effect on the shelf-life and quality of fresh-cuts:

- Ripe pear slices had a marketable shelf-life of about 2 days at 0°C. However, partially ripe and mature-green slices had a shelf-life of 8 days at 0°C, based on visual quality (Gorny et al., 2000). A study (Lopez-Galvez et al., 1997) of the effect of maturity of bell pepper (mature green, partially colored, or red fruit) from 24 different cultivars on shelf-life of diced and sliced products showed that fresh-cut pieces prepared from peppers that are ripening or are fully red generally have better shelf-life than pieces from mature green fruits.

- Because the activity of PAL can be used as an indicator of lettuce quality stability, the dynamics of PAL activity were monitored in relation to head maturity. The iceberg cultivar, Legacy, was harvested at different maturity stages (head firmness) and stored 2 weeks at 2.5°C before processing. The level of PAL activity at the mature or overmature stages was higher than that at the immature stage, indicating that mature or overmature lettuce shows oxidative browning discoloration more readily than immature lettuce after processing.

The effect of raw material conditions is not limited to these two factors. The firmness and size of raw materials, preprocessing conditions (such as fresh or storage), storage temperature, storage atmosphere, storage duration, and growing conditions also play a role in quality consistency of fresh-cut products:

- Pear fruit derived from smaller sized fruit generally have greater cut-surface discoloration and deteriorate more rapidly than slices derived from large fruit (Gorny et al., 2000). Gorny et al. (1999) found that 18- to 31-N (Newton) firmness was the optimal stage of ripeness for whole fruit to be processed into peach and nectarine slices. The flesh firmness of 44 to 58 N is optimal for fresh-cut pear slice processing (Gorny et al., 2000). Honeydew chunks with flesh firmness of 5 N had a shelf-life of less than 7 days. However, the shelf-life of fresh-cut honeydew was more than 7 days when the flesh firmness of honeydew raw material was 10 N (Table 12.4). There was significant reduction in the shelf-life of pear slices with increased storage duration of intact pears at –1°C in air or in a controlled atmosphere (Gorny et al., 1998; Gorny et al., 1999).
- Slices from McIntosh apples stored under 15 to 45% CO_2 and 1% O_2 for 6 weeks were firmer that those stored under 0% CO_2 and 1% O_2 (Amissah et al., 2001).
- Storage of whole lettuce heads before processing affected PAL activity, an indicator of browning discoloration formed from fresh-cut iceberg lettuce. Tissue from freshly harvested iceberg and romaine lettuces had a longer period of PAL induction than that of stored heads. The maximum PAL values deceased with increasing storage time of the heads. The initial rate of induction, however, increased with increasing storage time. In addition, higher cumulative levels of PAL were observed when higher amounts of organic fertilizer were applied to the iceberg lettuce. Different cooling methods of intact iceberg lettuce had no effect on maximum levels of PAL. However, the rate of induction of PAL in vacuum cooled samples reduced compared to noncooled and hydrocooled samples, regardless of maturity (Lopez-Galvez et al., 1996).
- Hong et al. (1998) conducted experiments to compare changes in the quality of slices of red tomato fruit from plants grown using black polyethylene or hairy vetch mulches under various foliar disease management systems (fungicide treatments). Slices from tomato fruit grown using hairy vetch mulch tended to have lower electrolyte leakage than those grown

with black polyethylene. Within fungicide treatment, slices from tomato fruit grown using hairy vetch showed less chilling injury (water-soaked areas) than those from tomatoes grown using black polyethylene. The percent of water-soaked areas of slices from tomato fruit grown using black polyethylene mulch under no fungicide application was more than eightfold that of slices from tomato fruit grown using hairy vetch under fungicide application.

BIOCHEMISTRY AND PHYSIOLOGY — BASIS FOR CONTROL OF QUALITY CONSISTENCY OF FRESH-CUT PRODUCE

Biochemistry and physiology of fresh-cut produce have been studied intensively (Solomos, 1994; Varoquaux and Wiley, 1994). However, additional basic information is needed to complement current information in order to develop improved methods for handling and storage of fresh-cuts without loss of quality.

For example, biochemistry and physiological function of the plant hormone ethylene were elucidated decades ago. Cutting plant tissues induces elevated ethylene production rate, and wound ethylene may accelerate deterioration and senescence in vegetative tissues and promote ripening of climacteric fruits. It has been found that ethylene produced by the physical action of fresh-cut processing is sufficient to accelerate softening of banana and kiwifruit, as well as chlorophyll loss in spinach, but does not have the same effects in broccoli (Abe and Watada, 1991). For tomato slices, however, formed ethylene increased ripening (Mencarelli et al., 1989) and reduced the chilling damage or water-soaking area in the package during storage (Hong and Gross, 2000). These results indicate the complexity of the effect of ethylene on quality of various fresh-cut products. More research is required before ethylene absorbents can be used for fresh-cut packages.

Chilling injuries have been observed in many tropical and subtropical vegetables and fruits when the whole produce is stored at the refrigerated temperature. The symptoms include softening or wet spots (decay), uneven color, and surface pitting. However, many chill-sensitive, fresh-cut produce commodities did not develop symptoms of chilling injury and the tissue appeared normal during refrigerated storage at 0 to 10°C (Watada et al., 1996). Zucchini slices survived better at 5°C than at 10°C (Izumi et al., 1996). However, chilling injury is a major quality issue with tomato slices stored at 5°C (Hong and Gross, 2000).

Oxidative browning at the cut surface is the limiting factor in storage of many fresh-cut fruits and vegetables. Oxidative browning occurs when the products of phenylpropanoid metabolism are oxidized in reactions catalyzed by phenolases, such as polyphenoloxidase (PPO) or peroxidase (Vamos-Vigyazo, 1981). The chemical reaction is basically the same in all plant tissue; however, the inhibition of this process by antibrowning agents depends on the commodities. Ascorbate is a very effective browning inhibitor for fresh-cut apples (Ponting et al., 1972), but it does not work well with lettuce (Zhuang and Barth, 1999). Acetate can inhibit pinking of lettuce ribs effectively (Castaner et al., 1996); sulfiting agents, although banned from use on fresh-cut produce, are still the best antibrowning treatment for processing cut potatoes and mushrooms (Sapers et al., 1989, 1994). Vacuum packaging inhibits browning discoloration of all of these products. However, it can result in off-odor formation during inappropriate package selection and storage temperature and thus reduce the shelf-life.

SENSORY QUALITY AND SENSORY TEST

Fresh-cut processors need a rapid, reliable, and objective sensory quality measurement for shelf-life validation, new product development, and evaluation of new technologies. Currently, most fresh-cut companies rely on a trained food technologist (either a research and development or quality assurance technician) for the sensory quality measurements; however, this method is subjective and results could be biased. The information generated by this method does not satisfy research and development needs, fresh-cut customers, and the providers of new technologies for fresh-cut business.

Many objective methods, including instrumental, physical, and chemical, have been developed to measure sensory quality changes of food during storage. Most of them require expensive equipment, such as the electronic nose, spectrophotometer, instron, and GC-MS. Currently, for most fresh-cut processors, the equipment costs too much to own and maintain. In addition, although much work is done in developing highly advanced instruments for quality detection of agricultural products, there is still a question as to the exact correlation of these measurements to sensory perceptions. Therefore, more attention must be paid to developing objective sensory evaluation methods in the fresh-cut industry.

REFERENCES

Abe, K. and Watada, A.E., 1991, Ethylene absorbent to maintain quality of lightly processed fruits and vegetables, *J. Food Sci.*, 56:1493–1496.

Amiot, M.J. et al., 1995, Influence of cultivar, maturity stage, and storage conditions on phenolic composition and enzymatic browning on pear fruit, *J. Agri. Food Chem.*, 43:1132–1137.

Amissah, J.G.N. et al., 2001, 2001 IFT Annual Meeting Book of Abstracts, June 23–27, 2001, New Orleans, LA, 230.

Anonymous, 2000, Cleaning up the image of fresh sprouts, *Fresh Cut*, 8.11:16–18.

ANZFA, 2001, Incidence of foodborne illness, http://www.anzfa.gov.au/mediareleasespublications/publications/foodsafetystandardscostsandbenefits/incidenceoffoodborne633.cfm.

Babic, I. and Watada, E.E., 1996, Microbial populations of fresh-cut spinach leaves affected by controlled atmospheres, *Postharv. Biol. Technol.,* 9:187–193.

Badawy, A.S., Gerba, C.P., and Kelley, L.M., 1985, Survival of rotavirus SA-11 on vegetables, *Food Microbiol.*, 2:199–205.

Banjongsinsiri, P. et al., 2000, 2000 IFT Annual Meeting Book of Abstracts, June 10–14, 2000, Dallas, TX, 134.

Barnard, R.J. and Jackson, G.J., 1984, The transfer of human infections by foods, in *Giardia and Giardiasis*, Erlandsen, S.L. and Mayer, E.A., Eds., New York: Plenum Press, 365–378.

Barry-Ryan, C. and O'Beirne, D., 1998, Quality and shelf-life of fresh-cut carrot slices as affected by slicing method, *J. Food. Sci.*, 63:851–856.

Bennik, M.H.J. et al., 1996, Microbiology of minimally processed, modified-atmosphere packaged chicory endive, *Postharv. Biol. Technol.*, 9:209–221.

Berrang, M.E., Brackett, R.E., and Beuchat, L.R., 1989, Growth of *Aeromonas hydrophila* on fresh vegetables stored under a controlled atmosphere, *Appl. Environ. Microbiol.*, 55:2176–2171.

Berrang, M.E., Brackett, R.E., and Beuchat, L.R., 1990, Microbial, color and textural qualities of fresh asparagus, broccoli and cauliflower stored under controlled atmosphere, *J. Food Prot.*, 53:391–395.

Beuchat, L.R. and Brackett, R.E., 1990, Growth of *listeria monocytogenes* on lettuce as influenced by shredding, chlorine treatment, modified atmosphere packaging, temperature and time, *Appl. Environ. Microbiol.*, 55:775–758.

Beuchat, L.R. and Brackett, R.E., 1991, Growth of *listeria monocytogenes* on tomatoes as influenced by shredding, chlorine treatment, modified atmosphere packaging, temperature and time, *Appl. Environ. Microbiol.*, 57:1367–1371.

Blanchard, M. et al., 1996, Modified atmosphere preservation of freshly prepared diced yellow onion, *Postharv. Biol. Technol.*, 9:173–185.

Blostein, J., 1993, An outbreak of *Salmonella javiana* associated with consumption of watermelon, *J. Environ. Health.*, 56:29–31.

Bolin, H.R. et al.,1977, Factors affecting the storage stability of shredded lettuce, *J. Food Sci.*, 42:1319–1321.

Brackett, R.E., 1987, Antimicrobial effect of chlorine on *Listeria monocytogenes*, *J. Food Prot.*, 50:999–1003.

Brackett, R.E., 1994, Microbiological spoilage and pathogens in minimally processed refrigerated fruits and vegetables, in *Minimally Processed Refrigerated Fruits and Vegetables*, Wiley, R.C., Ed., College Park, MD: Chapman & Hall, 269–312.

Brocklehurst, T.F. and B.M. Lund, B.M., 1981, Properties of *Pseudomonas* causing spoilage of vegetables stored at low temperatures, *J. Appl. Bacteriol.*, 50:259–266.

Brooks, B.W. et al., 1998, Specificity of monoclonal antibodies to *Campylobacter jejuni* lipopolysaccharide antigens, *Immunol. Invest.*, 27:257–265.

Castaner, M. et al., 1996, Inhibition of browning of harvested head lettuce, *J. Food Sci.*, 61:314–316.

CDC, 1979, *Salmonella oranienburg* gastroenteritis associated with consumption of precut watermelons, *Morbidity Mortality. Wkly. Rep.*, 28:522–523.

CDC, 1991, Multistate outbreak of *Salmonella poona* infections — U.S. and Canada, 1991, *Morbidity Mortality Wkly. Rep.*, 40:549–552.

CDC, 1993, Multistate outbreak of *Salmonella* serotype *montevideo* infections, EPI-AID, 93–79.

Draughon, F.A., Chen, S., and Mundt, J.O., 1988, Metabiotic association of *Fusarium*, *Alternaria*, and *Rhizoctonia* with *Clostridium botulinum* in fresh tomatoes, *J. Food Sci.*, 53:120–123.

Farber, J. M. et al., 1998, Changes in populations of *Listeria monocytogenes* inoculated on packaged fresh-cut vegetables, *J. Food Prot.*, 61:192–195.

Feldsine, P.T. et al., 1997, Assurance polyclonal enzyme immunoassay for detection of *Listeria monocytogenes* and related *Listeria* species in selected foods: collaborative study, *J. AOAC Int.*, 80:775–90.

Fratamico, P.M. and Strobaugh, T.P., 1998, Evaluation of an enzyme-linked immunosorbent assay, direct immunofluorescent filter technique, and multiplex polymerase chain reaction for detection of *Escherichia coli* O157:H7 seeded in beef carcass wash water, *J. Food Prot.*, 61:934–938.

Gangar, V. et al., 1998, LOCATE enzyme-linked immunosorbent assay for detection of *Salmonella* in food: collaborative study, *J. AOAC Int.*, 81:419–437.

Gayler, G.E. et al., 1955, An outbreak of salmonellosis traced to watermelon, *Publ. Health Rep.*, 70:311–313.

Golden, D.A., Rhodehamel, E.J., and Kautter, D.A., 1993, Growth of *Salmonella* spp. in cantaloupe, watermelon and honeydew melons, *J. Food Prot.*, 56:194–196.

Gorny, J.R. et al., 2000. Quality changes in fresh-cut pear slices as affected by cultivar, ripeness stage, fruit size, and storage regime, *J. Food Sci.*, 65:541–544.

Gorny, J.R., Hess-Pierce, B., and Kader, A.A., 1998, Effects of fruit ripeness and storage temperature on the deterioration rate of fresh-cut peach and nectarine slices, *Hort-Science*, 33:110–113.

Gorny, J.R., Hess-Pierce, B., and Kader, A.A., 1999, Stonefruit cultivars, storage atmospheres and chemical treatments affect the deterioration rate of fresh-cut peach and nectarine slices, *J. Food Sci.*, 64:429–432.

Hagenmaier, R.D. and Baker, R.A., 1998, A survey of the microbial population and ethanol content of bagged salad, *J. Food Prot.*, 61:357–359.

Hao, Y.Y. et al., 1999, Microbiological quality and production of botulinal toxin in film-packaged broccoli, carrots, and green beans, *J. Food Prot.*, 62:499–508.

Hauschild, A.H.W., 1989, *Clostridium botulinum* in *Foodborne Bacterial Pathogens*, Doyle, M.P., Ed., New York: Marcel Dekker, 111–189.

Hong, J.H. and Gross, K.C., 2000, Involvement of ethylene in development of chilling injury in fresh-cut tomato slices during cold storage, *J. Amer. Soc. Hort. Sci.*, 125:736–741.

Hong, J.H. et al., 1998, Program and Abstracts, Fresh Fruits and Vegetables: Quality and Food Safety, May 3–6, 1998, Beltsville, MD, 36.

Hurst, W.C., 1995, Sanitation of lightly processed fruits and vegetables, *HortScience*, 30:22–24.

Izumi, H., Watada, A.E., and Douglas, W., 1996, Low O_2 atmospheres affect storage quality of zucchini squash slices treated with calcium, *J. Food Sci.*, 119:540–545.

Jacxsens, L. et al., 1999, Behavior of *Listeria monocytogenes* and *Aeromonas* spp. on fresh-cut produce packaged under equilibrium-modified atmosphere, *J. Food Prot.*, 62:1128–1135.

Klepzig, I. et al., 1999, Microbiological state of precut mixed salad: shelf life, hygienic state and characteristics of some pathogens, *Archiv fuer Lebensmittelhygiene*, 50:95–104.

Konczal, J.B., Hughes, M.D., and Overby, K.T., 1992, Packaging technology for fresh precut produce, presentation at Packaging Machinery Manufacturers Institute, Chicago, IL.

Lin, C.M. and Wei, C.I., 1997, Transfer of *Salmonella montevideo* onto the interior surfaces of tomatoes by cutting, *J. Food Prot.*, 60:858–863.

Lopez-Galvez, G. et al., 1997, Quality of red and green fresh-cut peppers stored in controlled atmosphere, in CA '97 Proceedings Vol. 5: *Fresh-Cut Fruits and Vegetables and MAP*, Gorny, J.R., Ed., Davis, CA: University of California, 152–160.

Lopez-Galvez, G., Saltveit, M., and Cantwell, M., 1996, Wound-induced phenylalanine ammonia lyase activity: factors affecting its induction and correlation with the quality of minimally processed lettuces, *Postharv. Biol. Technol.*, 9:223–233.

Lund, B.M., 1982, The effect of bacteria on post-harvest quality of vegetables and fruits, with particular reference to spoilage, in *Bacteria and Plants*, Soc. Appl. Bacteriol. Symp. Ser. No. 10, Rhodes-Roberts, M.E. and Skinner, F.A., Eds., New York: Academic Press, 133–153.

Lund, B.M., 1983, Bacterial spoilage, in *Post-Harvest Pathology of Fruits and Vegetables*, Dennis, C., Ed., New York: Academic Press, 219–257.

Lyon, J.M. and Breidenbach, R.W., 1987, Chilling injury, in *Postharvest Physiology of Vegetables*, Weichmann, J., Ed., New York: Marcel Dekker, Inc., 305–326.

Mayers, P. and Couture, H., 1999, Bacterial foodborne illness in Canada: the problem, Canadian Food Inspection Agency website: http://cfia-acia.agr.ca/english/toc.html.

Mencarelli, F., Saltveit, M.E., Jr., and Massantini, R., 1989, Lightly processed foods: ripening of tomato fruit slices, *Acta Hort.*, 244:193–200.

Morris, G.K., 1984, *Shigella*, in *Compendium of Methods for the Microbiological Examination of Foods*, 2nd ed., Speck, M.L., Ed., Washington: American Public Health Association, 343–350.

Nguyen-The, C., 1991, Qualite microbiologique des vegetaux prets a l'emploi, *C.R. Acad. Agric. Fr.*, 77:7–13.

Nguyen-The, C. and Carlin, F., 1994, The microbiology of minimally processed fresh fruits and vegetables, *Crit. Rev. Food Sci. Nutr.*, 34:371–401.

Odumeru, J.A. et al., 1997, Assessment of the microbiological quality of ready-to-use vegetables for health-care food services, *J. Food Prot.,* 60:954–960.

Piagentini, A.M. et al., 1997, Survival and growth of *Salmonella hadar* on minimally processed cabbage as influenced by storage abuse conditions, *J. Food Sci.,* 62:616–618.

Portela, S.I. and Cantwell, M.I., 2001, 2001 IFT Annual Meeting Book of Abstracts, June 23–27, New Orleans, LA, 56.

Ponting, J.D., Jackson, R., and Watters, G., 1972, Refrigerated apple slices: preservative effects of ascorbic acid, calcium and sulfites, *J. Food Sci.*, 37:434–436.

Prakash, A. et al., 2000, Effects of low-dose gamma irradiation on the shelf life and quality characteristics of cut romaine lettuce packaged under modified atmosphere, *J. Food Sci.*, 65:549–553.

Ries, A.A. et al., 1990, A multistate outbreak of *Salmonella chester* linked to imported cantaloupe, *30th Intersci. Conf. Antimicrob. Agents Chemother.*, Am. Soc. Microbiol., Washington, D.C., 238.

Rodriguez, L.A. et al., 2001, 2001 IFT Annual Meeting Book of Abstracts, June 23–27, New Orleans, LA, 57.

SafetyAlerts, 2001a, Crunch pak apple packages recalled in 17 states, www.safetyalerts.com, March 26.

SafetyAlerts, 2001b, Boskovich Fresh-Cut recalls fresh, halved and seeded red bell peppers, www.safetyalerts.com, September 20.

Sapers, G.M. et al., 1989, Enzymatic browning in Atlantic potatoes and related cultivars, *J. Food Sci.*, 54:362–365.

Sapers, G.M. et al., 1994, Enzymatic browning control in minimally processed mushrooms, *J. Food Sci.*, 59:1042–1047.

Solomon, H.M. et al., 1990, Outgrowth of *Clostridium botulinum* in shredded cabbage at room temperature under a modified atmosphere, *J. Food Prot.*, 53:831–834.

Solomos, T., 1994, Some biological and physical principles underlying modified atmosphere packaging, in *Minimally Processed Refrigerated Fruits and Vegetables*, Wiley, R.C., Ed., College Park, MD: Chapman & Hall, 183–225.

Steinbruegge, E.G., Maxcy,R., and Liewen, M., 1988, Fate of *Listeria monocytogenes* on ready to serve lettuce, *J. Food Prot.*, 51: 596–599.

Sugiyama, H. and Yang, K.H., 1975, Growth potential of *Clostridium botulinum* in fresh mushrooms packaged in semi-permeable plastic film, *Appl. Microbiol.*, 30:964–969.

Sumner, S.S. and Peters, D.L., 1997, Microbiology of vegetables, in *Processing Vegetable Science and Technology*, Smith, D.S. et al., Eds., Lancaster, PA: Technomic Publishing Co., Inc., 87–106.

Todd, E.C.D., 1989, Preliminary estimates of costs of foodborne disease in the United States, *J. Food Prot.,* 52:595–601.

USDA-FSIS, 1998, *Microbiology Laboratory Guidebook*, 3rd ed., Vols. 1 & 2, Dev, B.P. and Lattuada, C.P., Eds., Washington, D.C.: U.S. Govt. Printing Office, Supt. Of Documents.

Vamos-Vigyazo, L., 1981, Polyphenol oxidase and peroxidase in fruits and vegetables, *Crit. Rev. Food Sci. Nutr.*, 15:49–127.

Varoquaux, P., Mazollier, J., and Albagnac, G., 1996, The influence of raw material characteristics on the storage life of fresh-cut butterhead lettuce, *Postharv. Biol. Technol.*, 9:127–139.

Varoquaux, P. and Wiley, R.C., 1994, Biological and biochemical changes in minimally processed refrigerated fruits and vegetables, in *Minimally Processed Refrigerated Fruits and Vegetables*, Wiley, R.C., Ed., College Park, MD: Chapman & Hall, 226–268.

Wang, R.F., Cao, W.W., and Cerniglia, C.E., 1997, A universal protocol for PCR detection of 13 species of foodborne pathogens in foods, *J. Appl. Microbiol.*, 83:727–736.

Watada, A.E., Ko, N.P., and Minott, D.A., 1996, Factors affecting quality of fresh-cut horticultural products, *Postharv. Biol. Technol.*, 9:115–125.

Wood, R.C., Hedberg, C., and White, K., 1991, A multistate outbreak of *Salmonella javiana* associated with raw tomatoes, Abst., Epidemic Intelligence Service 40th Annu. Conf. CDC, Atlanta, GA, 69.

Zhang, S. and Farber, J.M., 1996, The effects of various disinfectants against *Listeria monocytogenes* on fresh-cut vegetables, *Food Microbiol.*, 13:311–321.

Zhuang, H. and Barth, M.M., 1999, 1999 IFT Annual Meeting Book of Abstracts, July 24–28, Chicago, IL, 226.

Zhuang, R.-Y., Beuchat, L.R., and Angulo, F.J., 1995, Fate of *Salmonella montevideo* on and in raw tomatoes as affected by temperature and treatment with chlorine, *Appl. Environ. Microbiol.*, 61:2127–2131.

13 Irradiation of Fresh and Minimally Processed Fruits, Vegetables, and Juices*

Brendan A. Niemira

CONTENTS

INTRODUCTION

Vegetable products, including fresh produce, pre-cut salad vegetables, and fresh juices, typically receive only minimal processing. Often eaten raw or only lightly cooked, these commodities play an important dietary role in many societies.

* Mention of brand or firm name does not constitute an endorsement by the U. S. Department of Agriculture above others of a similar nature not mentioned.

1-58716-041-2/03/$0.00+$1.50

Although concerns related to the presence of pathogenic bacteria on food products have historically been associated with animal products such as meat, milk, and eggs, vegetable products have come under international investigation as sources of food-borne illness. This increased prevalence is due to changes in farm technology, processing and shipping practices, and changing consumption patterns associated with minimally processed fruits, vegetables, and juices (Hedberg et al., 1994; Tauxe et al., 1997; NACMCF, 1999). Beuchat (1996) discussed the incidence of human pathogens on fresh produce, citing contact with contaminated soil, water, compost, harvesting or processing equipment, or human handlers as means by which produce may become contaminated.

Application of conventional preservation and antimicrobial tools, such as heat or chemical wash treatments, to the special needs of produce is being widely investigated (Hotchkiss and Banco, 1992; Beuchat, 1998); among the alternative antimicrobial technologies is the use of ionizing radiation. Irradiation as a food treatment is endorsed by a variety of professional and governmental organizations such as the United States Food and Drug Administration (Anonymous, 2000b), U.S. General Accounting Office (Anonymous, 2000a), American Dietetic Association (Wood and Bruhn, 1999), and the U.N. World Health Organization (WHO, 1999).

The majority of food irradiation research efforts in recent decades have been devoted to treatment of meat products, and irradiation as a terminal control step is becoming more widespread for a variety of meat products in the U.S. (see Sommers, chap. 14, this volume). Low-dose irradiation may be implemented during postpackaging as a terminal control step and is therefore of interest to producers of nonthermally pasteurized (NTP) juices, fresh sprouts, pre-cut vegetables, prepared salad mixes, fruit salads, and other minimally processed vegetable products (NFPA, 2000), but applications are as yet limited.

Vegetables were among the first experimental subjects in studies of physiological response to irradiation (Guilleminot, 1908; Miege and Coupe, 1914). Produce, fruit juices, milk, eggs, and meat were all subjects of research to extend shelf-life with irradiation (Proctor et al., 1955; Thayer et al., 1996). Initial studies with irradiated produce typically used relatively high doses and were intended to achieve pasteurization-level reductions in spoilage bacteria and fungi (Diehl, 1995). Although this research helped to define the effective doses to eliminate spoilage organisms, the doses employed often exceeded the maximum radiation tolerances of vegetable commodities tested, resulting in loss of quality (Howard and Buescher, 1989; Prakash et al., 2000b). For this reason, irradiation was generally regarded as unsuitable for application to produce (Maxie and Abdel-Kader, 1966; Yu et al., 1996; Osterholm and Potter, 1997). However, more recent research with lower radiation doses, i.e., less than 3 kiloGray (10 kGy = 1 Mrad), has suggested a role for irradiation as one of several "hurdles" in fruit and vegetable processing (Tauxe et al., 1997; Thayer and Rajkowski, 1999; Anonymous, 2000a).

Preservation of fresh produce is limited to refrigeration and judicious use of modified atmosphere packaging (Sumner and Peters, 1997). Unlike meat, fresh produce is living tissue, able to conduct respiration, maintain water relations with its environment, and continue synthesis of defense compounds and secondary metabolites (Salisbury and Ross, 1984). Irradiation can alter these processes, leading to changes in firmness, aroma, color, or taste (Yu et al., 1996; Prakash et al., 2000b). Fresh and lightly processed vegetable products, as well as enzymatically active NTP juices, are relatively sensitive to storage conditions of temperature, humidity, light levels, and packaging atmosphere. The effects of irradiation on plant physiology and surface-associated microbial ecology are expressed during refrigerated handling and storage (Howard and Buescher, 1989; Al-Kahtani et al., 2000; Prakash et al., 2000a).

The purpose of this chapter is to discuss low-dose ionizing irradiation of minimally processed fruit, vegetable, and juice products to increase food safety. The discussion will draw on current research to describe pathogen prevalence and survival in various commodities; the applicability of the various irradiation technologies and their interaction with other antimicrobial measures; and factors that can influence pathogen sensitivity to radiation and postirradiation product quality. As cut vegetables, prepared salads, and other minimally processed vegetable products increase in popularity and global distribution, application of antimicrobial measures, including irradiation, will serve to protect the public from the threat of food-borne illness.

MICROFLORA OF MINIMALLY PROCESSED VEGETABLE PRODUCTS

Plant parts such as leaves, stems, fruits, and roots typically support 10^3 to 10^6 colony forming units (CFUs) per gram of plant tissue (Sumner and Peters, 1997; Mercier and Lindow, 2000). The vast majority of these microorganisms consist of nonpathogenic bacteria that interact with the plant and with each other, often forming biofilms as part of the phytoplane ecology (Carmichael et al., 1999; Fett, 2000). These organisms come from the environment in which the plants are grown, including the soil, water, air, and manure and compost, as well as from postharvest handling, processing, and shipping (Beuchat, 1996). Formation of bacterial biofilms on the inert surfaces of processing equipment such as steel, glass, rubber, and plastics is a well-known phenomenon (Costerton et al., 1995), but the widespread nature of the biofilm habitat on leaf and root surfaces has been illustrated only in recent years (Fett, 2000).

The biofilm habitat is a complex community of many bacterial species bound to the plant surface in a durable exopolysaccharide matrix. Many of the details regarding this communal existence are still being defined, but it is known that the sheltered matrix environment allows for interspecies exchange of nutrients and metabolites and protects the resident bacteria from desiccation and exposure to antimicrobial chemical rinses (Morris et al., 1997; Fett, 2000). The microbial populations on and in fruits and vegetables are carried into derivative juices or fresh-cut products during processing (Parish and Higgins, 1988; Sizer and Balasubramaniam, 1999). Fresh-cut produce cannot be heat pasteurized without loss of quality. Although the majority of fruit and vegetable juices sold in the U.S. are pasteurized and therefore present little risk of food-borne pathogens, fresh NTP juices command a premium price for flavor and aroma. NTP juices also have a relatively short shelf-life due to their microbial load (Buchanan et al., 1998).

The increasing significance of human pathogens on fresh fruits, vegetables, and juices has been recognized in recent years (Beuchat, 1996; Tauxe et al., 1997; Thayer and Rajkowski, 1999). Like nonpathogens, pathogenic organisms come from the environment in which the plants are grown as well as from postharvest handling, processing, and shipping (Beuchat, 1996). Because of repeated international outbreaks of salmonellosis and enterohemorrhagic *Escherichia coli* O157:H7 infections associated with raw sprouts (Taormina et al., 1999), in 1999 the U.S. Food and Drug Administration warned the public against consumption of this commodity. *Salmonella* and *E. coli* O157:H7 contamination of NTP juices (Cook et al., 1998; Buchanan et al., 1998) and contamination of widely consumed products such as lettuce, tomatoes, cabbage, and melons have led to an increased effort to address the problem of bacterial contamination of minimally processed fruits, vegetables, and juices.

IRRADIATION TECHNOLOGIES AND REGULATIONS

Diehl (1995) and Sommers (chap. 14, this volume) discuss the technologies used for irradiation of food products in detail. The advantages and disadvantages of the three commercial types of ionizing radiation are summarized in Table 13.1. It is important to note that each type of radiation ultimately leads to the introduction of high-energy electrons within the cells of the commodity and the resident microorganisms. Products are typically irradiated from two or more sides to ensure even coverage and maximum penetration. Ozone and heat are generated during the irradiation process for each of the technologies; proper ventilation and temperature control are therefore especially important for the irradiation of produce to avoid product degradation from these secondary effects (Maxie and Abdel-Kader, 1966).

Direct comparisons of the effects of various technologies on fruits, vegetables, or juice products are lacking, but in direct-comparison studies using meats, electron beam irradiation (1.25 or 3.0 kGy) was found to produce sensory and color changes in pepperoni very similar to those produced by thermal treatment (Johnson et al., 2000). However, while electron beam and gamma irradiation (1.5 or 3.0 kGy) were effective at eliminating *Salmonella typhimurium* from refrigerated beef steaks, electron beam was not as effective in the elimination of *Pseudomonas fluorescens*, a common spoilage bacterium found on meats and produce (Chung et al., 2000). Additional direct comparisons of the various irradiation methods are therefore warranted, including studies using fruit and vegetable products as the test subjects.

Under current U.S. regulations, dry or dehydrated vegetable-derived spices, seasonings, and flavorings as well as coloring agents (e.g., paprika) may be irradiated to 30.0 kGy (Anonymous, 2000b). The regulatory limit for fresh fruits and vegetables, however, is 1.0 kGy; furthermore, radiation is limited to the specific uses of disinfestation (i.e., inactivation of arthropod pests) and inhibition of produce growth and maturation (Anonymous, 2000b). In contrast, irradiation to control food-borne pathogens is permitted at higher doses for shell eggs (3.0 kGy), fresh and frozen poultry (3.0 kGy), fresh meats (4.5 kGy), and frozen meats (7.0 kGy). As of 2001, European regulators have approved only dry spices for irradiation (Gil-Robles and Funke, 1999).

In 1999, a coalition of U.S. food processors petitioned the U.S. Food and Drug Administration to amend U.S. regulations to allow doses of up to 4.5 kGy for a wide variety of refrigerated and ready-to-eat meat and vegetable products, including juices, with doses up to 10.0 kGy sought for frozen meat, vegetable, and juice

TABLE 13.1
Technologies for the Production of Ionizing Radiation

Factors	Electron Beam	X-ray	Gamma
Source	Electrons are generated on an emission coil and accelerated to high energy, 5 to 10 MeV.[a] Extensive cooling equipment required.	Created when high-energy electrons (up to 5 MeV) strike a metal plate (e.g., tungsten or tantalum alloys), typical conversion efficiency is 5 to 10%. Extensive cooling equipment required.	Radioactive decay of ^{60}cobalt (2.5 MeV) or ^{137}cesium (0.51 MeV). Cooling equipment required.
Mechanism	High-energy electrons cleave water molecules, creating oxygen and hydroxyl radicals that damage DNA, membranes. Direct cleavage of DNA also occurs.	High-energy photons stimulate atoms within target to release high-energy electrons, which cleave water molecules into radicals. Direct cleavage of DNA also occurs.	High-energy photons stimulate atoms within target to release high-energy electrons, which cleave water molecules into radicals. Direct cleavage of DNA also occurs.
Shielding required	During operation, >2 m concrete or <1 m steel/iron/lead. When source is powered off, no radiation is emitted.	During operation, >2 m concrete or <1 m steel/iron/lead. When source is powered off, no radiation is emitted.	>5 m water or >2 m concrete or <1 m steel/iron/lead. Source cannot be turned off, shielding of source must be the default position.
Speed[b]	Seconds	Seconds	Minutes (depending on source strength)
Penetrability[c]	6 to 8 cm, suitable for relatively thin or low density products	30 to 40 cm, suitable for all products	30 to 40 cm, suitable for all products

[a] MeV = million electron volts.
[b] Speed of dose delivery. The desired dose will vary, depending on the target organism and commodity irradiated.
[c] Penetrability in food, avg. density approximately 1 g/cm^3. This figure will vary for individual commodities due to localized variation in density associated with bone, voids, fibrous matter, etc.

products. Elimination of human pathogens is the primary goal of these requested dose limits; the potential for extension of shelf-life is regarded as a secondary goal (Anonymous, 2000c; NFPA, 2000). Therefore, as scientific data accumulate and commercial interest in irradiation increases, U.S. and international regulations concerning the irradiation of fresh produce and juices are expected to change to reflect increased understanding (Anonymous, 2000c; Gil-Robels and Funke, 1999).

IRRADIATION PHYSIOLOGY AND MICROBIAL ECOLOGY

The physiological effects of ionizing radiation on spoilage organisms and human pathogens have been recently reviewed (Monk et al., 1994; Diehl, 1995). Cells are damaged when high-energy electrons fracture molecules, typically the water molecules, into highly reactive radicals, which can damage DNA, proteins, and cell membranes. Bacterial DNA is generally more sensitive to damage from radicals during active replication or transcription (Snyder and Poland, 1995). Plasmid DNA may also be damaged, disrupting and inactivating plasmid-encoded genes; pathogenic bacteria such as *Yersinia enterocolitica*, which carry plasmid-encoded virulence genes, may therefore have a reduced virulence after irradiation (Sommers and Bhaduri, 2001). Under conditions of very limited free water, such as on dried or frozen products, fewer hydroxyl and oxygen radicals are produced, and their mobility is reduced; direct breakage of DNA therefore becomes the predominant means of radiation damage (Diehl, 1995). Because the DNA molecule represents a relatively small target for direct interaction with the high-energy electrons, radiation resistance of bacteria on dried or frozen products increases (Snyder and Poland, 1995; Thayer and Boyd, 1995; Nieto-Sandoval et al., 2000).

Fungi are typically more resistant to radiation than are bacteria (O'Connor and Mitchell, 1991; Monk et al., 1994). D_{10} values, i.e., the amount of radiation necessary to effect a 90% (1-log) reduction, have been reported for yeasts and molds in the range of 1 to 3 kGy (Lovell and Flick, 1966; Lescano et al., 1991; Narvaiz et al., 1992), as opposed to D_{10} of 0.3 to 0.7 kGy for pathogenic bacteria on produce and juices (Buchanan et al., 1998; Rajkowski and Thayer, 2000; Niemira et al., 2001). Human pathogenic viruses are typically more resistant to radiation than bacteria or fungi. Due to relatively high D_{10} for pathogenic viruses such as hepatitis and Norwalk (Ostetholm and Potter, 1997), the doses required to achieve meaningful virus population reductions are typically in excess of what most fresh produce products can withstand without unacceptable softening and loss in quality (Howard and Buescher, 1989; Monk et al., 1994; Yu et al., 1996). Although low-dose irradiation has been shown to suppress some plant pathogenic fungi responsible for storage losses (Thayer and Rajkowski, 1999), irradiation as a primary control step in a pathogen reduction program is best suited to control of bacterial pathogens rather than viruses or fungal pathogens.

Comparatively few studies have been conducted on the effect of low radiation doses on the microbial ecology of indigenous microflora with respect to inoculated pathogens. Loss of rare individuals within the community, effect on species present in low densities, or other undefined changes in community structure and function are typical concerns when considering the impact of antimicrobial measures (A. Matos, personal communication). A concern expressed by various authors (Thayer and Boyd, 2000) regarding irradiated foods is possible diminution of interspecies competition among the microorganisms; a suppression of the indigenous microflora by irradiation could lead to increased pathogen growth in storage.

Studies of irradiated meats have tended to support the position that, while irradiation of a product reduces the total population of bacteria on the sample, it does not lead to increased recolonization by inoculated pathogens (Licciardello et al., 1970; Tiwari and Maxcy, 1972; Thayer and Boyd, 2000). However, Matches and Liston (1968) found that *Salmonella* grew more rapidly on irradiated vs. nonirradiated fish fillets. Carlin et al. (1996) showed that the growth of *Listeria monocytogenes* inoculated onto previously disinfected leaves of endive (*Cichorium endiva*) was significantly enhanced, resulting in approximately 1.5 log units more pathogen on disinfected leaves vs. water-washed leaves within 4 days of storage at 10°C. This study relied on chemical disinfection rather than irradiation; however, the result indicates that the response of competitive phytoplane microorganisms to irradiation will be an important factor in establishing the appropriate use of this technology.

IRRADIATION OF FRUITS AND VEGETABLES

The Case for Irradiation Processing

With increasing consumption of fresh produce in the U.S. and increased globalization of the fresh produce market has come increasing concern over produce-associated food-borne illness (Beuchat, 1996; Tauxe et al, 1997; Taormina et al., 1999; Thayer and Rajkowski, 1999). Conventional antimicrobial procedures such as washing, chemical sanitization, thermal treatment, and modified atmosphere packaging have historically been developed and refined to suppress spoilage organisms effectively (Luh, 1997). Much research is dedicated to improving their efficacy against pathogenic bacteria such as *L. monocytogenes*, *Salmonella*, and *E. coli* (Beuchat, 1998; Jacxsens et al., 1999; Liao and Sapers, 2000).

It was long believed that pathogenic bacteria were not likely to be present in the interior of fruits and vegetables and that antimicrobial measures could reasonably be limited to surface disinfection (Anonymous, 1999). However, it is increasingly recognized that leaves, fruits, and seeds provide bacteria with numerous mechanisms to avoid these antimicrobial measures. As more recent research has shown, bacteria not only are likely to enter fruits and vegetables through natural openings (stomata, calyx, stem, stem scar, etc.), abiotic wounds, or phytopathogenic penetrations, but they also can survive within the produce for days or weeks (Anonymous, 1999; Buchanan et al., 1999; Fisher and Dolden, 1998; Takeuchi and Frank, 2000; Riordan et al., 2000). The consequences of internalization are reduction in the effectiveness of chemical treatments such as hot water, trisodium phosphate, chlorine, chlorine dioxide, and peroxyacetic acid (Zhuang and Beuchat, 1996; Beuchat, 1998; Pao and Davis, 1999; Fett, 2000).

The protective nature of bacterial biofilms also reduces the efficacy of antimicrobial measures (Morris et al., 1997; Fett, 2000). Biofilms can differ in size and species composition, depending on the host plant species, the position of the leaf within the plant canopy, or their location on an individual leaf due to variations in substrate availability or microclimate suitability (Mercier and Lindow, 2000; Fett, 2000). Human pathogens such as *E. coli* and *Salmonella* are known to form durable biofilms on industrial surfaces (Dewanti and Wong, 1995; Korber et al., 1997). The participation of enteric bacteria, as well as the psychrotrophs *L. monocytogenes* and *Y. enterocolitica*, in pre-existing phytoplane biofilms formed by nonpathogenic bacteria is not known (Fett, 2000; Liao and Sapers, 2000). By penetrating sheltered areas in the surface, subsurface, or interior of fruits and vegetables, ionizing radiation inactivates bacteria throughout the leaf.

Microbiology of Irradiated Produce

Ionizing radiation extends shelf-life by inactivating storage pathogens (Thomas, 1983; Willemot et al., 1996). The use of irradiation to reduce human pathogens on produce, rather than on meat, poultry, or shellfish, is a comparatively new area of study (Monk et al., 1994; Thayer and Rajkowski, 1999). Most studies to date have shown that irradiation of produce typically reduces resident microorganisms as effectively as irradiation of meats and poultry. Ongoing research is determining the doses required to achieve meaningful reductions while preserving the quality of produce.

Howard et al. (1995) reported that 1.0 kGy reduced total aerobic bacteria and lactic acid bacteria in *pico de gallo*, a cold salad made of tomatoes, yellow onions, and jalapeño peppers, over six weeks of refrigerated storage, thereby extending the shelf-life. A dose of 1.0 kGy reduced total aerobic plate count and *L. monocytogenes* by ~4 logs (99.99%) on pre-cut bell pepper; however, on peppers stored at 15 or 10°C, the pathogen regrew to initial levels within 4 days, while pathogen levels remained low on peppers stored at 5°C (Farkas et al., 1997). Also in that study, spoilage bacteria were reduced by 5 logs on carrot cubes by 1.0 kGy. The authors concluded that irradiation, when combined with good manufacturing practices, could effectively reduce pathogen levels throughout the useful shelf-life of the produce.

Irradiation of diced celery (Prakash et al., 2000b) showed that 1.0 kGy reduced *E. coli* and *L. monocytogenes* to undetectable levels (>5 logs). Total aerobic counts following irradiation and three conventional treatments, i.e., acidification, blanching, or chlorination, showed that acidification reduced initial counts to a degree comparable to that of 1.0 kGy by the end of the storage period (22 days), but the aerobic population following conventional treatments had regrown to equal (acidification) or exceed (blanching and chlorination) the untreated controls; that of irradiated (0.5 and 1.0 kGy) celery remained significantly lower throughout the study.

The D_{10} values of 14 phytopathogenic fungi were evaluated in irradiated (0.5 to 1.5 kGy) mango pulp (El-Samahay et al., 2000). These ranged from 0.39 for *Aspergillus sydowii* and *A. ustus* to 1.11 for *Penicillium oxalicum*, confirming the relatively high radiation resistance of fungi. The D_{10} values were also determined in saline solution and demonstrated some variance with the data obtained in mango pulp. The more complex medium caused a decrease in D_{10} for some fungi, e.g., *A. ustus* (55%), *Scopulariopsis brevicaulis* (55%), and *A. sydowii* (76%), and an increased D_{10} for others, e.g., *P. brevicompactium* (147%), *A. flavus* (159%), and *P.*

oxicalum (179%). The highest dose (1.5 kGy) reduced total aerobic bacteria count by ~1.8 logs, a difference that persisted throughout the study (50 days). Shelf-life in irradiated (1.0 kGy) fruits was extended from 25 to 50 days in refrigerated storage. Although the interiors of the fruit used in this study were sampled and found to be free of contamination, the variable effect of fruit pulp on the radiation resistance of resident microorganisms is clearly demonstrated. Pulp represents a combination of plant cell contents with extracellular fluid found in the intercellular spaces; nevertheless, this observation should be considered in light of evidence of internalization of pathogenic bacteria into intercellular spaces (Takeuchi and Frank, 2000).

Several types of salad sprouts were inoculated with a cocktail of *E. coli* O157:H7 strains or *Salmonella* strains and irradiated by Rajkowski and Thayer (2000). Gamma radiation effectively reduced the bacterial load, with *E. coli* O157:H7 D_{10} values of 0.34, 0.27, and 0.26 on radish, alfalfa, and broccoli sprouts, respectively. *Salmonella* D_{10} on radish sprouts was 0.54 for isolates derived from meat sources and 0.46 for isolates derived from vegetable sources.

On dry spices (<10% moisture), D_{10} values for resident microorganisms were considerably higher than on fresh fruits and vegetables. Irradiated (10 kGy) paprika showed a 4-log reduction in total mesophilic bacteria, and similar reductions occurred in total *Enterobacteriaceae* and coliforms without significant loss of color (Nieto-Sandoval et al., 2000).

Irradiation may be combined with modified atmosphere packaging (MAP) to achieve a hurdle-type control of spoilage and pathogens. This approach has shown success with irradiated packaged meats (Thayer and Boyd, 2000). After irradiation under MAP, total aerobic counts were reduced ~3 logs on iceberg lettuce (Hagenmaier and Baker, 1997) and ~1.5 logs on romaine lettuce (Prakash et al., 2000a) by doses of 0.19 kGy and 0.35 kGy, respectively, differences that persisted during storage. Rate of spoilage and shelf-life were reported to be unchanged for iceberg lettuce, but the shelf-life of romaine lettuce was extended from 14 to 18 days to the length of the study, 22 days. MAP reduced total aerobic plate counts of shredded carrot irradiated (0.45 kGy) by ~2 logs, a difference that persisted throughout a 9-day storage evaluation (Hagenmaier and Baker, 1998).

Ionizing radiation is demonstrably able to penetrate into protected areas of produce (surface, subsurface, and interior) to damage bacteria; however, the extent to which a biofilm habitat may influence the radiation sensitivity of bacteria, either native nonpathogens or pathogenic contaminants, is not well understood. As the biofilm-participatory bacteria are suspended in a water-saturated exopolysaccharide matrix, they may be more at risk from oxygen and hydroxyl radicals generated during the irradiation process, thereby demonstrating an increased sensitivity similar to that seen in solution studies. However, the exact composition of the matrix is not yet known and may include radical scavenging compounds, which would reduce the efficacy of the irradiation (Sommers et al., 2002). Close proximity of different bacterial species with different levels of resistance may lead to postirradiation competition for resources that places additional pressure on radiation-injured bacteria; conversely, the matrix may provide for additional recovery from radiation injury by absorbing the radicals in the chemical bonds of the matrix or by protecting the bacteria from desiccation or competition. The microbial ecology of irradiated biofilms is an area of active discussion, and development of competing hypotheses and will be an important area of research in the future.

Sensorial Properties of Irradiated Produce

Radiation-induced softening due to hydrolysis of pectins is a well-known phenomenon that is irradiation-dose dependent (Yu et al., 1996). Irradiated (1.0 kGy) celery was evaluated using analytical machinery and a trained sensory panel. The irradiated material had quantifiable texture, color, and aroma qualities equivalent or superior to untreated (controls), blanched, acidified, or chlorinated celery and maintained these qualities throughout the 22-day storage period of the study (Prakash et al., 2000b). Also, irradiated (1.0 kGy) celery had qualitative flavor, aroma quality, appearance, texture, and overall acceptability characteristics equal or superior to blanched, acidified, or chlorinated samples. The authors concluded that irradiation processing clearly results in a superior celery product vs. conventional processing means.

Iceberg lettuce (Hagenmaier and Baker, 1997) and romaine lettuce (Prakash et al., 2000a) were amenable to low-dose (<0.5 kGy) irradiation, with texture and shelf-life equal or superior to untreated controls. Strawberries softened following 1.0 or 2.0 kGy and chemical analysis showed an association with an increase in water-soluble pectin and a decrease in oxalate-soluble pectin (Yu et al., 1996). Cucumber pickles showed softening with increased dosage up to 1.0 kGy, but there was little effect on neutral sugars in the cell wall (Howard and Buescher, 1989). Potatoes showed softening following 0.2 kGy, and *in vitro* treatment of starches showed loss of viscosity postirradiation (Al-Kahtani et al., 2000).

Gibberellic acid–treated grapefruit tolerated 0.3 kGy with little loss of quality of the fruit, pulp, or juice, but after 0.6 kGy, pitting of the skin, softening of the fruit, and loss of juice quality rose to unacceptable levels (Miller and McDonald, 1996b). In a series of studies on blueberries, 0.5 kGy was identified as a dose that several cultivars would tolerate without loss of firmness and overall quality, but tolerance to 1.0 kGy was variable (Miller et al., 1994, 1995; Miller and McDonald, 1996a). Color, pH, decay rate, and other agronomic factors were reportedly not affected by dose level in these studies. Mango irradiated to 1.5 kGy were as acceptable to a sensory panel as nonirradiated controls immediately after treatment, based on odor, color, and overall appearance; irradiated fruit remained acceptable after 50 days of refrigerated storage, while nonirradiated controls became unacceptable after fewer than 30 days (El-Samahy et al., 2000).

Irradiation is known to oxidize a portion of the total ascorbic acid (vitamin C) to the dehydro form (Romani et al., 1963). Both of these forms of the vitamin are biologically active, suggesting minimal nutritional impact (Thayer, 1994); however, it is uncommon to see values for the dehydro form co-reported in analyses of ascorbic acid concentration.

Treatments up to 2.0 kGy did not produce any negative impact on radish sprouts (Rajkowski and Thayer, 2000). A dose of 1.0 kGy caused a 12% loss of ascorbic acid content in pre-cut bell pepper and a slight increase in beta-carotene content of cubed carrots, and the dose extended the shelf-life of both products (Farkas et al., 1997). Treatment with 1.0 kGy did not affect the sensorial properties of *pico de gallo* but did result in loss of measurable ascorbic acid (Howard et al., 1995). The authors of that study discussed the high antioxidant concentration of the component vegetables (tomato, onion, jalapeño) as one possible mechanism for reducing the development of off-flavors and odors that can result from high-dose irradiation. Antioxidants can absorb oxygen and hydroxyl radicals formed during irradiation (Sommers et al., 2002) and may therefore be expected to protect plant tissues and cellular structures; this protection would presumably not extend to surface-bound microorganisms.

Although antioxidants added to nutrient solutions are known to protect suspended bacteria from the effects of ionizing radiation (Sommers et al., 2002), the role of natural antioxidants in determining radiation tolerance of fruits and vegetables has not been addressed directly. In reviews by Kader (1986) and Dupont et al. (1997), a variety of fruits and vegetables was grouped according to their tolerance for irradiation. More radiation tolerant products are, in some cases, also reported to be higher in antioxidant concentration, and lower radiation tolerance was similarly associated with low antioxidants (Cao et al., 1996; Wang et al., 1996); however, methodological differences in how the produce was evaluated in the various studies prevent definitive conclusions. Complications may arise through the localization or compartmentalization of the native antioxidants within the plant tissues, the effect of homogenization of tissues as part of the chemical analysis, heat degradation during sample preparation, or some other artifact of analysis.

A full discussion on methods to detect prior irradiation of fruits, vegetables, and juices is beyond the scope of this chapter. Many of the methods used for detection of irradiated meats, poultry, and shellfish may be adapted for these products; see reviews by Schreiber et al. (1993) and Sommers (chap. 14, this volume). Detection of radiation-induced hydrocarbons by gas chromatography ormass spectrometry has been successfully employed in irradiated (1.0 kGy) perilla (*Perilla frutescens*) seeds (Lee et al., 2000). Thermoluminescence has shown promise in detecting irradiation treatment of spices (Schreiber et al., 1993) and potatoes (Al-Kahtani et al., 2000). Measurement of electron paramagnetic resonance (EPR) is commonly used as a means of dosimetry, with alanine pellets as the dosimetry sample. EPR can detect prior irradiation of alfalfa (*Medicago sativa*) seeds (D.W. Thayer, personal communication). It should be noted that *ex post facto* dosimetry is difficult: although these and other detection methods may be able to indicate that a given product was irradiated, determining when the product was so treated, or the dose delivered, is unlikely (Diehl, 1995).

Influence of Plant Variety on Microbiological and Sensorial Response

Variety-specific response to irradiation is an important consideration, not only in determining the suitability of irradiation for treatment of modern varieties, but also in evaluating the applicability of information acquired in previous studies of irradiated produce. The most economically significant varieties of grains, vegetables, and fruits (and therefore the most likely subjects of irradiation research and implementation) change from year to year (Thelen, 1999; Wiersema and Leon, 1999). In studies with inoculated meat, pathogenic bacteria demonstrate varying radiation sensitivity, depending on the specific type of meat evaluated (Thayer et al., 1995). Evidence suggests that, within a given crop plant species, the variety can also lead to varying radiation sensitivity.

Reviews of produce radiation sensitivity often combine different varieties under a single heading, e.g., "leafy vegetables" or "grape" (Kader, 1986; Dupont et al., 1997), a practice that does not recognize evidence of varietal differences. In separate studies of iceberg lettuce, *Lactuca sativa* var. Raleigh (Hagenmaier and Baker, 1997), and romaine lettuce, *L. sativa* var. Longifola (Prakash et al., 2000a), the texture, headspace gas composition, and microbiology of irradiated produce were assessed. Although irradiated (0.45 kGy) iceberg lettuce showed respiration 33% greater than nonirradiated controls, irradiated (0.35 kGy) romaine lettuce showed decreased respiration, based on oxygen content of the headspace gas after 1 day in storage. At relatively high doses (0.81 kGy), iceberg lettuce lost firmness, while at lower doses (0.2 and 0.5 kGy) the shear force of irradiated and nonirradiated leaves was virtually the same, with no apparent differences in texture. In contrast, loss of firmness of romaine lettuce irradiated at low doses (0.15 and 0.35 kGy) was readily apparent. Irradiation reduced the aerobic microbial population in iceberg lettuce (0.2 kGy) by approximately 3 log units, but in romaine lettuce irradiated at a higher dose (0.35 kGy), the reduction was only 1.5 log units.

Several varieties of sprouts were inoculated with a cocktail of *E. coli* O157:H7 strains and irradiated by Rajkowski and Thayer (2000). D_{10} values obtained on radish, alfalfa, and broccoli sprouts were 0.34, 0.27, and 0.26, respectively. Direct statistical comparisons were not reported, but the standard errors presented in that study were small, suggesting a variety-specific influence on the radiation resistance of the inoculated bacteria.

A comparison of the potato (*Solanum tuberosum*) varieties Ajax and Diamant (Al-Kahtani et al., 2000) showed differences in the postirradiation (0.20 kGy) viscosity of starches, whether isolated from irradiated tubers or irradiated *in vitro*. This difference was ascribed to unspecified varietal differences in chemical profile related to pectins, cellulose, and other tissue constituents. The blueberry (*Vaccinium* spp.) cultivars Brightwell and Tifblue did not differ in their response to gamma radiation, showing softening and similar loss of quality following 1.0 kGy but no negative impact after 0.5 kGy (Miller and McDonald, 1996a). However, although the flavor and texture of Sharpblue blueberries were considered acceptable following electron beam irradiation (1.0 kGy) (Miller et al., 1995), flavor and texture of Climax blueberries similarly irradiated (1.0 kGy) (Miller et al., 1994) declined significantly. Color, pH, decay rate, and other agronomic factors were not affected by dose level in these studies.

The nature of varietal specificity can be most clearly considered when data for different varieties are co-reported, but meaningful inferences are still possible when separate studies are considered. Care must be taken, however, when comparing the results from different single-variety studies conducted under widely disparate experimental conditions that perhaps include different means of irradiation, e.g., gamma vs. electron beam.

The mechanisms behind varietal specificity lie in as yet undefined differences in physiology and biochemistry related to irradiation response. Different levels of starch, pectin, and cellulose may lead to variable changes in viscosity, turgidity, or softening as various chemical bonds are hydrolyzed (Howard and Buescher, 1989; Yu et al., 1996; Al-Kahtani et al., 2000). Levels of natural antioxidants may also play a role, by absorbing the oxygen and hydroxyl radicals produced during the irradiation process before they can damage tissue structures (Urbain, 1986). Also, the state of maturity of the vegetable at the time of irradiation may influence the efficacy of irradiation treatment and/or effects on product quality (Howard et al., 1995).

Differential microbial response to irradiation on different varieties may have multiple mechanisms. Varietal differences in surface topography, cuticle thickness, secondary metabolite profile, and other anatomical and biochemical factors are known to influence the development of plant pathogens (Agrios, 1988) as well as the efficacy of surface disinfectants applied to produce (Adams et al., 1989; Korber et al., 1997; Beuchat, 1998). By providing microhabitats with a unique microflora or biofilm community or a localized complement of secondary metabolites, plant surfaces with more varied topography may result in increased survival of pathogens in protective niches. Bacteria are known to penetrate into the tissues of lettuce (Takeuchi and Frank, 2000) and apple (Buchanan et al., 1999; Liao and Sapers, 2000), making them inaccessible to chemical disinfectants. The anatomical structures of most fruits and vegetables do not present a significant barrier to ionizing photons (gamma and x-ray), although the depth of penetration of ionizing electrons may be a factor for unusually thick or dense products (Table 13.1). A given mix of *Salmonella* isolates was shown to be approximately twice as resistant to gamma radiation when irradiated on seeds used for sprouting as when irradiated on the sprouts themselves (Thayer et al., 1999). The level of free water on a comparatively dry product such as seeds may have been a factor in this study; however, the authors ascribe the result to the presence of unidentified physiochemical factors associated with the topology of the seeds.

Fruits and vegetables present a complex substrate for studies of irradiation, with significant complicating factors of maturity, anatomy, topology, surface and internal chemistry, and native microflora, which may compete with or succor contaminating pathogenic bacteria. Furthermore, plant varieties wax and wane in commercial popularity and, as new combination products are created, e.g., salads, salsas, etc., the complexity of this interaction increases. Although a full discussion of packaging materials *vis–à-vis* irradiation is beyond the scope of this chapter, it should be noted that relatively few plastics are approved for use in irradiation processing out of concern for potential generation of volatiles and secondary food additives (Anonymous, 2000b). The effect of these factors, singly and in combination, on the efficacy of irradiation, as well as of chemical sanitizers and other treatments, will determine what doses will be most effective at eliminating contaminating bacteria and best tolerated by the fruit or vegetable product of interest.

IRRADIATION OF JUICES

The Case for Irradiation Processing

Conventional heat pasteurization is used for the overwhelming majority of juices sold in the U.S.; this pasteurization process results in the loss of essential oils and other juice components, changing the flavor of the resulting juice. Although fresh, NTP juices are prized for premium flavor and aroma, these products have also been responsible for outbreaks of salmonellosis, enterohemorrhagic *E. coli* infection, hemolytic uremic syndrome, and other illnesses that have led to numerous hospitalizations and deaths (Luedtke and Powell, 2000). In response to these food-borne illnesses, the U.S. Food and Drug Administration has implemented a policy (Anonymous, 2001) requiring processors of NTP juices to implement control plans and technologies that will provide 5-log (99.999%) reductions in human pathogen load. These regulations will take effect for all processors by 2004; within this time frame, nonthermal antimicrobial measures must be brought to commercial standards of reliability and cost effectiveness if NTP juices are to be sold in the U.S.

A variety of nonthermal means of reducing the microbial load of NTP juices was recently discussed (Sizer and Balasubramaniam, 1999). These include pulsed electric fields, minimal thermal processing, high-pressure processing, and ultraviolet radiation. Regulatory approval for the use of ionizing radiation is currently being sought (NFPA, 2000). These technologies are being researched to determine the most applicable means of reducing the risk of pathogen contamination of NTP juices while preserving the flavor and aroma characteristics that allow these products to command premium prices.

Microbiology of Irradiated Juices

The majority of research efforts related to irradiation of juices has targeted spoilage organisms such as yeasts and molds rather than bacterial pathogens (Monk et al., 1994). Niemira (2001) reported a D_{10} value of 0.35 kGy for *S. enteritidis* irradiated in reconstituted orange juice. Buchanan et al. (1998) showed that irradiation (1.0 kGy) effectively inactivated *E. coli* O157:H7 in inoculated commercial apple juices, with D_{10} values of 0.12, 0.16, and 0.21 kGy for the three isolates tested. Niemira et al. (2001) found that the resistance of four *Salmonella* isolates irradiated in orange juice also varied, with D_{10} values of 0.71 kGy (*S. anatum*), 0.48 kGy (*S. newport*), 0.35 kGy (*S. infantis*), and 0.38 kGy (*S. stanley*). For the most resistant isolate (*S. anatum*), 3.5 kGy was indicated as the dose required to achieve a 5-log reduction. Similarly, *Salmonella hartford* was reduced to undetectable levels in irradiated (3.0 kGy) orange juice (Pickett et al., 2000). In that study, it was possible to recover the pathogen through selective enrichment.

The antimicrobial efficacy of irradiation is influenced by several factors. Native variation in pathogen resistance among different isolates has been noted (Buchanan et al., 1998; Niemira et al., 2001). Buchanan et al. (1998) showed that the radiation sensitivity of three strains of *E. coli* O157:H7 irradiated in apple juice was reduced 54 to 67% by previous growth of the cultures in acid environments. Also, the sensitivity of one test strain decreased in more turbid juices. This difference was ascribed to the antioxidant power of the suspended solids, although data for the antioxidant power of the juices were not presented. Although *L. monocytogenes* irradiated *in vitro* was increasingly protected in solutions of increasing antioxidant power (Sommers et al., 2001), *S. enteritidis* suspended in commercial citrus juices of varying composition (orange vs. orange/tangerine, extra pulp vs. pulp free, regular vs. calcium enriched, etc.) and varying antioxidant power did not differ in D_{10} value following gamma irradiation (Niemira, 2001). The role of natural and artificial antioxidants is therefore an area to be addressed more fully.

Sensorial Properties of Irradiated Juices

As in whole fruits, irradiation oxidizes a portion of juice ascorbic acid to the dehydro form; as in fruits, this conversion is expected to have little nutritional impact (Romani et al., 1963; Thayer, 1994). In a wide-ranging study, Fetter et al. (1969) irradiated a variety of commercially pasteurized juices. Taste panel evaluation found a reduction of flavor quality in irradiated (5.0 kGy) grape juice, but similarly irradiated juices of orange, guava, tomato, red currant, black currant, apricot, peach, pear, and apple were not affected. Chachin and Ogata (1969) treated grape, apple, and orange juices with sterilizing (2 to 80 kGy) doses of gamma radiation. Loss of grape juice anthocyanin and orange juice beta carotene was evident after 10 kGy and was dose dependent up to 80 kGy. Browning was reported in apple juice after 5 kGy and ascorbic acid was reduced in apple and orange juices. Dehydroascorbic acid was not reported. Although 10 kGy reduced the acceptability of orange and apple juice, 0.01% propyl gallate reduced the loss of quality of orange juice thus irradiated.

In a later study, high-dose irradiation (10 kGy) of orange juice resulted in a similarly unacceptable degree of flavor degradation and browning; however, the addition of 0.1% sorbic acid before irradiation effectively eliminated loss of flavor and reduced browning (Thakur and Singh, 1993). Spoto et al. (1997) irradiated orange juice concentrate and determined the effect of storage time and temperature on juice quality and acceptability. In that study, the highest dose (up to 5.0 kGy) combined with 25°C storage resulted in loss of "orange" flavor and an increase in "bitter," "medicinal," and "cooked" ratings by sensory panelists. Lower dose (2.5 kGy) and cooler storage (0 or 5°C) were proposed as an acceptable processing regimen.

Niemira et al. (2001) found no evident changes in the appearance or aroma of reconstituted orange juice irradiated to 2.5 kGy. Pasteurized orange juice irradiated to doses as high as 5.0 kGy demonstrated little reduction in product color or concentration of key aroma and flavor compounds (X. Fan, personal communication). Unpasteurized apple cider irradiated to 3.0 kGy was identifiable, although not unacceptable, to sensory panelists evaluating aroma; similar results were obtained with reconstituted and fresh orange juices (B.A. Niemira and X. Fan, unpublished data). In contrast, Miller and McDonald (1996b) reported that the juice from irradiated grapefruit was acceptable after 0.3 kGy but declined in quality at 0.6 kGy. Pickett et al. (2000) reported increasing off-flavor in unpasteurized orange juice irradiated to 3.0 kGy, rendering the juice unpalatable.

Although research into the organoleptic qualities of irradiated juices continues, the factors that may lead to undesirable flavors, aromas, or changes in appearance are complex. While extensive studies into the effect of the variety and maturity of the source fruit on the radiation sensitivity of the resulting juice are lacking, it is known that juice composition is variety dependent (Huggart et al., 1975). In light of the evidence for varietal specificity in irradiated whole produce, the possibility of varietal specificity in juices with regard to negative irradiation effects cannot be discounted. Temperature control during the irradiation process, as well as the possibility of ozone generation in the headspace gas of sealed containers, should also be carefully considered as potential secondary factors in juice irradiation.

CONCLUSIONS

With regard to food safety concerns, irradiation is, first and foremost, an antimicrobial treatment. As with other antimicrobial and preservation treatments such as chlorination, blanching, MAP, and ozonation, the specific details of how best to apply the treatment are critical. In the case of low-dose irradiation, many of these details are still being determined. A great many factors that influence what doses are necessary to inactivate bacteria while preserving produce and juice quality have yet to be explored in a systematic way, including:

1. Pathogens living in a biofilm habitat
2. Multicomponent salads and vegetable mixes
3. Varietal specificity in radiation tolerance and native microbial load
4. Internalization of pathogens; variation in concentrations of native antioxidants
5. Different combination protocols with other treatments in a hurdle approach
6. The specific type of irradiation employed, i.e., gamma vs. x-ray vs. e-beam
7. Other, as yet unidentified, factors

These factors will play a part in determining where, how, and whether fresh fruits, vegetables, and juices are irradiated. Irradiation holds a great deal of promise as a minimally invasive, zero-contact, postpackaging terminal control step. The application of the technology awaits the results of research to address the concepts and concerns presented in this chapter.

ACKNOWLEDGMENTS

The author would like to thank J.L. Alford, W.W. Kirk, and A. Matos for their thoughtful reviews, M.E. Niemira and C.H. Sommers for valuable discussion; and K. Lonczynski for technical assistance in the preparation of this manuscript.

REFERENCES

Adams, M.R., Hartley, A.D., and Cox, L.J., 1989, Factors affecting the efficacy of washing procedures used in the production of prepared salads, *Food Microbiol.*, 6:69–77.

Agrios, G.N., 1988, *Plant Pathology*, 3rd ed., San Diego, CA: Academic Press.

Al-Kahtani, H.A. et al., 2000, Irradiation and storage effects on some properties of potato starch and use of thermoluminescence for identification of irradiated tubers, *Am. J. Potato Res.*, 77:245–259.

Anonymous, 1999, Potential for infiltration, survival and growth of human pathogens within fruits and vegetables, U.S. Food and Drug Administration, Center for Food Safety and Applied Nutrition, http://vm.cfsan.fda.gov/~comm/juicback.html.

Anonymous, 2000a, Food irradiation: available research indicates that benefits outweigh risks, U.S. General Accounting Office, GAO/RCED-00–217.

Anonymous, 2000b, Irradiation in the production, processing and handling of food, Code of Federal Regulations, 21CFR179.

Anonymous, 2001, FDA publishes final rule to increase safety of fruit and vegetable juices, P01–03, http://www.fda.gov/bbs/topics/NEWS/2001/NEW00749.html.

Beuchat, L.R., 1996, Pathogenic microorganisms associated with fresh produce, *J. Food Prot.*, 59(2):204–216.

Beuchat, L.R., 1998, Surface decontamination of fruits and vegetables eaten raw: a review, Food Safety Unit, World Health Organization, WHO/FSF/FOS 98.2.

Buchanan, R.L. et al., 1998, Inactivation of *Escherichia coli* O157:H7 in apple juice by irradiation, *Appl. Envir. Microbiol.*, 64(11):4533–4535.

Buchanan, R.L. et al., 1999, Contamination of intact apples after immersion in an aqueous environment containing *Escherichia coli* O157:H7, *J. Food Prot.*, 62(5):444–450.

Cao, G., Sofic, E., and Prior, R.L., 1996, Antioxidant capacity of tea and common vegetables, *J. Agric. Food Chem.*, 44:3426–3431.

Carlin, F., Nguyen-The, C., and Morris, C.E., 1996, Influence of background microflora on *Listeria monocytogenes* on minimally processed fresh broad-leaved endive (*Chicorium endiva* var. latifolia), *J. Food Prot.*, 59(7):698–703.

Carmichael, I. et al., 1999, Bacterial colonization and biofilm development on minimally processed vegetables, *J. Appl. Microbiol.*, Symp. Suppl., 85:45S-51S.

Chachin, K. and Ogata, K., 1969, Changes of chemical constituents and quality in some juice irradiated with the sterilizing dose level of gamma rays, *Food Irradiation,* 4(1):85–90.

Chung, M.-S., Ko, Y.-T., and Kim, W.-S., 2000, Survival of *Pseudomonas fluorescens* and *Salmonella Typhimurium* after electron beam and gamma irradiation of refrigerated beef, *J. Food Prot.*, 63(2):162–166.

Cook, K.A. et al., 1998, Outbreak of *Salmonella* serotype Hartford infections associated with unpasteurized orange juice, *JAMA*, 280:1504–1509.

Costerton, J.W. et al., 1995, Microbial biofilms, *Annu. Rev. Microbiol.*, 49:711–745.

Dewanti, R. and Wong, A.C.L., 1995, Influence of culture conditions on biofilm formation by *Escherichia coli* O157:H7, *Int. J. Food Microbiol.*, 26:147–164.

Diehl, J.F., 1995, *Safety of Irradiated Foods*, 2nd ed., New York: Marcel Dekker, Inc.

Dupont, J.L. et al., 1997, Nuclear technology in agriculture and nutrition, *Nucl. News*, 40(7):58–62.

El-Samahay, S.K. et al., 2000, Microbiological and chemical properties of irradiated mango, *J. Food Safety*, 20:139–156.

Farkas, J. et al., 1997, Effects of low-dose gamma radiation on shelf-life and microbiological safety of pre-cut/prepared vegetables, *Adv. Food Sci.*, 19(3/4):111–119.

Fett, W.F., 2000, Naturally occurring biofilms on alfalfa and other types of sprouts, *J. Food Prot.*, 63(5):625–632.

Fetter, F. et al., 1969, Das flavourverhalten einger gamma-bestrahlter fruchtsaefte, Mitteilungen: -rube, -wein, -obstbau und fruechteverwertung, 19:140–151.

Fisher, T.L. and Dolden, D.A., 1998, Fate of *Escherichia coli* O157:H7 in ground apples used in cider production, *J. Food Prot.*, 61(10):1372–1374.

Gil-Robels, J.M. and Funke, K.H., 1999, European Parliament Directive 1992/3/EC (Document A4–0008/1999), On the establishment of a community list of foods and food ingredients treated with ionising radiation.

Guilleminot, H., 1908, Comparative effects of radium and x-rays on vegetable cells, *Compt. Rend.*, 145:798–801.

Hagenmaier R.D. and Baker, R.A., 1997, Low-dose irradiation of cut iceberg lettuce in modified atmosphere packaging, *J. Agric. Food Chem.*, 45:2864–2868.

Hagenmaier R.D. and Baker, R.A., 1998, Microbial population of shredded carrot in modified atmosphere packaging as related to irradiation treatment, *J. Food Sci.*, 63(1):162–164.

Hedberg, C.W., MacDonald, K.L., and Osterholm, M.T., 1994, Changing epidemiology of food-borne disease: a Minnesota perspective, *Clin. Infect. Dis.*, 18:671–682.

Hotchkiss, J.H. and Banco, M.J., 1992, Influence of new packaging technologies on the growth of microorganisms in produce, *J. Food Prot.*, 55(10):815–820.

Howard, L.R. and Buescher, R.W., 1989, Cell wall characteristics of gamma-radiated refrigerated cucumber pickles, *J. Food Sci.*, 54(5):1266–1268.

Howard, L.R., Miller, G.H., and Wagner, A.B., 1995, Microbiological, chemical and sensory changes in irradiated *pico de gallo*, *J. Food Sci.*, 60(3):461–464.

Huggart, R.L., Rouse, A.H., and Moore, E.L., 1975, Effect of maturity, variety and processing on color, cloud, pectin and water-insoluble solids of orange juice, *Fla. State Hort. Soc. Proc.*, 88:342.

Jacxsens, L. et al., 1999, Behavior of *Listeria monocytogenes* and *Aeomonas* spp. on fresh-cut produce packaged under equilibrium-modified atmosphere, *J. Food Prot.*, 62(10):1128–1135.

Johnson, S.C. et al., 2000, Irradiation in contrast to thermal processing of pepperoni for control of pathogens: effects of quality indicators, *J. Food Sci.*, 65(7):1260–1265.

Kader, A.A., 1986, Potential application of ionizing radiation in postharvest handling of fresh fruits and vegetables, *Food Technol.*, 40:117–121.

Korber, D.R. et al., 1997, Substratum topography influences susceptibility of *Salmonella enteritidis* biofilms to trisodium phosphate, *Appl. Environ. Microbiol.*, 63:3352–3358.

Lee, H.-J., Byun, W.-W., and Kim, K.-S., 2000, Detection of radiation-induced hydrocarbons and 2-alkylcyclobutanones in irradiated perilla seeds, *J. Food Prot.*, 63(11):1563–1569.

Lescano, G. et al., 1991, Effect of chicken breast irradiation on microbiological, chemical and organoleptic quality, *Lebensm. Wiss. Technol.*, 24:130–134.

Liao, C.-H. and Sapers, G.M., 2000, Attachment and growth of *Salmonella Chester* on apple fruits and *in vivo* response of attached bacteria to sanitizer treatments, *J. Food Prot.*, 63(7):876–883.

Licciardello, J.J., Nickerson, J.T.R., and Goldblith, S.A., 1970, Inactivation of *Salmonella* in poultry with gamma radiation, *Poultry Sci.*, 49:663–675.

Lovell, R.T. and Flick, G.J., 1966, Irradiation of Gulf Coast area strawberries, *Food Technol.*, 20:949–952.

Luedtke, A. and Powell, D., 2000, A Timeline of Fresh Juice Outbreaks. Food Safety Risk Management and Communications Project Fact Sheet, U. of Guelph, Guelph, Ontario, Canada, http://www.plant.uoguelph.ca/safefood/micro-haz/juice-outbreaks.htm.

Luh, B.S., 1997, Principles and applications of vegetable processing, in *Processing Vegetables: Science and Technology*, Smith, D.S., et al., Eds., Lancaster, PA: Technomic Publishing, Inc., Ch. 1.

Matches, J.R. and Liston, J., 1968, Growth of Salmonellae on irradiated and nonirradiated seafoods, *J. Food Sci.*, 33:406–410.

Maxie, E.C. and Abdel-Kader, A., 1966, Food irradiation — physiology of fruits as related to feasibility of the technology, *Adv. Food Res.*, 15:105–138.

Mercier, J. and Lindow, S.E., 2000, Role of leaf surface sugars in colonization of plants by bacterial epiphytes, *Appl. Environ. Microbiol.*, 66(1):369–374.

Miege, E. and Coupe, H., 1914, Influence of x-rays on plants, *Compt. Rend.*, 159:338–340.

Miller, W.R. et al., 1994, Quality of "Climax" blueberries after low dosage electron beam irradiation, *J. Food Safety*, 17(1):71–79.

Miller, W.R., McDonald, R.E., and Smittle, B.J., 1995, Quality of "Sharpblue" blueberries after electron beam irradiation, *HortScience*, 30(2):306–308.

Miller, W.R. and McDonald, R.E., 1996a, Quality of "Brightwell" and "Tifblue" blueberries after gamma irradiation for quarantine treatment, *HortScience*, 31(7):1234.

Miller, W.R. and McDonald, R.E., 1996b, Postharvest quality of GA-treated Florida grapefruit after gamma irradiation with TBZ and storage, *Postharvest Biol. Technol.*, 7:253–260.

Monk, J.D., Beuchat, L.R., and Doyle, M.P., 1994, Irradiation inactivation of food-borne microorganisms, *J. Food Prot.*, 58(2):197–208.

Morris, C.E., Monier, J., and Jaques, M., 1997, Methods for observing microbial biofilms directly on leaf surfaces and recovering them for isolation of culturable microorganisms, *Appl. Environ. Microbiol.*, 63(4):1570–1576.

Narvaiz, P., Lescano, G., and Kairiyama, E., 1992, Irradiation of almonds and cashew nuts, *Lebensm. Wiss. Technol.*, 25:232–235.

National Food Processors Association (NFPA), 2000, Petition to amend 21CFR179 (Irradiation in the production, processing and handling of food), http://www.nfpa-food.org/petition/petition.pdf.

National Advisory Committee on Microbiological Criteria for Foods (NACMCF), 1999, Microbiological safety evaluation and recommendations on fresh produces, *Food Control*, 10:117–134.

Niemira, B.A., 2001, Citrus juice composition does not influence radiation sensitivity of *Salmonella enteritidis, J. Food Prot.*, 64(6):869–872.

Niemira, B.A., Sommers, C.H., and Boyd, G., 2001, Irradiation inactivation of four *Salmonella* species in orange juices with varying turbidity, *J. Food Prot.*, 64(5):614–617.

Nieto-Sandoval, J.M. et al., 2000, Effect of electron beam irradiation on color and microbial burden of red paprika, *J. Food Prot.*, 63(5):633–637.

O'Connor, R.E. and Mitchell, G.E., 1991, Effect of irradiation on microorganisms in strawberries, *Int. J. Food. Microbiol.*, 12:247–256.

Osterholm, M.T. and Potter, M.E., 1997, Irradiation Pasteurization of Solid Foods: Taking Food Safety to the Next Level. Emerging Infectious Disease, Vol. 3, No. 4, http://www.cdc.gov/ncidod/eid/vol3no4/osterhol.htm.

Pao, S. and Davis, C.L., 1999, Enhancing microbiological safety of fresh orange juice by fruit immersion in hot water and chemical sanitizers, *J. Food Prot.*, 62(7):756–760.

Parish, M.E. and Higgins, D.P., 1988, Isolation and identification of lactic acid bacteria from samples of citrus molasses and unpasteurized orange juices, *J. Food Sci.*, 53:645–646.

Pickett, K.T. et al., 2000, Effects of low-dose irradiation on the microbial counts, sensory attributes and ascorbic acid content of unpasteurized orange juice, Abstract 86H-10, Institute of Food Technologists Annual Meeting, 2000, http://ift.confex.com/ift/2000/techprogram/paper_3867.htm.

Prakash, A. et al., 2000a, Effects of low-dose gamma irradiation on the shelf-life and quality characteristics of cut Romaine lettuce packaged under modified atmosphere, *J. Food Sci.*, 65(3):549–553.

Prakash, A. et al., 2000b, Effects of low-dose gamma irradiation and conventional treatments on shelf-life and quality characteristics of diced celery, *J. Food Sci.*, 65(6):1070–1075.

Proctor, B.E. et al., 1955, Extension of food storage life by irradiation, *Food Technol.*, 9:523.

Rajkowski, K.T. and Thayer, D.W., 2000, Reduction of *Salmonella* spp. and strains of *Escherichia coli* O157:H7 by gamma radiation of inoculated sprouts, *J. Food Prot.*, 63(7):871–875.

Riordan, D.C.R., Sapers, G.M., and Annous, B.A., 2000, The survival of *Escherichia coli* O157:H7 in the presence of *Penicillium expansum* and *Glomerella cingulata* in wounds on apple surfaces, *J. Food Prot.*, 63(12):1637–1642.

Romani, R.J. et al., 1963, Radiation physiology of fruit — ascorbic acid, sulfhydryl and soluble nitrogen content of irradiated citrus, *Radiat. Bot.*, 3:58.

Salisbury, F.B. and Ross, C.W., 1984, *Plant Physiology*, 3rd ed., Belmont, CA: Wadsworth Publishing Co.

Schreiber, G.A. et al., 1993, Luminescence techniques to identify the treatment of food by ionizing radiation, *Food Struct.*, 12:385–396.

Sizer, C.E. and Balasubramaniam, V.M., 1999, New intervention processes for minimally processed juices, *Food Technol.*, 53(10):64–67.

Snyder, O.P. and Poland, D.M., 1995, Food Irradiation Today, Hospitality Institute of Technology and Management, St. Paul, MN, http://www.hi-tm.com/documents.irrad. html.

Sommers, C.H. and Bhaduri, S., 2001, Loss of crystal violet binding in *Yersinia enterocolitica* following gamma irradiation, *Food Microbiol.*, 18:367–374.

Sommers, C.H., Handel, A.P., and Niemira, B.A., 2001, Radiation resistance of *Listeria monocytogenes* on the presence or absence of sodium erythorbate, *J. Food Sci.*, in press.

Sommers, C.H., chap. 14, this volume.

Spoto, M.H.F. et al., 1997, Sensory evaluation of orange juice concentrate as affected by irradiation and storage, *J. Food Process. Preserv.*, 21(3):179–191.

Sumner, S.S. and Peters, D.L., 1997, Microbiology of vegetables, in *Processing Vegetables: Science and Technology*, Smith, D.S. et al., Eds., Lancaster, PA: Technomic Publishing Inc., Ch. 4.

Taormina, P.J., Beuchat, L.R., and Slutsker, L., 1999, Infections associated with eating seed sprouts: an international concern, *Emerging Infect. Dis.*, 5(5):626–634.

Tauxe, R. et al., 1997, Microbial hazards and emerging issues associated with produce: a preliminary report to the National Advisory Committee on Microbiological Criteria for Foods, *J. Food Prot.*, 60(11):1400–1408.

Takeuchi, K. and Frank, J.F., 2000, Penetration of *Escherichia coli* O157:H7 into lettuce tissues as affected by inoculum size and temperature and the effect of chlorine treatment on cell viability, *J. Food Prot.*, 63(4):434–440.

Thakur B.R. and Singh, S.A., 1993, Effect of sorbic acid on irradiation-induced sensory and chemical changes in sweetened orange juice and mango pulp, *Int. J. Food Sci. Technol.*, 28(4):371–376.

Thayer, D.W., 1994, Wholesomeness of irradiated foods, *Food Technol.*, 48(5):132–136.

Thayer, D.W. and Boyd, G., 1995, Radiation sensitivity of *Listeria monocytogenes* on beef as affected by temperature, *J. Food Sci.*, 60:237–240.

Thayer, D.W. et al., 1995, Variations in radiation sensitivity of foodborne pathogens associated with the suspending meat, *J. Food Sci.*, 60(1):63–67.

Thayer, D.W. et al., 1996, Radiation pasteurization of food, Council for Agricultural Science and Technology, Issue paper number 7, April 1996, http://www.cast-science.org/past_ip.pdf.

Thayer, D.W. and Rajkowski, K.T., 1999, Developments in irradiation of fresh fruits and vegetables, *Food Technol.*, 53(11):62–65.

Thayer, D.W. et al., 1999, Elimination of *E. coli* O157:H7 and control of *Salmonella* on sprouts and alfalfa seed by gamma irradiation and sanitation, Abs., Sprout Summit — Best Practices and Recommendations for the Production of Safer Sprouts from Seeds, National Center for Food Safety and Technology, Nov 15–16, 1999, Summit-Argo, IL.

Thayer, D.W. and Boyd, G., 2000, Reduction of normal flora by irradiation and its effect on the ability of *Listeria monocytogenes* to multiply on ground turkey stored at 7°C when packaged under a modified atmosphere, *J. Food Prot.*, 63(12):1702–1706.

Thelen, K., 1999, Michigan State University Field Crop Advisory Team Alert, Vol. 14, No. 18.

Thomas, P., 1983, Radiation preservation of foods of plant origin. V. Temperate fruits: pome fruits, stone fruits and berries, *CRC Crit. Rev. Food Sci. Nutr.*, 24:357–400.

Tiwari, N.P. and Maxcy, R.B., 1972, Comparative growth of salmonellae, coliforms, and other members of the microflora of raw and radurized ground beef, *J. Milk Food Technol.*, 35:455–460.

Urbain, W.M., 1986, *Food Irradiation*, New York: Academic Press, Inc.

Wang, H., Cao, G., and Prior, R.L., 1996, Total antioxidant capacity of fruits, *J. Agric. Food Chem.*, 44:701–705.

Wiersema, J.H. and Leon, B. 1999, *World Economic Plants: A Standard Reference*, Boca Raton, FL: CRC Press.

Willemot, C.M., Marcotte, M., and Deschenes, L., 1996, Ionizing radiation for preservation of fruits, in *Processing Fruits: Science and Technology, Vol 1. Biology, Principles and Applications*, Somogyi, L.P. and Ramaswamy, H.P., Eds., Lancaster, PA: Technomic Publishing, Co., Ch. 9, 221–260.

Wood, O.B. and Bruhn, C.H., 1999, Food irradiation — position of the ADA (American Dietetic Association), http://www.eatright.org/airradi.html.

World Health Organization (WHO), 1999, High-dose irradiation: wholesomeness of food irradiated with doses above 10 KGy, Joint FAO/IAEA/WHO Study Group on High-Dose Irradiation, Geneva, September 15–20, 1997, WHO Technical Report, Series 890.

Yu, L., Reitmeier, C.A., and Love, M.H., 1996, Strawberry texture and pectin content as affected by electron beam irradiation, *J. Food Sci.*, 61(4):844–846.

Zhuang, R.Y. and Beuchat, L.R., 1996, Effectiveness of trisodium phosphate for killing *Salmonella montevideo* on tomatoes, *Lett. Appl. Microbiol.*, 22:97–100.

14 Irradiation of Minimally Processed Meats*

Christopher H. Sommers

CONTENTS

INTRODUCTION

Meats are an ideal medium for the proliferation of microorganisms, and the preservation of meat products has been an ongoing challenge driven by the need for extended shelf-life and safety. As humans have increased in population, formed

* Reference to a brand or firm name does not constitute endorsement by the US Department of Agriculture over others of a similar nature not mentioned.

1-58716-041-2/03/$0.00+$1.50

communities, built cities, and strained the planet's resources, this need has become critical. What is produced on one continent may be consumed on another the following week. The global economy includes the packaging and shipping of minimally processed meat products on a massive scale. The U.S. exported 2607 million pounds of beef, 1374 million pounds of pork, and 5300 million pounds of chicken in 2001 alone (Nunes, 2001a,b; Cooper, 2001).

The technologies for preservation of meats include refrigeration, cooking, salting, curing, smoking, canning, and the use of chemical preservatives and modified atmosphere packaging. Other processes, including pulse electric fields and high-pressure technologies, are in various stages of development. Despite advances in food processing technology, an estimated 76,000,000 annual cases of food-borne-related illnesses in the U.S. result in approximately 5000 deaths (Mead et al., 1999). In the year 2000 alone, approximately 2.4 million pounds of beef were recalled due to possible contamination with *Escherichia coli* O157:H7. In the last 5 years, more than 50 million pounds of ready-to-eat meat products have been recalled in the U.S. due to contamination with *Listeria monocytogenes*. *Homo sapiens*, even with the aid of advanced technologies, has not beaten the microbe.

Another weapon in the food safety arsenal is ionizing radiation. The use of ionizing radiation for the sterilization or pasteurization of meats has been a topic of research for over 40 years. Proctor et al. (1943) used x-rays to preserve hamburger meat, extend its shelf-life, and maintain its organoleptic qualities. It is arguably the most studied and tested technology in the history of modern food processing (WHO, 1994; Diehl, 1995). It has been used extensively for the sterilization of medical devices and pharmaceuticals, elimination of microorganisms from spices, de-infestation of agricultural commodities, and elimination of parasites.

The U.S. FDA has approved the use of low-dose ionizing radiation to control food-borne pathogens on meat and poultry products and to extend product shelf-life (Federal Register, 1997). Refrigerated red meats and meat by-products may be irradiated to a dose of 4.5 kGy, while frozen meat and meat by-products may be irradiated to a dose of 7.0 kGy. The approval includes meats of bovine, ovine, porcine, and equine sources. Ionizing radiation doses ranging from 1.5 to 3.0 kGy may be used for the decontamination of poultry. The World Health Organization has approved treatment of foodstuffs with ionizing radiation up to doses of 10 kGy (WHO, 1994). More than 40 countries have approved the use of irradiation of foods and at least 30 are currently irradiating at least one food type for commercial purposes (Louharanou, 2001). Because overall coordinated approvals for irradiation of minimally processed meats do not exist for the European Economic Community as a trade group, this chapter focuses on irradiation of meat products in the North American market.

This chapter concerns ionizing radiation doses of less than 10 kGy, commonly referred to as radurization, radiation pasteurization, or cold pasteurization for the treatment of minimally processed meats. In this review, minimally processed meats include meat or poultry cuts, whole poultry carcasses, or ground meat. Doses of less than 10 kGy vary significantly in their ability to eliminate pathogens and reduce microorganisms associated with spoilage, depending on the species of microorganism, product composition, temperature during irradiation, and inclusion of modified atmosphere. The author tries to provide information about the effects of these variables as pertains to the radiation pasteurization of minimally processed meats, including variables that would affect processing in a commercial environment.

RADIATION-ABSORBED DOSE

Another chapter in this book addresses the use of ionizing radiation for the treatment of fruits and vegetables; this section describes the scientific terminology and technology commonly associated with ionizing radiation treatment of foodstuffs for both chapters. The SI unit for radiation-absorbed dose is the Gray (Gy). One Gy is equal to 100 rads, 1 kGy is equal to 100 krads, and 10 kGy is equal to 1 Mrad. Older texts often use the electron volt (eV) as a unit of absorbed energy, where 1 rad = 100 erg $gram^{-1}$ = 6.24×10^{13} eV $gram^{-1}$ (Pryor, 1977). This difference in use of measurement units has produced considerable confusion for those unfamiliar with ionizing radiation processing of meats and other foods.

Guidelines pertaining to electrons produced at 5 MeV and 10 MeV by x-ray and electron beam machines, respectively, refer to the energies at which the x-rays or electrons are produced, not a unit of radiation-absorbed dose. Many texts refer to the energy of photons, in MeV, released from isotopes such as ^{60}Co or ^{137}Cs. Again, those values are not related to absorbed dose. Radiation-absorbed dose produced by machine, as with isotope-produced gamma radiation, is measured in Gy as determined by appropriate dosimetry (see below). Pasteurization of meats with ionizing radiation is often referred to as cold pasteurization because of the low rate of heat transfer, 0.23°C/kGy, which occurs as a result of the process.

TYPES AND SOURCES OF IONIZING RADIATION

Ionizing radiation refers to forms of radiant energy that possess sufficient energy to create negatively and positively charged ions in the target substrate. The three forms of ionizing radiation approved for pasteurization of meat products include gamma radiation, beta radiation, and x-rays (WHO, 1994; Federal Register, 1997).

Gamma rays are produced by the natural decay of isotopes (either ^{60}Co or ^{137}Cs). Most, but not all, commercial gamma irradiators use ^{60}Co as the source material. The advantage of gamma radiation is the high penetrability of the photons produced, allowing irradiation of high bulk density materials. ^{60}Co releases photons having energies of 1.33 MeV and 1.17 MeV, while ^{137}Cs releases photons with an energy of 0.66 MeV (Pryor, 1977). The approximate penetration of ^{60}Co and ^{137}Cs photons in a layer of water sufficient to reduce signal intensity by 90% is 37, 35, and 25 cm for 1.33, 1.17, and 0.66 MeV, respectively. Because of the 5.27-year half-life of ^{60}Co, approximately 12% of the source material must be replaced annually to maintain the original radiation dose rate (WHO, 1994). The half-life of ^{137}Cs is 30 years.

Beta radiation is produced by electron beam machine. Electrons produced at a maximum energy of 10 MeV may be used for food irradiation. The advantages of electron beam machines as the radiation source include:

1. The machine can be turned on and off as needed.
2. There is no source that needs replenishing.

3. The radiation dose is delivered quickly.
4. Because of their high through-put, electron beam machines may be placed at the end of existing manufacturing lines.

The disadvantages of electron beam include the low penetrability of electrons (10 MeV) — approximately 5 cm in meat products — and the large requirements for electricity, cooling, and maintenance (Diehl, 1995; WHO, 1994).

Machine-generated x-rays with a maximum energy of 5 MeV may be used to pasteurize meat products (WHO, 1994). X-rays offer the advantages of gamma rays (penetration into products) combined with the advantages of electron beam machines, including the ability to turn the source on and off as needed. However, the energy efficiency of x-ray units is low, with most of the energy put into the production of x-rays lost in the form of heat.

DOSIMETRY

Dosimetry standards for ionizing radiation pasteurization of foods, including minimally processed meat products, are often overlooked in discussions of the technology. The American Society for Testing and Materials (ASTM) Subcommittee E10.01 has prepared 24 standards for radiation processing of materials, which include those for minimally processed meats, fruits, and vegetables. These standards include: (1) E1204–97, Practice for Dosimetry in Gamma Irradiation Facilities for Food Processing, (2) E1431–98, Practice for Dosimetry in Electron and Bremsstrahlung Irradiation Facilities for Food Processing, and (3) E1900–97, Guide for Dosimetry in Radiation Research on Food and Agricultural Products (ASTM, 2000). The majority of ASTM standards have also been adopted by the International Standards Organization (ISO), with the equivalent ISO standards listed as the first footnote in the ASTM Standard (ASTM, 2000).

HOW IONIZING RADIATION KILLS MICROORGANISMS

Ionizing radiation kills microorganisms by damaging the bacterial chromosome (Ward, 1991). It induces DNA strand breaks, transition mutations, transversion mutations, frameshift mutations, and deletion mutations (Glickman et al., 1980; Raha and Hutchison, 1991; Sargentini and Smith, 1994; Wijker et al., 1996). The loss of large plasmids containing genes required for pathogen virulence can be induced by exposure to ionizing irradiation (Sommers and Bhaduri, 2001). Ionizing radiation damages DNA via two mechanisms: direct action against the bacterial chromosome by photon-induced breakage of the DNA phosphodiester backbone or indirect damage to the DNA by the radiolysis products of water, primarily hydroxyl radicals. Indirect damage accounts for approximately 70% of the DNA damage induced by ionizing radiation.

Ionizing radiation also disrupts the microorganism cell membrane. Naidu et al. (1998) found that the lipopolysaccharide (LPS) cell membrane component isolated from *Salmonella typhimurium* was depolymerized following exposure to ionizing radiation. The irradiated LPS and lipid A were less toxic and mitogenic in animal studies. Ionizing radiation disrupts the cellular membrane–DNA complexes required for plasmid partitioning and active sites for the DNA repair process (Watkins, 1980; Khare et al., 1982). Although it is theoretically possible to produce virulence from a previously nonvirulent microorganism, no such occurrence has been reported in the 50 years following the introduction of ionizing radiation as a food safety technology (WHO, 1994).

Incidence of Food-borne Illness

The primary function of ionizing radiation pasteurization of meats is to eliminate pathogenic bacteria to provide safer meat products to consumers and to extend product shelf-life through the reduction of spoilage microorganisms. Mead et al. (1999) identified a number of pathogens responsible for food-related illnesses in humans and listed estimated illnesses, food-borne transmission frequencies, hospitalizations, and fatality rates in the U.S. on an annual basis. The infectious agents were classified as bacterial, parasitic, or viral in origin. Bacterial pathogens accounted for 60% of hospitalizations, parasites for 5%, and viruses for 34% (Mead et al., 1999). The radiation resistances of agents in those three classes are discussed in this section. Other information is provided by reports from the Food-borne Diseases Active Surveillance Database (Shallow et al., 1997, 1999; Wallace et al., 1997).

Determination of Radiation Resistance

The D_{10} value, the estimate of the ionizing radiation dose required to eliminate 90% of viable organisms, varies from agent to agent and even within species of the same organism. In general, meat processors, in an attempt to comply with hazard analysis and critical control point (HACCP) guidelines, call for elimination of 5 $\log_{10}$, or 99.999%, of a given pathogen. A radiation dose of 2.5 kGy would be utilized to achieve a 99.999% reduction of a pathogen if the D_{10} value were 0.5 kGy. In the majority of peer-reviewed scientific publications, the D_{10} value was determined by inoculation of pathogens into or onto a meat matrix, treatment of the sample with multiple doses of ionizing radiation, and determination of the number of viable microorganisms on specific or nonspecific growth media. The D_{10} value is commonly calculated as the reciprocal of the slope (Ley, 1983).

A number of factors affects the radiation resistance of bacteria, viruses, and parasites. Thayer (2000) reviewed a number of parameters that affect D_{10} value of the pathogen *E. coli* O157:H7. These include (1) isolates chosen for D_{10} determination, (2) growth state of the organism (logarithmic or stationary phase), (3) past history, including development of acid tolerance, (4) presence of biofilms, (5) sample size and packing conditions, (6) culture media used, (7) colony counting methodology, (8) radiation-absorbed dose calculation and measurement, and (9) method of D_{10} calculation. Processors should consider D_{10} variability and uncertainty when selecting radiation doses required for elimination of specific pathogens.

D_{10} Values of Specific Organisms

Salmonella and *Campylobacter*

Tauxe (1999) identified five bacterial pathogens of concern responsible for food-borne illnesses and the incidences per 100,000 cases. Those pathogens included *Campylobacter, Salmonellae, E. coli* O157:H7, *L. monocytogenes*, and *Yersinia enterocolitica.* The following sections describe the radiation resistances of those organisms when suspended in various meat matrixes.

Campylobacter spp. and *Salmonella* spp. cause an estimated 2.45 million cases of food-borne illness annually (Mead et al., 1999). In addition to this role, both have been identified as inducing Rieter's Syndrome, a reactive arthritis that can develop following infection by those organisms (Smith, 1994). Certain serotypes of *Campylobacter* have been associated with the autoimmune disease Gullain–Barre Syndrome, an acute demyelating peripheral neuropathy that can be induced by bacteria and viruses (Allos et al., 1998).

Salmonella spp. and *Campylobacter* spp. can be readily isolated from retail meat products. Duffy et al. (2001) investigated the incidence of microbial contamination of retail pork products sold in the U.S.. An average incidence rate of 9.6% *Salmonella* spp.–positive products was found for four types of retail pork products. *C. jejuni* and *C. coli* were found to have the lowest incidence among the food-borne pathogens on pork, with an overall incidence rate of 1.3%. In contrast, Berrang et al. (2001) found *Campylobacter* spp. to be readily detectable on both skinless and skin-present chicken parts and concluded that consumers should not expect to lower the number of microorganisms significantly by removing the skin. Differences in the incidence rate can be attributed to the microbial ecology of the host animal.

Lambert and Maxcy (1984) estimated the D_{10} of *C. jejuni*, the most common pathogen found on poultry products, to be approximately 0.16 kGy. Clavero et al. (1994) found that *C. jejuni* inoculated into ground beef and irradiated at refrigerated temperature had D_{10} values ranging from 0.175 to 0.199 kGy. Collins et al. (1996) found a D_{10} of 0.15 kGy for *C. jejeuni* suspended in vacuum-packaged ground pork. The radiation resistance of *Campylobacter* varies depending on species and even within species. Using minced poultry irradiated at refrigeration temperatures, Patterson (1995) found the D_{10} of *Campylobacter* spp. to be 0.13 to 0.19 kGy for three strains of *C. jejuni* and 0.12 to 0.25 kGy for three strains of *C. coli.* The D_{10} of single strains of *C. fetus* and *C. lari* were 0.14 kGy (Patterson, 1995).

Thayer et al. (1990) found that *Salmonella* spp. inoculated into aerobically packed, refrigerated, and mechanically deboned chicken meat had D_{10} ranging from 0.36 to 0.77 kGy. Patterson (1988) determined the D_{10} value for *S. typhimurium* to be 0.50 kGy on aerobically packed chicken irradiated at 10°C. Clavero et al. (1994) found the D_{10} of *S. typhimurium* suspended in low-fat and high-fat ground beef at refrigeration temperature to range from 0.62 to 0.66 kGy. In a study on the effect of the meat species as a factor affecting D_{10} value, the radiation resistance of *Salmonella* spp. ranged from 0.51 to 0.71 kGy. The individual D_{10} values were 0.70, 0.67, 0.51, 0.71, and 0.71 kGy when suspended in beef, lamb, pork, turkey breast, and turkey leg meat, respectively (Thayer et al., 1995). In that study, the inoculated meats were vacuum-packed, irradiated (4°C), and plated under identical conditions. Abu-Tarbush et al. (1997) irradiated naturally contaminated whole chickens (4°C) to doses of 2.5, 5.0, 7.5, and 10.0 kGy and concluded that a radiation dose of 2.5 kGy was sufficient to destroy *Salmonella*, *Campylobacter*, and *Yersinia* spp. without affecting product quality.

E. Coli O157:H7

E. coli O157:H7 causes an estimated 73,000 illnesses and 52 deaths annually (Mead et al., 1999). Because of its high mortality rate and ability to induce hemolytic uremic syndrome (HUS), *E. coli* O157:H7 has been designated as an adulterant in meats and other food products. However, it should be noted that non-O157 shiga-toxin-producing *E. coli* are the cause of an estimated 7 to 20% of HUS cases and 36,000 illnesses annually in the U.S. (Mead et al., 1999). Enterotoxigenic and other diarrheogenic *E. coli* cause an estimated 79,000 illnesses each on an annual basis in the U.S.

Thayer and Boyd (1993) obtained a D_{10} value of 0.28 kGy for *E. coli* O157:H7 suspended in refrigerated ground beef. Clavero et al. (1994) obtained a D_{10} of 0.24 and 0.25 kGy for *E.coli* O157:H7 inoculated onto aerobically packed low-fat and high-fat ground beef, respectively. Thayer et al. (1995) found D_{10}'s of 0.30, 0.32, 0.30, 0.30, and 0.29 kGy for *E. coli* O157:H7 suspended in vacuum-packed refrigerated beef, lamb, pork, turkey breast, and turkey leg meat, respectively; the species of meat did not affect the radiation resistance of *E. coli* O157:H7. Patterson (1988) obtained D_{10} values of 0.35 to 0.39 kGy and 0.27 kGy for *E. coli* suspended in refrigerated (4°C) aerobically packed and vacuum-packed minced chicken meat, respectively. Results from these studies indicate that a radiation dose of 1.6 kGy should eliminate 99.999% of *E. coli* O157:H7 in beef, a 5-$\log_{10}$ reduction.

L. Monocytogenes

L. monocytogenes causes an estimated 2500 cases of food-borne illness annually, with a mortality rate of approximately 20% in immunocompromised or other susceptible individuals (Mead et al., 1999). It is capable of growth at refrigeration temperatures, can survive in high-salt environments, and is readily isolated from retail meat products (Smith, 1996). It should be noted, however, that Listeriosis is most often associated with postprocess contamination of ready-to-eat foods that are refrigerated for long periods as opposed to raw meat products. Duffy et al. (2001) found *L. monocytogenes* to have an incidence (detection) rate of approximately 19.8% in retail pork products. Miettinen et al. (2001) detected *L. monocytogenes*, representing 3 serotypes and 14 different pulse field gel electrophoresis (PFGE) types, on 62% of retail poultry parts produced by two abattoirs and processing plants.

The D_{10} for *L. monocytogenes* suspended in refrigerated vacuum-packed beef, pork, lamb, or turkey meat ranged from 0.45 to 0.50 kGy (Thayer et al., 1995). Unlike *Salmonella* spp., the meat substrate had no effect on *L. monocytogenes* radiation resistance. Monk et al. (1994) obtained a D_{10} of 0.58 to 0.59 kGy and 0.51 to 0.57 kGy in refrigerated low-fat and high-fat ground beef, respectively. The D_{10}

of *L. monocytogenes* on raw turkey breast was found to be 0.56 kGy, while the D_{10} on cooked turkey breast was found to be 0.69 kGy (Thayer et al., 1998). Patterson (1989) found D_{10} values for *L. monocytogenes* ranging from 0.42 to 0.55 kGy on poultry meat. Shamsuzzaman et al. (1992) obtained a D_{10} of 0.59 for *L. monocytogenes* inoculated onto uncooked chicken breast meat. A radiation dose of approximately 3.0 kGy would be sufficient to eliminate 99.999% of *L. monocytogenes* from refrigerated minimally processed meat and poultry.

Y. Enterocolitica

The emerging pathogen *Y. enterocolitica* causes an estimated 96,000 illnesses annually. Like *Campylobacter* and *Salmonella*, it can cause the autoimmune disorder Rieter's Syndrome (Smith, 1994). *Y. enterocolitica* can grow at refrigeration temperatures and is a frequent contaminant found on retail pork products. Because of the ability of swine to act as a reservoir for *Y. enterocolitica,* this organism is considered a pathogen of concern for the pork industry (Davies, 1997). Duffy et al. (2001) found that 19.8% of retail whole-muscle pork and 11.5% of retail ground pork products were contaminated with *Y. enterocolitica* and that most contamination occurred appeared to occur after the product left the processing plant. In-line irradiation of packaged retail pork products would benefit the pork processing industry.

Sommers and Bhaduri (2001) found that D_{10} ranged from 0.17 to 0.21 kGy for *Y. enterocolitica* that was suspended in phosphate buffer or raw ground pork and contained a virulence plasmid. In addition, ionizing radiation induced loss of the virulence plasmid tenfold over that observed in unirradiated controls, regardless of the suspending matrix. Not only did ionizing radiation eliminate *Y. enterocolitica* from raw pork, but the surviving population was less, not more, virulent. That finding has specific relevance to cases in which large numbers of food-borne pathogens might be deliberately inoculated into a food product in an act of bioterrorism.

Irradiation attenuates bacteria that survive the irradiation process. Kamat et al. (1997) found D_{10} of 0.25 kGy for two *Y. enterocolitica* strains inoculated into raw pork. Sommers and Novak (2002) determined D_{10} values ranging from 0.15 to 0.23 kGy for eight different virulence plasmid-containing and virulence plasmid-lacking *Y. enterocolitica* strains suspended in vacuum-packaged raw ground pork. Tarkowski et al. (1984) found that the radiation resistance of *Y. enterococolitica* varied by isolate. In general, a radiation dose of 1.5 kGy should be adequate to produce a 5-$\log_{10}$ reduction of *Y. enterocolitica* in raw meats.

Spore Formers and Spores

Doses of ionizing radiation up 10 kGy are sufficient to eliminate most vegetative food-borne pathogens; however, they are not sufficient to produce a 5-$\log_{10}$ reduction in spores of organisms such as *Clostridium botulinum*, *Bacillus cereus*, or *Clostridium perfringens*. *C. botulinum*, *C. perfringens*, and *B. cereus* are responsible for an estimated 58 and 27,000 cases of food-borne illness annually in the U.S. (Mead et al., 1999). Because of recent events, elimination and attenuation of spore-forming bacteria from food and non-foodstuffs have regained importance for the health and safety of the American people.

Spores are more resistant to ionizing radiation than vegetative cells of the same species, possibly due to reduced water activity in the spores, with a subsequent decrease in indirect DNA damage as a result of the hydrolysis of water. *C. botulinum* spores have relatively large D_{10} values that range from 1.29 to 3.34 kGy (Anellis and Koch, 1962). Spores of the spoilage bacterium *Clostridium sporogenes* have been reported to have D_{10} values of 6.3, 7.8, and 10.1 kGy in beef, pork, and chicken fat, respectively (Shamsuzzaman and Lucht, 1993). *C. perfringens* vegetative cells had D_{10} values ranging from 0.347 to 0.826 kGy when inoculated onto a variety of food matrices (Kamat et al., 1989). Harewood et al. (1994) reported a D_{10} value of 2.71 kGy for *C. perfringens* spores inoculated onto clams. Irradiation to a dose of 10 kGy would therefore be insufficient to produce a 5-$\log_{10}$ reduction in spore number (Roberts and Ingram, 1965).

Bacillus cereus spores suspended in phosphate buffer were reported to have D_{10} values of 2.5 to 4.0 kGy, as opposed to vegetative cells that had D_{10} values of 0.30 to 0.65 kGy (Kamat et al., 1989). Thayer and Boyd (1994) tested the radiation resistance of endospores and logarithmic and stationary-phase *B. cereus* suspended in vacuum-packed mechanically deboned chicken meat (MDCM). D_{10} values of 0.187, 0.446, and 2.67 kGy were obtained for logarithmic phase, stationary phase, and endospore *B. cereus*, respectively. D_{10} values of *B. cereus* endospores ranged from 1.92 to 2.78 kGy when suspended in MDCM, beef, pork, turkey breast, or beef gravy. The D_{10} value for lyophilized *B. cereus* spores irradiated in air was 3.4 kGy vs. 2.3 kGy on plate count broth (Ma and Maxcy, 1981).

Viruses

Ionizing radiation pasteurization is not an effective method for elimination of viruses from meat products. Although reduced in titer, polio virus could be recovered from hydrated and dehydrated beef pot roast, pork, or chicken following ionizing radiation doses up to 6.0 kGy (Heidelbaugh and Giron, 1969). Sullivan et al. (1971) determined the D_{10} values for 30 viruses of public health significance. D_{10} values ranged from 3.8 to 5.0 kGy when viruses were suspended in Eagle's Minimal Medium.

Pirtle et al. (1997) investigated the combined effects of ionizing radiation (4.4 and 5.27 kGy) and heat treatment on the destruction of two RNA and DNA viruses suspended in either saline or raw ground pork. The authors concluded that ionizing radiation did not increase the sensitivity of the viruses to heat inactivation and that the method has no practical application toward the goal of virus removal in meat products above that of heat treatment alone. Foot-and-mouth disease virus suspended in bovine tissues can be eliminated at radiation doses of 15.0 or 25.0 kGy in combination with heat treatment (Lasta et al., 1992). However, those doses are well above the 4.5 or 7.0 kGy doses allowed for pasteurization of refrigerated and frozen meats and therefore of no practical significance in regard to ionizing radiation pasteurization of meats.

Parasites

Parasites commonly found in raw meats include *Trichenella spiralis*, *Taenia solium*, *Taenia saginata*, and *Toxoplasma gondii*. Parasite cysts and their larvae are fairly sensitive to ionizing radiation because they are unable to develop into their larval form following exposure to ionizing radiation. Brake et al. (1985) found that 0.15 to 0.30 kGy prevented maturation of ingested larvae in the gut of the host animals and the production of parasite progeny. Split hog carcasses irradiated to 0.3 kGy for elimination of *T. spiralis* were acceptable to the market in quality. Verster et al. (1977) were able to eradicate *T. solium* and *T. saginata*, pork and beef tapeworms, from pork and beef carcasses using radiation doses from 0.2 to 0.6 kGy with no effect on carcass quality factors. Tolgay et al. (1972) found that a radiation dose of 0.4 kGy was sufficient to prevent maturation of *T. saginata* cysts in beef but that more than 1.0 kGy was needed for elimination of adult tapeworm. Dubey et al. (1998) investigated the ability of ionizing radiation to eliminate sporulated and unsporulated *T. gondii*. Unsporulated *T. gondii* oocysts irradiated to doses of 0.4 to 0.8 kGy sporulated but were not infective to mice. Sporulated oocysts irradiated to 0.4 kGy were infective but unable to reproduce. Even minimum ionizing radiation doses used for elimination of bacterial pathogens from meats would eliminate most parasites.

TSE and BSE

A key topic in food safety has been the emergence, incidence, and transmission of transmissible spongiform encephalopathies (TSEs), primarily bovine spongiform encephalopathy (BSE) and its human counterpart, Creutzfeld–Jakob disease (CJD). The infectious agent responsible for TSE has been termed "prion," which Prusiner et al. (1982) defined as "a proteinaceous infectious particle — a small infectious agent consisting largely or solely of protein" (Schrueder, 1994). Alper (1966) first described attempts to inactivate prions using ionizing radiation. In those experiments, brains of mice showing advanced signs of the TSE "Scrapie" were isolated, irradiated to a dose of 25 kGy, diluted, and injected into healthy mice. Subsequently using the LD_{50} of the treated mice, Alper (1966) calculated the radiation dose required to provide a 1-$\log_{10}$ reduction in prion biological activity to be 43.0 kGy. Ionizing radiation, even at sterilization doses, cannot inactivate prions in meat products and therefore would be ineffective against BSE as an infectious agent.

SPOILAGE ORGANISMS AND IRRADIATION

Gram-negative *pseudomonads* tend to dominate the background microflora of non-irradiated raw meats and are more sensitive to ionizing radiation than are Gram-positive lactic acid bacteria and micrococci (Patterson, 1988; Ingram, 1975). Therefore, lactic acid bacteria, which are more resistant to ionizing radiation than the *pseudomonads*, are the most common spoilage organisms in radiation pasteurized meats (Urbain, 1978). Grant and Patterson (1991) found that *Lactobacilli* dominated the background microflora of refrigerated pork irradiated to a dose of 1.75 kGy. Hastings et al. (1986) were able to isolate multiple *Lactobacillus* spp., including *L. sake*, *L. curvatus*, and *L. farciminis*, from beef pasteurized to a dose of 5.0 kGy. Unlike bacterial food-borne pathogens, the D_{10} values of *Lactobacillus* spp. were found to be higher in the exponential phase (log) of growth than in the stationary phase (Hastings et al., 1986). Ehioba et al. (1987, 1988) found that the background microflora of unirradiated ground pork consisted primarily of Gram-negative *Pseudomonas* and *Enterobacter* species but was primarily *Lactobacilli* following irradiation and refrigerated storage. Background microflora of pork loins irradiated to a dose of 3.0 kGy, unlike nonirradiated controls, consisted of Gram-positive bacteria following extended refrigerated storage (Lebepe et al, 1990).

THE EFFECT OF TEMPERATURE ON RADIATION RESISTANCE OF MICROORGANISMS

Minimally processed red meats may be irradiated to doses of 4.5 kGy for fresh refrigerated products or to 7.0 kGy for frozen products. Depending on water activity, salt content, and other variables, the initial freezing point of raw meat can vary between 0 and –5°C. Raw meats become hard frozen (greater than 80% of the available water frozen) between –5 and –10°C, with ice content increasing as temperature decreases. A number of studies have addressed the parameters affecting the freezing and thawing of raw meats (Miles et al., 1997; Succar and Harakawa, 1990; Chang and Tao, 1981).

Irradiation of meats at subfreezing temperatures decreases lipid oxidation, minimizes vitamin loss, and decreases the probability of chemical changes leading to off-flavors and odors. However, maintaining product temperature at subfreezing levels during irradiation increases the radiation resistance of food-borne pathogens. As early as 1956, the relationship between temperature and radiation resistance of microorganisms was noted (Cain et al., 1956). The increased radiation resistance of microorganisms at subfreezing temperatures is attributed to decreased a_w and hydroxyl radical mobility following radiolysis of water (Bruns and Maxcy, 1979; Taub et al., 1979). However, most studies concerning irradiation of meats, including the majority of those conducted in the 1980s and 1990s, did not address the effects of temperature on D_{10} value.

Significant effects of product temperature have been observed for Gram-negative and Gram-positive organisms. The radiation resistance of *Staphylococcus aureus* is inversely related to temperature (Thayer and Boyd, 1992). Thayer and Boyd (1995) reported a D_{10} value for *L. monocytogenes* of 0.45 kGy on ground beef when the temperature was maintained at 5°C during the irradiation process. Ground beef maintained at a temperature of –20°C obtained a D_{10} value of 1.21 kGy. Lopez-Gonzalez et al. (1999) found that *E. coli* O157:H7 suspended in beef had a D_{10} of 0.62 kGy when irradiated at –15°C, as opposed to a D_{10} of 0.41 kGy at 5°C. *S. typhimurium* suspended in mechanically deboned chicken meat is more resistant to ionizing radiation at subfreezing temperatures than at above freezing temperatures (Thayer and Boyd, 1990). *Shewanella putrefaciens* (*Pseudomonas putrefaciens*) is associated with the spoilage of red meats; the radiation resistance of *S. putrefaciens* was also inversely related to temperature (Thayer and Boyd, 1996). Thayer and Boyd noted a similar inverse relationship between radiation resistance and temperature for stationary phase *B. cereus* (Thayer and Boyd, 1994).

Many irradiation facilities lack the ability to control temperature during the process. Some processors wrap product in insulating blankets to maintain constant temperature during pasteurization. In order to maintain meat products in the frozen state during processing, conveyor totes often contain dry ice as a method of temperature control. However, little is known concerning the effect of deep-freezing temperatures (–76°C) on the radiation resistance of microorganisms. Thayer and Boyd (2001) found D_{10} for *E. coli* O157:H7 and *S. aureus*, suspended in ground beef and maintained at –76°C, to be 1.1 and 0.8 kGy, respectively. This is much higher that the D_{10} values described earlier for the same organisms. Sommers et al. (in press) found D_{10} for *Y. enterocolitica* suspended in raw ground pork of 0.18, 0.40, and 0.55 kGy at temperatures of 5, –20, and –76°C, respectively. These data indicate that dry ice packing of meats during irradiation can increase D_{10} values significantly and that processors must account for these differences when selecting a dose of ionizing radiation sufficient to achieve pasteurization while simultaneously maintaining product quality factors.

Modified Atmosphere Packaging and Irradiation

The shelf-life of fresh meat products can be extended in the presence of modified atmosphere, which can inhibit the growth of spoilage organisms and pathogens. A gas volume–product volume ratio ranging from 1:1 to 3:1 is recommended, with product requiring a longer shelf-life receiving the larger gas–product volume ratio (Farber and Dodds, 1995). The modified atmospheres typically used vary with the meat product (Farber and Dodds, 1995; Day, 1992; AFDO, 1990). Red meats may be packed in modified atmospheres containing oxygen for the purpose of maintaining the bright red color associated with myoglobin in its oxygenated form (oxymyoglobin). However, the modified atmospheres rich in oxygen commonly used for red meats, which can range from 60 to 80% in retail and bulk packs (Day, 1992), may not be appropriate for use with irradiation due to ionizing radiation–induced lipid oxidation in the presence of oxygen. Poultry is typically packed under atmospheres containing CO_2 and N_2 (Day, 1992). Regulations implemented by the U.S. Food and Drug Administration allow irradiation of meats only in oxygen-permeable packaging.

Studies on the modified atmosphere packaging (MAP) of minimally processed meats in combination with ionizing radiation have yielded inconsistent results. Much of this may be due to the natural biological diversity of microorganisms belonging to different species or to variation in experimental conditions between studies. It is clear that a more systematic study of the effects of MAP in combination with ionizing radiation under actual meat industry processing conditions is required (Lee et al., 1996). The ability of pathogens to proliferate on minimally processed meats during long-term refrigerated storage is as important as the D_{10} values obtained. Grant and Patterson (1991) observed that pork irradiated to 1.75 kGy and incubated for 9 days at 10°C had lower counts of *S. tyhphimurium*, *Y. enterocolitica*, and *L. monocytogenes* than in nonirradiated MAP (25% CO_2: 75% N_2) pork. Zhao et al. (1996) observed that the postirradiation growth of *Salmonella* spp. over 15 days refrigerated storage was CO_2-dependent. Thayer and Boyd (1999) found that *L. monocytogenes* suspended in turkey meat was more sensitive to ionizing radiation in 100% CO_2 as opposed to 100% N_2 and that the postirradiation growth of *L. monocytogenes* could be inhibited by CO_2 in a concentration-dependent manner.

The radiation resistance of specific spoilage organisms in the presence of MAP has also been examined. Hastings et al. (1986) found that *Lactobacillus* spp. were more sensitive to ionizing radiation in the presence of a 100% CO_2 atmosphere than in the presence of air, vacuum, or N_2. The D_{10} of a mixture of three *Lactobacillus* reference strains were 1.15, 0.87, 1.18, and 1.00 kGy when irradiated in the presence of N_2, CO_2, vacuum, and air, respectively. Patterson (1988) determined the radiation resistances of a number of food-borne pathogens and spoilage organisms in the presence of modified atmospheres. *Pseudomonas putida* suspended in minced chicken meat had D_{10} of 0.11 kGy in the presence of CO_2, 0.08 kGy in the presence of N_2, and 0.06 kGy under vacuum. As with aerobically packed or vacuum-packed meat products, Gram-positive spoilage bacteria are more resistant to ionizing radiation than the Gram-negative pseudomonads.

CONCLUSIONS

Radomyski et al. (1994) concluded that an ionizing radiation dose of 3.0 kGy should be sufficient to eliminate more than 99% of *E. coli* O157:H7, *Salmonella* spp., *L. monocytogenes*, *Y. enterocolitica*, *C. jejuni*, and parasites from refrigerated minimally processed meats. In contrast, based on results obtained by Thayer et al. (1998) with *L. monocytogenes*, a radiation dose of 5.5 kGy would be required to produce a 5-$\log_{10}$ reduction of common vegetative bacterial pathogens in frozen meats. These values are well within the 4.5- and 7.0-kGy limits approved for use by the U.S. FDA for refrigerated and frozen meat products; however, treatment of minimally processed meats with ionizing radiation does not replace good manufacturing and handling procedures. Roberts and Weese (1998) observed that irradiation of refrigerated high aerobic plate count ground beef patties did not result in effective shelf-life extension. Consideration should be given to product temperature, modified atmosphere, packing configuration, and the radiation-absorbed dose, as determined by routine dosimetry, to ensure a 5-$\log_{10}$ reduction (99.999%) in pathogen number.

Further research into methodologies for elimination of viruses and spore-forming bacteria from minimally processed meats is needed. Approximately one third of food-borne illnesses are due to viral, as opposed to bacterial, contamination of food products (Mead et al., 1999). Although illnesses due to spore forming bacteria are relatively low in the U.S., recent events merit a reexamination of methods for elimination and attenuation of spores in food matrices. The use of combined industrial processes has the potential to further reduce microbial populations over individual processing technologies (Lacroix and Ouattara, 2000). Other technologies include the use of antimicrobial packaging, ozonation, surface treatment with sanitizers or acids, ultra-high pressure or hydrodynamic pressure technologies, surface thermal applications, and pulsed electric fields.

ACKNOWLEDGMENTS

I would like to thank Drs. Connie Briggs, William Mackay, and Donald Thayer, and Glenn Boyd and Laren Melenski, for review of the chapter.

REFERENCES

Abu-Tarboush, H.M. et al., 1997, Sensory and microbial quality of chicken as affected by irradiation and postirradiation storage at 4°C, *J. Food Prot.*, 60:761–770.

AFDO, 1990, Retail guidelines, refrigerated foods in reduced oxygen packages, Association of Food and Drug Officials, Washington, D.C.

Allos, B.M. et al., 1998, *Campylobacter jejuni* strains from patients with Guillain–Barre, *Emerging Infect. Dis.*, 4:263–268.

Alper, T., 1966, The exceptionally small size of the scrapie agent, *Biochem. Biophys. Res. Commn.*, 22:278–284.

Anellis, A. and Koch, R., 1962, Comparative resistance of strains of *Clostridium botulinum* to gamma rays, *Appl. Microbiol.*, 10:326–330.

ASTM, 2000, ASTM Dosimetry Standards for Radiation Processing, 4th Int. Workshop on Dosimetry for Radiation Process., American Society for Testing and Materials, West-conshohocken, PA.

Berrang, M.E., Ladely, S.R., and Buhr, R.J., 2001, Presence and level of *Campylobacter, Coliforms, Escherichia coli*, and total aerobic bacteria recovered from broiler parts with and without skin, *J. Food Prot.*, 64:184–188.

Brake, R.J. et al., 1985, Destruction of *Trichenella spiralis* by low dose irradiation of pork, *J. Food Safety*, 7:127–143.

Bruns, M.W. and Maxcy, R.B., 1979, Effect of irradiation temperature and drying on survival of highly radiation resistant bacteria in complex menstrua, *J. Food Sci.*, 44:1743–1746.

Cain, R.F., Bubl, E.C., and Anderson, A.W., 1956, Effect of intermittent radiations and concomitant increase in temperature during radiation on the acceptability of ground beef, *Food Technol.*, 10:537–540.

Chang, H.D. and Tao, L.C., 1981, Correlation of enthalpies of food systems, *J. Food Sci.*, 46:1493.

Clavero, M.C. et al., 1994, Inactivation of *Escherichia coli* O157:H7, *Salmonellae*, and *Campylobacter jejuni* in raw ground beef by irradiation, *Appl. Environ. Microbiol.*, 60:2069–2075.

Collins, C.I., Murano, E.A., and Wesley, I.V., 1996, Survival of *Arcobacter butzleri* and *Campylobacter jejuni* after irradiation treatment in vacuum-packaged ground pork, *J. Food Prot.*, 59:1164–1166.

Cooper, C., 2001, Chicken: keeping eggs in all the baskets, *Meat Poultry*, September, 2001:39–42.

Davies, P., 1997, Food safety and its impact on domestic and export markets, *Swine Health Prod.*, 5:13–20.

Day, B.P.F., 1992, Guidelines for the good manufacturing and handling of modified atmosphere packed food products, Campden Food and Drink Research Association, Technical Manual No. 34, Chipping Campden, Gloucestershire, U.K.

Diehl, J.F., 1995, Safety of Irradiated Food, 2nd ed., New York: Marcel Dekker, Inc.

Dubey, J.P. et al., 1998, Effect of gamma irradiation on unsporulated and sporulated *Toxiplasma gondii* oocysts, *Int. J. Parasit.*, 28:369–375.

Duffy, E.A. et al., 2001, Extent of microbial contamination in United States pork retail products, *J. Food Prot.*, 64:172–178.

Ehioba, R.M. et al., 1987, Effect of low-dose (100 krad) gamma irradiation on the microflora of vacuum-packaged ground pork with and without added sodium phosphates, *J. Food Sci.*, 52:1477.

Ehioba, R.M. et al., 1988, Identification of microbial isolates from vacuum-packaged ground pork irradiated at 1 kGy, *J. Food Prot.*, 53:278.

Farber, J. and Dodds, K., 1995, in *Principles of Modified-Atmosphere and Sous-Vide Product Packaging*, Lancaster, PA: Technomic Publishing, Inc.

Federal Register, 1997, Irradiation in the production, processing, and handling of food, Vol. 62, No. 232, 64107–64121.

Glickman, B.W., Rietveld, K., and Aaron, C.S., 1980, γ-Ray induced mutational spectrum in the *lacI* gene of *Escherichia coli, Mutat. Res.*, 69:1–12.

Grant, I.R. and Patterson, M.F., 1991, Effect of irradiation and modified atmosphere packaging on the microbiological safety of minced pork stored under temperature abuse conditions, *Int. J. Food Sci. Technol.*, 26:521–533.

Harewood, P., Rippley, S., and Montesalvo, M., 1994, Effect of gamma-irradiation on shelf life and bacterial and viral loads in hard-shelled clams, *Appl. Environ. Microbiol.*, 60:2666–2670.

Hastings, J.W., Holzapfel, W.H., and Niemand, J.G., 1986, Radiation resistance of *Lactobacilli* isolated from radurized meat relative to growth and environment, *Appl. Environ. Microbiol.*, 52:898–901.

Heidelbaugh, N.D. and Giron, D.J., 1969, Effect of processing on recovery of polio virus from inoculated foods, *J. Food Sci.*, 34:239–241.

Ingram, M., 1975, Microbiology of foods pasteurized by ionizing radiation, Technical Report Series, Report IFIP-R33, International Project in the Field of Food Irradiation, Institut fur Strahlentechnologie, Karlsruke, Federal Republic of Germany.

Kamat, A.S., Nerkar, D.P., and Nair, P.M., 1989, *Bacillus cereus* in some Indian foods: incidence and antibiotic, heat, and radiation resistance, *J. Food Safety*, 10:31–41.

Kamat, A.S. et al., 1997, Control of *Yersinia enterocolitica* in raw pork and pork products by γ irradiation, *Int. J. Food Microbiol.*, 36:69–76.

Khare, S. et al., 1982, Radiation effects on membranes, *Radiat. Res.*, 90:233–243.

Lacroix, M. and Outtara, B., 2000, Combined industrial processes with irradiation to assure innocuity and preservation of food products — a review, *Food Res. Int.*, 33:719–724.

Lambert, J.B. and Maxcy, R.B., 1984, Effect of gamma radiation on *Campylobacter jejeuni*, *J. Food Sci.,* 49:665–667.

Lasta, J. et al., 1992, Combined treatments of heat, irradiation, and pH effects on infectivity of foot-and-mouth disease virus in bovine tissues, *J. Food Sci.*, 57:36–39.

Lebepe, S. et al., 1990, Changes in microflora and other characteristics of vacuum-packaged pork loins irradiated at 3.0 kGy, *J. Food Sci.*, 55:918.

Ley, F.J., 1983, New interest in the use of irradiation in the food industry, in *Food Microbiology: Advances and Prospects*, Roberts, T.A. and Skinner, F. A., Eds., London: The Society for Applied Bacteriology Symposium Series No. 11, 113–129.

Lopez-Gonzalez, V. et al., 1999, Influence of various commercial packaging conditions on survival of *Escherichia coli* O157:H7 to irradiation by electron beam vs. gamma rays, *J. Food Prot.*, 62:10–15.

Louharanou, P., 2001, Review of status of global acceptance of food irradiation, Report of FAO/IAEA Research Coordinated Meeting, St. Hyacinthe, Canada.

Ma, K. and Maxcy, R.B., 1981, Factors influencing radiation resistance of vegetative bacteria and spores associated with radappertization of meat, *J. Food Sci.*, 46:612–616.

Mead, G., Hudson, W., and Ariffin, R., 1990, Survival and growth of *Listeria monocytogenes* on irradiated poultry carcasses, *Lancet*, 335:1036.

Mead, P.S. et al., 1999, Food-related illness and death in the United States, *Emerging Infect. Dis.*, 5:607–625.

Miettenin, M.K. et al., 2001, Prevalence of *Listeria monocytogenes* in broilers at the abattoir, processing plant, and retail level, *J. Food Prot.*, 64:2001.

Miles, C.A. et al., 1997, Estimating the initial freezing point of foods from composition data, *Int. J. Food Sci. Technol.*, 32:389–400.

Monk, J.D. et al., 1995, Irradiation inactivation of *Listeria monocytogenes* and *Staphylococcus aureus* in low and high fat frozen and refrigerated ground beef, *J. Food. Prot.*, 51:969–974.

Naidu, M., Chander, R., and Nair, P.M., 1998, Effect of gamma irradiation on chemical and biological properties of lipopolysaccharide from *Salmonella typhimurium*, *Indian J. of Exp. Biol.*, 36:588–592.

Nunes, K., 2001a, Beef: as the cattle cycle turns, *Meat Poultry*, September 2001, 26–28.

Nunes, K., 2001b, Pork: balancing supply, demand and slaughter capacity, *Meat Poultry*, September 2001, 31–36.

Patterson, M.F., 1988, Sensitivity of bacteria to irradiation on poultry meat under various atmospheres, *Lett. Appl. Microbiol.*, 7:55.

Patterson, M., 1989, Sensitivity of *Listeria monocytogenes* to irradiation on poultry meat and in phosphate-buffered saline, *Lett. Appl. Microbiol.*, 8:181–184.

Patterson, M.F., 1995, Sensitivity of *Campylobacter* spp. to irradiation in poultry meat, *Lett. Appl. Mircobiol.*, 20:338–340.

Pirtle, E.C., Proescholdt, T.A., and Beran, G.W., 1997, Trial of heat inactivation of selected viruses following irradiation, *J. Food Prot.*, 60:426–429.

Proctor, B.E., van der Graaff, R.J., and Fram, H., 1943, Reports on quartermaster contract projects, July 1942, by the Food Technology Laboratories, MIT, in *Historical Development of Food Irradiation*, Goldblith, S.A., Ed., *Proc. Int. Symp. Food Irradat.*, Vienna, 1966, 217.

Prusiner, S.B. et al., 1982, Further purification and characterization of scrapie prions, *Biochemistry*, 21:6942–6950.

Pryor, W.A., 1977, in *Free Radicals in Biology*, Vol. III, New York: Academic Press, 10–24.

Radomyski, T. et al., 1994, Elimination of pathogens of significance in food by low-dose irradiation: a review, *J. Food Prot.*, 57:73–86.

Raha, M. and Hutchison, F., 1991, Deletions induced by gamma rays in the genome of *Escherichia coli*, *J. Mol. Biol.*, 220:193–198.

Roberts, T.A. and Ingram, M., 1965, Radiation resistance of spores of *Clostridium* species in aqueous solution, *J. Food Sci.*, 30:879–885.

Roberts, W.T. and Weese, J.O., 1998, Shelf-life of ground beef patties treated by gamma radiation, *J. Food Prot.*, 61:1387–1389.

Sargentini, N.J. and Smith, K.C., 1994, DNA sequence analysis of γ-radiation (anoxic)–induced and spontaneous *lacI^d^* mutations in *Escherichia coli* K-12, *Mutat. Res.*, 309:147–163.

Schrueder, B.E.C., 1994, BSE agent hypothesis, *Livestock Prod. Sci.*, 38:23–33.

Shallow, S. et al., 1999, Preliminary FoodNet data on the incidence of foodborne illnesses — selected sites, United States, *Morbidity Mortality Wkly. Rep.*, 49:201–205.

Shallow, S. et al., 1997, Foodborne diseases active surveillance network, 1996, *Morbidity Mortality Wkly. Rep.*, 46:258–261.

Shamsuzzaman, K. et al., 1992, Microbiological and other characteristics of chicken breast meat following electron-beam and *sous-vide* treatments, *J. Food Prot.*, 55:528–533.

Shamsuzzaman, K. and Lucht, L., 1993, Resistance of *Clostridium sporogenes* spores to radiation and heat in various nonaqeous suspension media, *J. Food Prot.*, 56:10–12.

Smith, J.L., 1994, Arthritis and food-borne bacteria, *J. Food Prot.*, 57:935–941.

Smith, L.T., 1996, Role of osmolytes in adaptation of osmotically stressed and chill stressed *Listeria monocytogenes* grown in liquid media and on processed meat surfaces, *Appl. Environ. Microbiol.*, 62:3088–3093.

Sommers, C.H. and Bhaduri, S., 2001, Loss of crystal violet binding activity in stationary phase *Yersinia entrocolitica* following gamma irradiation, *Food Microbiol.*, 18:367–374.

Sommers, C.H. et al., Effect of temperature on the radiation resistance of virulent *Yersinia enterocolitica, Meat Sci.,* 61:323–328.

Sommers, C.H. and Novak, J.S., 2002, Radiation resistance of virulence plasmid-containing and virulence plasmid-less *Yersinia enterocolitica, J. Food Prot.,* 65:556–559.

Succar, J. and Hayakawa, K., 1990, A method to determine initial freezing point of foods, *J. Food Sci.*, 55:1711–1713.

Sullivan, R. et al., 1971, Inactivation of thirty viruses by gamma radiation, *Appl. Microbiol.*, 22:61–65.

Tarkowski, J.A. et al., 1984, Low dose gamma irradiation of raw meat. I. Bacteriological and sensory quality effects in artificially contaminated samples, *Int. J. Food Microbiol.*, 1:13–23.

Taub, I.A., Halliday, J.W., and Sevilla, M.D., 1979, Chemical reactions in proteins irradiated at subfreezing temperatures, *Adv. Chem. Serol.*, 180:109–140, 167.

Tauxe, R., 1999, Foodborne disease: public health challenges and the need for new prevention technologies, Meeting Proceedings: Food Irradiation 99, May 12–14, 1999, Washington, D.C.

Thayer, D. et al., 1990, Radiation resistance of *Salmonella, J. Ind. Microbiol.*, 5:373–390.

Thayer, D.W. and Boyd, G., 1990, Survival of *Salmonella typhimurium* ATCC 14028 on the surface of chicken legs or in mechanically deboned chicken meat gamma-irradiated in air or vacuum at temperatures –20 to +20°C, *Poultry Sci.,* 70:1026–1033.

Thayer, D.W. and Boyd, G., 1992, Gamma ray processing to destroy *Staphylococcus aureus* in mechanically deboned chicken meat, *J. Food Sci.*, 60:237–240.

Thayer, D.W. and Boyd, G., 1993, Elimination of *Escherichia coli* O157:H7 in meats by gamma irradiation, *Appl. Environ. Microbiol.*, 59:1030–1034.

Thayer, D.W. and Boyd, G., 1994, Control of enterotoxic *Bacillus cereus* on poultry or red meats and in beef gravy by gamma irradiation, *J. Food Prot.*, 57:758–764.

Thayer, D.W. et al., 1995, Variations in radiation sensitivity of food-borne pathogens associated with the suspending meat, *J. Food Sci.*, 60:63–67.

Thayer, D.W. and Boyd, G., 1996, Inactivation of *Shewenella putrafaciens* by gamma irradiation of red meat and poultry, *J. Food Safety*, 16:151–160.

Thayer, D.W. and Boyd, G., 1999, Irradiation and modified atmosphere packaging for the control of *Listeria monocytogenes* on turkey meat, *J. Food Prot.*, 62:1136–1142.

Thayer, D.W. et al., 1998, Fate of gamma-irradiated *Listeria monocytogenes* during refrigerated storage on raw or cooked turkey breast meat, *J. Food Prot.,* 61:979–987.

Thayer, D.W., 2000, Sources of variation and uncertainty in the estimation of radiation D-10 values for foodborne pathogenes, *ORACBA News*, 5:10–14.

Thayer, D. and Boyd, G., 2000, Reduction of normal flora by irradiation and its effect on the ability of *Listeria monocytogenes* to multiply on ground turkey stored at 7°C when packaged under a modified atmosphere, *J. Food Prot.,* 63:1702–1706.

Thayer, D.W. and Boyd, G., 2001, Effect of irradiation temperature on inactivation of *E. coli* O157:H7 and *Staphylococcus aureus, J. Food Prot.*, 64:1624–1626.

Tolgay, Z. et al., 1972, Investigations on invasion capacity and destruction of *Cysticercus bovis* in beef treated by ionizing radiation (gamma rays from Co-60), *Tuerk. Veteriner. Hekimleri. Dernergi. Dergisi.*, 42:13–29, 38.

Urbain, W.M., 1978, Irradiation of meats and poultry, *Food Irradiat. Inf.*, 8:14–30.

Verster, A., Plessis, T., and van den Heever, L.W., 1977, The eradication of tapeworms in pork and beef carcasses by irradiation, in *Irradiation Processing*, London: Elsevier, 769–773.

Wallace, D.J. et al., 1997, Incidence of foodborne illnesses reported by the Foodborne Diseases Active Surveillance Network (FoodNet), *J. Food Prot.*, 63:807–809.

Ward, J.F., 1991, Mechanisms of radiation action on DNA in model systems — their relevance to cellular DNA, in *The Early Effects of Radiation on DNA*, Fielden, E.M. and O'Neill, P., Eds., New York: Springer-Verlag, 1–16.

Watkins, D.K., 1980, Stimulation of DNA synthesis in bacterial DNA-membrane complexes after low doses of ionizing radiation, *Int. J. Radiat. Biol. Relat. Chem. Med.*, 38:247–256.

WHO, 1994, in *Safety and Nutritional Adequacy of Irradiated Food,* World Health Organization, Geneva.

Wijker, C.A. et al., 1996, γ-radiation-induced mutation spectrum in the episomal *lacI* gene of *Escherichia coli* under oxic conditions, *Mutat. Res.*, 349, 229–239.

Zhao, Y. et al., 1996, Bacteriological, physiochemical, and sensory quality of fresh pork chops with low-dose irradiation and modified-atmosphere packaging, *J. Food Prot.,* 59:493–501.

15 Biological Control of Minimally Processed Fruits and Vegetables

Britta Leverentz, Wojciech Janisiewicz, and William S. Conway

CONTENTS

INTRODUCTION

Today's consumers prefer lightly processed products without chemical preservatives that are therefore more natural and safe (Stiles, 1996; Gould, 1996; Leistner and Gorris, 1995). To meet this demand, the number of convenience food items such as minimally processed, ready-to-use (RTU) fruits and vegetables has been increasing on the shelves of supermarkets (Reyes, 1996). Minimally processed food includes food destined for the fresh-cut market but also food that is going to be processed further, like peeled potatoes (Doan and Davidson, 2000) or produce that will be mixed in various types of salads (Thomas and O'Beirne, 2000; Huxoll and Bolin, 1989). To minimally process fruits and vegetables means that they may be sorted, graded, cleaned, peeled, trimmed, cut, sliced, shredded, washed, or disinfected (Doan and Davidson, 2000; Reyes, 1996; Brackett, 1994). They may also be mixed and

1-58716-041-2/03/$0.00+$1.50

combined with cooked vegetables, meats, pasta, and salad dressings (Francis et al., 1999; Brackett, 1987). Minimally processed fruits and vegetables are usually packaged and preserved by refrigeration (Francis et al., 1999; Kelly et al., 1998).

In general, less contamination is reported on fruits than on vegetables (Beuchat, 1996). This may be due partly to the lower pH of most fruits in comparison to vegetables (Brackett, 1994). However, recent outbreaks suggest that fresh-cut fruits, melons in particular, can be an important reservoir for *Salmonella* (Beuchat, 1996; Asplund and Nurmi, 1991; Gayler et al., 1955) (news report May 25, 2000; Sacramento, CA (SafetyAlerts) — California State Health Director Diana M. Bont, R.N., Dr. P.H.).

Biopreservation is a method to extend the storage life and safety of foods through the use of natural microflora or their bactericidal products (Stiles, 1996). This is accomplished by 1) the application of bacterial strains that grow rapidly or produce antagonistic substances, 2) adding these purified substances, 3) adding fermentation liquids from antagonistic organisms, or 4) adding mesophilic lactic acid bacteria (LAB) as a protectant in instances of temperature abuse (Torriani et al., 1997; Stiles, 1996).

EFFECT OF ENVIRONMENTAL PARAMETERS ON THE BACKGROUND MICROFLORA

INDIGENOUS MICROFLORA

The indigenous microflora on produce consists of large and diverse populations of microorganisms that may reach 10^2 to 10^9 CFU/g (Breidt and Fleming, 1997). Of these, 80 to 90% are Gram-negative rods like pseudomonads, *Enterobacter*, or *Erwinia* species (Francis et al., 1999; Bennik et al., 1998; Breidt and Fleming, 1997; Nguyen-The and Carlin, 1994). Pseudomonads may account for more than half of the bacterial population, while soft rotting species like *Erwinia* are seldom found on unprocessed vegetables (Nguyen-The and Carlin, 1994). Fluorescent pseudomonads and LAB such as *Lactobacillus* and *Leuconostoc* are part of the normal vegetable flora, while coliforms, yeasts, and molds can be present or may be introduced during processing (Nguyen-The and Carlin, 1994; Lund, 1992). Antilisterial lactococci are also found on fresh vegetables as part of the normal microflora (Cai et al., 1997).

The presence of epiphytic microorganisms on the intact produce determined by preharvest practices may influence resistance to postharvest diseases (Baldwin and Baker, 1992) but may also influence the microbiology of minimally processed produce (Nguyen-The and Carlin, 1994; Brackett, 1994). Microbial species that prevail on minimally processed produce are commonly found on plants in the field or after harvest (Nguyen-The and Carlin, 1994). Therefore, it is possible to reduce contamination by food-borne pathogens in storage before processing. For example, colonization of apple wounds by food-borne human pathogens such as *Escherichia coli* O157:H7 may be prevented by the application of biocontrol agents such as *Pseudomonas syringae* (Janisiewicz et al., 1999b). However, the more serious problem may be the potential for contamination of fresh-cut surfaces during processing (Francis et al., 1999; Breidt and Fleming, 1997).

Biocontrol methods applied to minimally processed produce are based on understanding the unique characteristics of the microorganisms present on that produce. The growth of microorganisms on produce is affected by various factors, including chemical composition, type of produce, presence of naturally occurring and added antimicrobial compounds and biocontrol agents, composition of the natural microflora, maturity of the produce, respiratory behavior (climacteric or nonclimacteric), environmental conditions (temperature, relative humidity, pH, redox potential, atmospheric composition) and handling during production, and processing and transport of the produce (Heard, 1999; Leistner, 1992; Brackett, 1987). The ability to manipulate and control these factors will determine the success of microflora in maintaining the quality and safety of the produce.

A pathogen present on minimally processed produce must successfully overcome several obstacles or hurdles before it can grow (Breidt and Fleming, 1997). The hurdle concept describes the use of multiple preservative factors, termed "hurdles," that are strategically combined to prevent the growth of pathogens and ensure the quality and microbial stability of food (Leistner, 1992). The major hurdles during processing are sanitation, modified atmosphere packaging (MAP), refrigeration, and competitive interactions of the bacteria. Additional hurdles may be needed to prevent the growth of food-borne pathogens and spoilage organisms. The use of beneficial microorganisms as protectants can be a very effective hurdle, but it only complements and does not replace good manufacturing practices (Breidt and Fleming, 1997; Holzapfel, 1995). If indigenous microflora is utilized as a hurdle, parameters influencing it must be considered.

Refrigeration

Human food-borne pathogens can survive on minimally processed fruits and vegetables and grow rapidly under ideal conditions (Conway et al., 2000; Janisiewicz et al., 1999a; Piagentini et al., 1997; Nguyen-The and Carlin, 1994). Because minimally processed produce is usually preserved by refrigeration, those organisms able to grow at low temperatures are of particular concern (Bennik et al., 1995; Nguyen-The and Carlin, 1994). They include *Listeria monocytogenes, Aeromonas* sp., and *Yersinia enterocolita* (Reyes, 1996; Lund, 1992). However, other food-borne pathogens cannot be excluded due to the abusive temperatures that may be reached during processing, marketing, and transport of the product to and from the retail outlet (Nguyen-The and Carlin, 1994; Bauman, 1991).

Beneficial LAB are a group of bacteria that thrive under low-temperature conditions, as well (Breidt and Fleming, 1998; Stiles, 1994); this may be the reason that long-term storage does not necessarily decrease product safety (Thomas and O'Beirne, 2000). Their growth at abusive temperatures may also successfully inhibit pathogenic growth (Gombas, 1989). Populations of beneficial LAB present on produce after prolonged storage at 3°C may proliferate at temperatures around 12°C, whereas the growth of *Listeria* and aerobic mesophilic (indigenous) populations is reduced (Thomas and O'Beirne, 2000).

Fluorescent pseudomonads and erwinias are considered spoilage organisms and can grow at refrigeration temperatures (Nguyen-The and Carlin, 1994; Brackett,

1994; Lund, 1992). These psychrotrophic organisms may have a competitive advantage over most pathogens, causing food spoilage before food-borne pathogens reach dangerous levels (Brackett, 1994).

Disinfection

Contamination of the produce with food-borne pathogens by soil, machines, or humans is possible, and the more a product is processed, the more the final microbial population will reflect that of its processing environment (Francis et al., 1999). Moisture and exudates on cut surfaces provide excellent growth conditions (Francis et al., 1999) and can lead to a six- to sevenfold increase in microbial populations. Therefore, washing seems to be the most important step after processing (Huxoll and Bolin, 1989).

Even though washing can be accomplished with water alone (Huxoll and Bolin, 1989), often either 100 ppm chlorine or 1% citric or ascorbic acid is added (Francis et al., 1999). However, the growth of food-borne pathogens may be greater when the background microflora has been reduced by chemical disinfection (Francis et al., 1999; Bennik et al., 1996; Carlin et al., 1996). Washing potatoes with chlorine is not feasible since Gram-negative microorganisms, such as *Pseudomonas fluorescens* I and *Vibrio fluvialis,* grow even better after the treatment (Gunez et al., 1997). In general, disinfection with organic acids seems to be more successful than disinfection with chlorine (Gunez et al., 1997). However, chemical disinfection does not guarantee that a product will be free of food-borne pathogens. The total elimination of *L. monocytogenes* from the surface of produce, for example, may be limited or unpredictable (Nguyen-The and Carlin, 1994).

The microbial background population present on the produce plays a role in suppressing populations of food-borne pathogens that may be added through contamination. For example, after chemical decontamination of the produce, the growth of food-borne pathogens that may arrive on the surface, together with those still present, may be favored in comparison to pathogens on untreated produce (Francis et al., 1999; Breidt and Fleming, 1998; Carlin et al., 1996). It is important to avoid recontamination of the produce after disinfection (Bennik et al., 1996), although that may be difficult to achieve. The risk of eliminating the inhibitory potential of the background flora that may prevent food-borne pathogens from spreading must be weighed against the risk of food-borne pathogens possibly remaining on the produce by avoiding disinfection. However, this risk may possibly be reduced when disinfection is followed by the application of a mixture of biocontrol agents.

Modified Atmosphere Packaging

MAP increases the duration at which fresh-cut produce can be stored, but it may not have a significant effect on the number and growth of food-borne pathogens (Bennik et al., 1996; Lund, 1992; Brackett, 1988). An increase in CO_2 reduces the maximum specific growth rate of organisms but not the maximum population density reached (Kim et al., 2000). However, MAP may cause shifts in the natural background microflora of beneficial organisms that can then lead to an actual increase of pathogen populations such as *L. monocytogenes* (Francis and O'Beirne, 1998a; Brackett, 1987).

The changes in the predominance of microbial species caused by MAP are specific to the produce (Heard, 1999). Increased CO_2 or reduced O_2 may favor LAB in minimally processed vegetables (Nguyen-The and Carlin, 1994); higher moisture in packaged fresh-cut produce also is conducive to microbial growth. Elevated CO_2 and humidity levels in packages generally increase populations of aerobic bacteria, yeasts, and molds (Brackett, 1987, 1988, 1990; Deak et al., 1987; Deak, 1984) and shift bacterial populations from Gram-negative to Gram-positive (Sinell, 1980). Depletion of oxygen in airtight-packaged minimally processed fresh fruits, vegetables, or mushrooms also favors toxin production by *Clostridium botulinum* (Nguyen-The and Carlin, 1994; Lund, 1992).

The growth of food-borne pathogens often cannot be detected by product odor or appearance (Piagentini et al., 1997), and total microbial counts at the end of the storage period may be unrelated to sample quality (Nguyen-The and Carlin, 1994). This is the reason that extended shelf-life may unfavorably influence the safety index (Breidt and Fleming, 1997; Hintlian and Hotchkiss, 1987). This index is defined as the ratio of spoilage bacteria to pathogenic bacteria in foods.

BIOCONTROL

Research has been done on adding bacterial strains as a biocontrol to minimally processed produce (Bennik, 1999; Torriani et al., 1997; Vescovo et al., 1995). Many such strains used on vegetables are selected bacteriocin-producing LAB or mixtures of these organisms (Heard, 1999). *Lactobacillus* strains inhibit enterococci and coliforms (Heard, 1999). Also, a complex bacterial population extracted from endive leaves the growth of *L. monocytogenes in vitro* completely inhibited (Carlin et al., 1996). On processed lettuce, competition with *Listeria innocua* is mostly based on *Enterobacter* spp. and LAB and not on pseudomonads (Francis and O'Beirne, 1998b). Further possibilities for biocontrol are the application of bacteriophages and of yeasts.

Lactic Acid Bacteria

Lactic acid bacteria (LAB) are a very important group in many foods, even though they have been used mostly in food fermentation, where the changes that these bacteria induce in the product preclude the growth of pathogens (Stiles, 1996; Fleming, 1990; Gombas, 1989). LAB are safe to consume, dominate the natural microflora of many foods (Stiles, 1996), and can dramatically reduce populations of *Enterobacteriaceae* present on vegetables (Vescovo et al., 1995). They are not pathogenic, are not putrefactive and, therefore, are acceptable to the human senses (Saleh and Ordal, 1955). Lactobacilli also are generally recognized as safe (GRAS) (Breidt and Fleming, 1998; Stiles, 1996; Holzapfel, 1995).

Saleh and Ordal were among the first to use LAB as biocontrol agents and demonstrated their effectiveness in controlling *C. botulinum* on frozen chicken (Saleh and Ordal, 1955). Later, *Lactobacillus* strains were used to control *L. monocytogenes* in pickle products (Romick, 1994), or they were applied to the naturally occurring microflora of ready-to-use vegetables to control coliforms and enterococci (Vescovo et al., 1995). The same group reported that LAB cultures or a strain of *Lactobacillus casei* and its culture permeate controlled *Aeromonas hydrophilia, L. monocytogenes, Salmonella typhimurium*, and *Staphylococcus aureus* on vegetables (Torriani et al., 1997; Vescovo et al., 1996). A similar inhibition was noted when the culture permeate of *L. casei* was tested against several food-borne pathogens (Torriani et al., 1997).

In experiments with cucumber juice and other cucumber products, the addition of *L. lactis*, at an initial concentration that was two log units lower than that of *L. monocytogenes,* still resulted in the suppression of the pathogen (Breidt and Fleming, 1997). However, when *L. monocytogenes* was studied in competition with *Lactococcus lactis*, *Leuconostac mesenteroides*, or *Lactobacillus plantarum*, the pathogen was primarily inhibited by the naturally occurring microflora rather than the added LAB strains (Breidt and Fleming, 1997). This shows that laboratory strains of LAB may be less suited for competition than indigenous LAB populations on produce (Breidt and Fleming, 1997). If laboratory strains of bacteriocinogenic LAB are used, they may be less stable on certain foods; relatively high population levels ($>10^7$ LAB/g) may be necessary to use them successfully as protectants against pathogens (Abee et al., 1995; Gombas, 1989).

A good source for LAB is sprouted seeds (Kelly et al., 1998). Cai et al. (1997) isolated antilisterial *Lactococcus* and *Enterococcus* isolates from bean sprouts. The optimum strategy to develop effective biocontrol strains may be the isolation of these strains from the specific products under the conditions (refrigerated) in which the biocontrol will be used. Naturally, these organisms would be most adapted to the growth on that product (Breidt and Fleming, 1997; Vescovo et al., 1995).

Mechanism of LAB Antagonism

Besides competition for space and nutrients, LAB may suppress food-borne pathogens by two major mechanisms of biocontrol. First, many LAB produce bacteriocins such as nisin, lacticin, pediocin, sakacin, leucocin, leuconocin, helveticin, caseicin, plantaricin (Holzapfel, 1995; Vandenbergh, 1993), enterocin (Abee et al., 1995), and mundticin (Bennik, 1999) that are active against other Gram-positive bacteria including *L. monocytogenes, Bacillus cereus, S. aureus*, and *C. botulinum* (Holzapfel, 1995). Some of the bacteriocins produced by LAB are broad spectrum, such as nisin, lacticin 481, pediocin AcH, sakacin A, and leucocin UAL 187; some are medium spectrum, such as leuconocin S and pediocin SJ–1; and others are narrow spectrum, like helveticin J and caseicin 80 (Holzapfel, 1995). Most are active against *L. monocytogenes* and *S. aureus,* and a few are active against *B. cereus*, *C. botulinum*, and *Clostridium perfringens*. Nisin is more effective at low pH due to its solubility below pH 5 (Stiles, 1994; Vandenbergh, 1993). This characteristic makes it ideal for use on fruits, which are in that pH range. Pediocin is useful over a variety of temperature and pH ranges when the initial inoculum of the food-borne pathogen is low (Vandenbergh, 1993). Mundticin, a bacteriocin produced by *Enterococcus mundtii* on mungbean sprouts, may also have potential, even though the inhibition of *L. monocytogenes* occurred on sterile vegetable medium and not on fresh produce (Bennik, 1999).

The control of Gram-negative pathogens such as *Salmonella* species and *E. coli* may be accomplished by combining nisin with EDTA or other iron chelators (Stevens et al., 1991, 1992). Of the many bacteriocins produced by LAB, nisin is the only compound approved as a preservative in 50 countries and the only purified bacteriocin that is commercially available (Thomas et al., 2000; Gould, 1996). In the U.S., it currently has GRAS (generally recognized as safe) status as listed in the code of federal regulations (FDA, 1999). It is used on pasteurized and processed cheese spreads (plain and with added vegetables, fruit, meats); sauces and nonstandard salad dressings; pasteurized, chilled soups; and, with permission from the USDA, liquid egg products (Danisco Cultor, 2001). A disadvantage of using nisin alone is the potential for developing resistance or growth conditions that are favorable for Gram-negative bacteria by eliminating indigenous Gram-positive bacteria (Stiles, 1996). However, the actual nisin-producing bacterial strains are also used as biocontrol agents to control Gram-positive bacteria, including spore formers (Abee et al., 1995; Vescovo et al., 1995; Stiles, 1994). A mixture of strains could provide a plethora of different compounds (Holzapfel, 1995).

The second major biocontrol mechanism is the acidification of the environment by LAB that counteracts spoilage organisms such as pseudomonads and *Enterobacteriacea* (Vescovo et al., 1995; Vandenbergh, 1993) and prevents the growth of food-borne pathogens (Breidt and Fleming, 1997; Gombas, 1989). The reduction in pH, due to organic acids such as lactic acid, results in the inhibition or destruction of bacteria such as clostridia, pseudomonads, *Salmonella, Listeria*, *Staphylococcus*, *A. hydrophila,* and *B. cereus*. Also, undissociated acids produced by LAB can reduce the pH inside the cell by diffusion, which inhibits *E. coli* at a pH of 5.1 rather than a pH of 4.5 (Torriani et al., 1997; Holzapfel, 1995; Vandenbergh, 1993; Beuchat, 1992).

In addition to these two major mechanisms, LAB also produce hydrogen peroxide, which can inhibit more sensitive spoilage organisms and pathogens such as (fluorescent) pseudomonads and *S. aureus* (Holzapfel, 1995; Vandenbergh, 1993).

Lytic Bacteriophages

Lytic bacteriophages (phages) may provide an attractive alternative for decontaminating fresh-cut fruits and vegetables that may contain various bacterial pathogens. Phages are natural, safe, and ubiquitous in the environment. For example, they are naturally present at concentrations of 2.5×10^8/ml in uncontaminated freshwater (Beuchat, 1996). In addition, it is believed that phages will not compromise food quality (Alisky et al., 1998). Unlike chemical sanitizers, phages are an environmentally safe biocontrol alternative. They are highly specific in that they will lyse only selected pathogenic bacteria and have no effect on the desirable microflora present on the produce. This may also reduce the likelihood of recolonization of the produce by food-borne pathogens.

Phages were used therapeutically in humans in the 1930s and 1940s, and although their use in the United States and Western Europe was curtailed after the use of antibiotics became widespread, phages are still used in the former Soviet Union and Eastern Europe (Alisky et al., 1998). The concern of bacterial resistance developing may be addressed by using a mixture of different phages. If the bacteria develop resistance to one phage, the probability of escaping the other phages would be very small. A mixture has the added advantage that a specific host range can be targeted by using phages with a narrow host range or phages with both narrow and broad host ranges that may complement each other.

There are reports on phages isolated for treatment of soft rot erwinias (Eayre et al., 1995) and plant pathogenic bacteria on fruit trees (Zaccardelli et al., 1992; Mesquita et al., 1983). Phages have been applied against a fish pathogen (Park et al., 2000) and against the growth of spoilage pseudomonads on beef (Greer, 1986). Recently, phages have been successfully applied as a biocontrol of *Salmonella* on fresh-cut fruits (Leverentz et al., 2001), and work is in progress to evaluate the effectiveness of phage treatment on vegetables.

The efficacy of phage treatment seems to be favored by a pH close to neutral. For example, in the therapeutic use of phages, neutralizing the pH of the stomach acid was necessary in order for the phage treatment to remain effective (Alisky et al., 1998). On fruit with an acidic pH around 4, phages will die off rapidly within 24 to 48 h (Leverentz et al., 2001). However, on fruit with a higher pH of 5.5 to 6.0, phages may be able to suppress further increases in the targeted bacterial populations over an extended time period. *Salmonella* populations on honeydew can be reduced by phage treatment by up to an unprecedented 3.5 log units (Leverentz et al., 2001).

Phage treatment may be a useful method for reducing populations of food-borne pathogens on fresh-cut produce. The effectiveness may be increased by combining the treatment with conventional methods such as washing the fresh-cut produce with water followed by storage at refrigeration temperatures. Phage treatment has the potential for use as an effective sanitizer and possibly as a short-term protectant to prevent the buildup of populations of food-borne pathogens like *Salmonella* to hazardous levels on minimally processed produce. Because the treatment is so specific, it can easily be combined with other biocontrol agents. At present an experimental use permit application is pending for use of phages in poultry plants and on nonfood-contact surfaces (personal communication, A. Sulakvelidze, 2001).

Yeasts

The epidermis of fruits and vegetables is an effective physical barrier to infection. However, wounds in the epidermis expose the fleshy inner tissues to decay by fungi. This situation closely resembles that of fresh-cut produce in which the cut surface must be protected against colonization by food-borne human pathogens. In developing protection of fresh-cut produce against food-borne pathogens, much can be learned from the successful use of yeast and bacterial antagonists in controlling wound-invading postharvest pathogens of fruits (Droby et al., 1998; Janisiewicz, 1998; Stack, 1998; Janisiewicz and Bors, 1995; Wilson and Wisniewski, 1989).

Yeasts are common on fresh produce, with populations ranging from 10^2 to 10^6 cells/g (Deak and Beuchat, 1996). They often occur simultaneously with LAB in natural food habitats because they have many common ecological determinants (Deak and Beuchat, 1996). The yeast genera commonly isolated from cabbage, tomatoes, bell pepper, and corn are nonfermenting *Cryptococcus* and *Rhodotorula* and fermenting *Candida* and *Kloeckera* (Golden et al., 1987; Deak et al., 1987; Geeson, 1979). The yeast genera commonly isolated from apple are *Rhodotorula, Sporobolomyces, Cryptococcus, Metschnikowia, Kloeckera, Hanseniaspora, Torulopsis, Aureobasidium,* and *Debaryomyces* (Davenport, 1976). The microflora on produce may change significantly, especially after contamination of damaged tissue during harvest, processing, and storage. Populations of microorganisms generally increase greatly during storage, but this increase has little relationship to the quality and safety of produce (Brackett and Splittstoesser, 1992).

Many of the yeast genera occurring on apple have been shown to have biocontrol activity against postharvest decay pathogens of fruits, namely *Penicillium* spp. and *Botrytis cinerea* (Ippolito et al., 2000; Chand-Goyal and Spotts, 1997; Filonow et al., 1996; Janisiewicz et al., 1994; Jijakli and Lepoivre, 1993; Beuchat, 1992; Roberts, 1990; Janisiewicz, 1987). They prevent decay development by preemptive, rapid colonization of fruit wounds, the main site of pathogen entry. Although preemptive exclusion is probably the most important mechanism of biocontrol, other mechanisms, such as induced resistance or production of lytic enzymes and siderophores, may also be involved (Clavente et al., 1999; Filonow, 1998; Grevesse et al., 1998; Filonow et al., 1996; Jijakli and Lepoivre, 1993; Wisniewski et al., 1991; Droby et al., 1989). However, reduction of fungal pathogens may be more important in connection with produce destined to be further processed as juice than with that sold as fresh-cut. Some fungal pathogens have the ability to increase the pH of the more acidic fruit tissue, which increases the possibility of the growth of food-borne pathogens (Conway et al., 2000; Wells and Butterfield, 1997).

Some yeasts may produce antibacterial factors against both Gram-positive and Gram-negative bacteria (Deak and Beuchat, 1996). These killer yeasts produce toxic polypeptides that also kill other yeasts within the same genus and even eukaryotic organisms such as fungi. Killer strains can make up 30% of yeast communities in natural habitats. Approximately 18 species include killer yeasts, but most are found in *Saccharomyces cerevisiae*. However, additional research is needed to determine the true potential of these yeasts for control of food-borne pathogens on minimally processed produce (Deak and Beuchat, 1996).

CONCLUSION

Biological control of food-borne pathogens on minimally processed fruits and vegetables can be achieved in various ways. The indigenous microflora, which differs by commodity, can be influenced by handling and processing conditions such as refrigeration, disinfection, and modified atmosphere packaging to prevent the growth of food-borne pathogens while preserving the quality of the produce. In addition, selected biocontrol agents or mixtures of these may be applied to minimally processed produce. This can include the application of strains of LAB that produce bacteriocins and organic acids to lower the pH. Further, specific bacteriophage mixtures can be effectively applied as a disinfectant or protectant, which can be combined with other biocontrol agents. Another approach may be to influence the microflora on the produce before processing through augmentation of selected bacterial or yeast biocontrol agents.

These interactions are complex; additional research is necessary to determine the optimal combination of parameters for effective control of food-borne pathogens while providing the consumer with safe, more "natural" minimally processed produce. A variety of possibilities for biocontrol on minimally processed produce is just beginning to be investigated, and the potential exists to develop very effective treatment combinations. However, further research is necessary to develop these treatments for commercial application.

REFERENCES

Abee, T., Krockel, L., and Hill, C., 1995, Bacteriocins: mode of action and potentials in food preservation and control of food poisoning, *Intl. J. Food Microbiol.*, 28:169–185.

Alisky, J. et al., 1998, Bacteriophages show promise as antimicrobial agents, *J. Infect.*, 36: 5–15.

Asplund, K. and Nurmi, E., 1991, The growth of salmonellae in tomatoes, *Int. J. Food Microbiol.*, 13:177–182.

Baldwin, E.A. and Baker, R.A., 1992, Natural enzyme and bicontrol methods for improving fruits and fruit quality, Chapter 3, in *Molecular Approaches to Improving Food Quality and Safety*, Bhatnagar, D. and Cleveland, T.E., Eds., New York: Van Nostrand Reinhold, 61–82.

Bauman, H., 1991, Safety and regulatory aspects, in *Food Product Development from Concept to the Marketplace,* Graf, E. and Saguy, I., Eds., New York: Van Nostrand Reinhold.

Bennik, M.H.J., 1999, Biopreservation in modified atmosphere stored mungbean sprouts: the use of vegetable-associated bacteriocinogenic lactic acid bacteria to control the growth of *Listeria monocytogenes*, *Lett. Appl. Microbiol.*, 28:226–232.

Bennik, M.H.J. et al., 1996, Microbiology of minimally processed, modified-atmosphere packaged chicory endive, *Postharvest Biol. Technol.*, 9:209–221.

Bennik, M.H.J. et al., 1995, Growth of psychrotrophic foodborne pathogens in a solid surface model system under the influence of carbon dioxide and oxygen, *Food Microbiol.*, 12:509–519.

Bennik, M.H.J. et al., 1998, The influence of oxygen and carbon dioxide on the growth of prevalent *Enterobacteriaceae* and *Pseudomonas* species isolated from fresh and controlled-atmosphere-stored vegetables, *Food Microbiol.*, 15:459–469.

Beuchat, L.R., 1992, Surface disinfection of raw produce, *Dairy, Food Environ. Sanit.*, 12:6–9.

Beuchat, L.R., 1996, Pathogenic microorganisms associated with fresh produce, *J. Food Prot.*, 59:204–216.

Brackett, R.E., 1987, Microbial consequences of minimally processed fruits and vegetables, *J. Food Qual.*, 10:195–206.

Brackett, R.E., 1988, Changes in the microflora of packaged fresh tomatoes, *J Food Qual.*, 11:89–105.

Brackett, R.E., 1990, Influence of modified atmosphere packaging on the microflora and quality of fresh bell peppers, *J. Food Prot.*, 53:255–257.

Brackett, R.E., 1994, Microbiological spoilage and pathogens in minimally processed refrigerated fruits and vegetables, in *Minimally Processed Refrigerated Fruits and Vegetables*, New York: Chapman & Hall, 269–312.

Brackett, R.E. and Splittstoesser, D.L., 1992, Fruits and vegetables, in *Compendium for the Microbiological Examination of Foods*, Washington, D.C.: American Public Health Association, 287–293.

Breidt, F. and Fleming, H.P., 1997, Using lactic acid bacteria to improve the safety of minimally processed fruits and vegetables, *Food Technol.*, 51:44–51.

Breidt, F. and Fleming, H.P., 1998, Modeling of the competitive growth of *Listeria monocytogenes* and *Lactococcus lactis* in vegetable broth, *Appl. Environ. Microbiol.*, 64:3159–3165.

Cai, Y., Ng, L.-K., and Farber, J.M., 1997, Isolation and characterization of nisin-producing *Lactococcus lactis* subsp. *lactis* from bean sprouts, *J. Appl. Microbiol.*, 83:499–507.

Carlin, F., Nguyen-The, C., and Morris, C.E., 1996, Influence of background microflora on *Listeria monocytogenes* on minimally processed fresh broad-leaved endive (*Cichorium endivia* var. *latifolia*), *J. Food Prot.*, 59:698–703.

Chand-Goyal, T. and Spotts, R.A., 1997, Biological control of postharvest diseases of apple and pear under semi-commercial and commercial conditions using three saprophytic yeasts, *Biol. Control*, 10:199–206.

Clavente, V., Benuzzi, D., and de Tosetti, M.I.S., 1999, Antagonistic action of siderophores from *Rhodotorula glutinis* upon the postharvest pathogen *Penicillium expansum*, *Int. Biodeterior. Biodegrad.*, 43:167–172.

Conway, W.S. et al., 2000, Survival and growth of *Listeria monocytogenes* on fresh-cut apple slices and its interaction with *Glomerella cingulata* and *Penicillium expansum*, *Plant Dis.*, 84:177–181.

Danisco Cultor, 2001, International acceptance of nisin as a food preservative, Technical Memorandum, TM 1–1e SPEC, Brabrand, Denmark.

Davenport, R.R., 1976, Distribution of yeast and yeast like organisms from aerial surfaces of developing apples and grapes, in *Microbiology of Aerial Plant Surfaces,* Dickinson, C.H. and Preece, T.F., Eds., London: Academic Press, 325–374.

Deak, T., 1984, Microbial ecological principles in controlled atmosphere storage of fruits and vegetables, in *Microbial Associations and Interactions in Food,* Kiss, I., Deak, T., and Incze, K., Eds., Budapest: Hungarian Academy of Science, 9–22.

Deak, T. and Beuchat, L., 1996, *Handbook of Food Spoilage Yeasts,* New York: CRC Press.

Deak, T. et al., 1987, Extending the shelf life of fresh sweet corn by shrink wrapping, refrigeration and irradiation, *J. Food Sci.*, 52:1625–1631.

Doan, C.H. and Davidson, P.M., 2000, Microbiology of potatoes and potato products: a review, *J. Food Prot.*, 63:668–683.

Droby, S. et al., 1989, Characterization of the biocontrol activity of *Debaryomyces hansenii* in the control of *Penicillium digitatum* on grapefruit, *Can. J. Microbiol.*, 35:794–800.

Droby, S. et al., 1998, Commercial testing of Aspire yeast preparation for the biological control of postharvest decay of citrus, *Biol. Control*, 12:97–101.

Eayre, C.G., Bartz, J.A., and Concelmo, D.E., 1995, Bacteriophages of *Erwinia carotovora* and *Erwinia ananas* isolated from freshwater lakes, *Plant Dis.*, 79:801–804.

FDA, 1999, Code of Federal Regulations, Food and Drug Administration, Dept. of Health and Human Services, Title 21 Food and Drugs, Chapter I, Part 184, Subpart B, Section 184.1539, Nisin Preparation, 507–508, National Archives and Records Administration.

Filonow, A.B., 1998, Role of competition for sugars by yeasts in the biocontrol of gray mold of apple, *Biocontrol Sci. Technol.*, 8:243–256.

Filonow, A.B. et al., 1996, Biological control of *Botrytis cinerea* in apple by yeasts from various habitats and their putative mechanisms of antagonism, *Biol. Control*, 7:212–220.

Fleming, H.P., 1990, Mixed cultures in vegetable fermentations, in *Mixed Cultures in Biotechnology. The McGraw-Hill Environmental Biotechnology Series*, Zeikus, G.J. and Johnson, E.A., Eds., 69–103.

Francis, G.A. and O'Beirne, D., 1998a, Effects of storage atmosphere on *Listeria monocytogenes* and competing microflora using a surface model system, *Int. J. Food Sci. Technol.*, 33:465–476.

Francis, G.A. and O'Beirne, D., 1998b, Effects of the indigenous microflora of minimally processed lettuce on the survival and growth of *Listeria innocua*, *Int. J. Food Sci. Technol.*, 33:477–488.

Francis, G.A., Thomas, C., and O'Beirne, D., 1999, The microbial safety of minimally processed vegetables, *Int. J. Food Sci. Technol.*, 34:1–22.

Gayler, G.E. et al., 1955, An outbreak of salmonellosis traced to watermelon, *Public Health Rep.*, 70:311–313.

Geeson, J.D., 1979, The fungal and bacterial flora of stored white cabbage, *J. Appl. Bacteriol.*, 46:189–193.

Golden, D.A., Heaton, E.K., and Beuchat, L.R., 1987, Effect of chemical treatments on microbiological, sensory and physical qualities of individually shrinkwrapped produce, *J. Food Prot.*, 50:673–680.

Gombas, D.E., 1989, Biological competition as a preserving mechanism, *J. Food Safety*, 10:107–117.

Gould, G.W., 1996, Industry perspectives on the use of natural antimicrobials and inhibitors for food applications, *J. Food Prot.,* Suppl., 8286.

Greer, G.G., 1986, Homologous bacteriophage control of *Pseudomonas* growth and beef spoilage, *J. Food Prot.*, 49:104–109.

Grevesse, C., Jijakli, M.H., and Lepoivre, P., 1998, Study of exo-β-1,3-glucanase activity production by the yeast *Pichia anomala* in relation to its antagonistic properties against *Botrytis cinerea* on postharvest apples, in *Twelfth Forum for Applied Biotechnology*, Med. Fac. Landbouw. Univ. Gent., 1685–1692.

Gunez, G., Splittstoesser, D.F., and Lee, C.Y., 1997, Microbial quality of fresh potatoes: effect of minimal processing, *J. Food Prot.*, 60:863–866.

Heard, G., 1999, Microbial safety of ready-to-eat salads and minimally processed vegetables and fruits, *Food Aust.*, 51:414–420.

Hintlian, C. and Hotchkiss, J., 1987, Comparative growth of spoilage and pathogenic organisms on modified atmosphere packaged cooked beef, *J. Food Prot.*, 50:218–223.

Holzapfel, W.H., 1995, Biological preservation of foods with reference to protective cultures, bacteriocins and food-grade enzymes, *Int. J. Food Microbiol.*, 24:343–362.

Huxoll, C.C. and Bolin, H.R., 1989, Processing and distribution alternatives for minimally processed fruits and vegetables, *Food Technol.*, 43:124–128.

Ippolito, A. et al., 2000, Control of postharvest decay of apple fruit by *Aureobasidium pullulans* and induction of defense responses, *Postharvest Biol. Technol.*, 19:265–272.

Janisiewicz, W.J., 1987, Postharvest biological control of blue mold on apples, *Phytopathology*, 77:481–485.

Janisiewicz, W.J., 1998, Biocontrol of postharvest diseases of temperate fruits, in *Plant–Microbe Interactions and Biological Control,* Boland, G.J. and Kuykendall, L.D., Eds., New York: Marcel Dekker, 171–198.

Janisiewicz, W.J. and Bors, B., 1995, Development of a microbial community of bacterial and yeast antagonists to control wound-invading postharvest pathogens of fruit, *Appl. Environ. Microbiol.*, 61:3261–3267.

Janisiewicz, W.J. et al., 1999a, Fate of *Escherichia coli* O157:H7 on fresh-cut apple tissue and its potential for transmission by fruit flies, *Appl. Environ. Microbiol.*, 65:1–5.

Janisiewicz, W.J., Conway, W.S., and Leverentz, B., 1999b, Biological control of postharvest decays of apple can prevent growth of *Escherichia coli* O157:H7 in apple wounds, *J. Food Prot.*, 62:1372–1375.

Janisiewicz, W.J., Peterson, D.L., and Bors, B., 1994, Control of storage decay of apples with *Sporobolomyces roseus*, *Plant Dis.*, 78:466–470.

Jijakli, M.H. and Lepoivre, P., 1993, Biological control of postharvest *Botrytis cinerea* and *Penicillium expansum* on apples, in *IOBC/WPRS Bulletin: Biological Control of Foliar and Post-Harvest Diseases*, 16th ed., 106–110.

Kelly, W.J., Davey, G.P., and Ward, L.J.H., 1998, Characterization of lactococci isolated from minimally processed fresh fruit and vegetables, *Int. J. Food Microbiol.*, 45:85–92.

Kim, K.K. et al., 2000, Isolation and identification of antifungal N-butylbenzenesulphonamide produced by *Pseudomonas* sp AB2, *J. Antibiot.*, 53:131–136.

Leistner, L., 1992, Food preservation by combined methods, *Food Res. Int.*, 25:151–158.

Leistner, L. and Gorris, G., 1995, Food preservation by hurdle technology, *Trends Food Sci. Technol.*, 6:41–46.

Leverentz, B. et al., 2001, Examination of bacteriophage as a biocontrol method for *Salmonella* on fresh-cut fruit — a model study, *J. Food Prot.*, 64:1116–1121.

Lund, B.M., 1992, Ecosystems in vegetable foods, *J. Appl. Bacteriol. Symp. Suppl.*, 73:115S-126S.

Mesquita, M.M.F., Souto, M.L.A., and Martins, J.M.S., 1983, Sensitivity to bacteriophages and bacteriocins among epiphytic fluorescent pseudomonads, *Revista de Biologia*, 12:529–538.

Nguyen-The, C. and Carlin, F., 1994, The microbiology of minimally processed fresh fruits and vegetables, *Crit. Rev. Food Sci. Nutr.*, 34:371–401.

Park, S.C. et al., 2000, Isolation of bacteriophages specific to a fish pathogen, *Pseudomonas plectoglossicida*, as a candidate for disease control, *Appl. Environ. Microbiol.*, 66:1416–1422.

Piagentini, A.M. et al., 1997, Survival and growth of *Salmonella hadar* on minimally processed cabbage as influenced by storage abuse conditions, *J. Food Sci.*, 62:616–618, 631.

Reyes, V.G., 1996, Improved preservation systems for minimally processed vegetables, *Food Aust.*, 48:87–90.

Roberts, R.G., 1990, Postharvest biological control of gray mold of apple by *Cryptococcus laurentii*, *Phytopathology*, 80:526–530.

Romick, T.L., 1994, Biocontrol of *Listeria monocytogenes*, a psychrotrophic pathogen model, in low salt, nonacidified, refrigerated vegetable products, Dissertation, North Carolina State University, Raleigh.

Saleh, M.A. and Ordal, Z.J., 1955, Studies on growth and toxin production of *Clostridium botulinum* in a precooked frozen food. II. Inhibition by lactic acid bacteria, *Food Res.*, 20:340–350.

Sinell, H.J., 1980, Packaging, in *Microbial Ecology of Foods*, New York: Academic Press, 193–204.

Stack, J.P., 1998, Postharvest biological control: commercial successes and a model for public and private sector cooperation, 7th Int. Congr. Plant Pathol., Edinburgh, Scotland. 1998.

Stevens, K. et al., 1991, Nisin treatment for inactivation of *Salmonella* species and other gram-negative bacteria, *Appl. Environ. Microbiol.*, 57:3613–3615.

Stevens, K. et al., 1992, Effect of treatment conditions on nisin inactivation of gram-negative bacteria, *J. Food Prot.*, 55:763–766.

Stiles, M.E., 1994, Potential for biological control of agents of foodborne disease, *Food Res. Int.*, 27:245–250.

Stiles, M.E., 1996, Biopreservation by lactic acid bacteria, *Antonie van Leeuwenhoek*, 70:331–345.

Thomas, C. and O'Beirne, D., 2000, Evaluation of the impact of short-term temperature abuse on the microbiology and shelf-life of a model ready-to-use vegetable combination product, *Int. J. Food Microbiol.*, 59:47–57.

Thomas, L.V., Clarkson, M.R., and Delves-Broughton, J., 2000, Nisin, in *Natural Food Antimicrobial Systems*, Naidu, A.S., Ed., Boca Raton, FL: CRC Press, 463–524.

Torriani, S., Orsi, C., and Vescovo, M., 1997, Potential of *Lactobacillus casei*, culture permeate, and lactic acid to control microorganisms in the ready-to-use vegetables, *J. Food Prot.*, 60:1564–1567.

Vandenbergh, P.A., 1993, Lactic acid bacteria, their metabolic products and interference with microbial growth, *FEMS Microbiol. Rev.*, 12:221–238.

Vescovo, M. et al., 1995, Inhibitory effect of selected lactic acid bacteria on microflora associated with ready-to-use vegetables, *Lett. Appl. Microbiol.*, 21:121–125.

Vescovo, M. et al., 1996, Application of antimicrobial-producing lactic acid bacteria to control pathogens in ready-to-use vegetables, *J. Appl. Bacteriol.*, 81:113–119.

Wells, J.M. and Butterfield, J.E., 1997, *Salmonella* contamination associated with bacterial soft rot of fresh fruits and vegetables in the marketplace, *Plant Dis.*, 81:867–872.

Wilson, C.L. and Wisniewski, M.E., 1989, Biological control of postharvest diseases of fruits and vegetables: an emerging technology, *Annu. Rev. Phytopathol.*, 27:425–441.

Wisniewski, M. et al., 1991, Mode of action of postharvest biocontrol yeast, *Pichia guilliermondii*. Characterization of attachment to *Botrytis cinera, Physiol. Mol. Plant Pathol.*, 39:245–258.

Zaccardelli, M. et al., 1992,, *Xanthomonas campestris* pv. *pruni* bacteriophages on peach trees and their potential use for biological control, *Phytopathologia Mediterranea*, 31:133–140.

Index

C

D

E

F

G

H

I

J

Q

R

S

T

U

V

W

X

Y

Z

For Product Safety Concerns and Information please contact our EU representative GPSR@taylorandfrancis.com Taylor & Francis Verlag GmbH, Kaufingerstraße 24, 80331 München, Germany

Batch number: 10725823

Printed by Printforce, the Netherlands